MSP430-based Robot Applications

MSP430-based Robot Applications

A Guide to Developing Embedded Systems

Dan Harres

AMSTERDAM • BOSTON • HEIDELBERG • LONDON
NEW YORK • OXFORD • PARIS • SAN DIEGO
SAN FRANCISCO • SINGAPORE • SYDNEY • TOKYO
Newnes is an imprint of Elsevier

Newnes is an imprint of Elsevier
225 Wyman Street, Waltham, MA 02451, USA

First edition 2013

British Library Cataloguing-in-Publication Data
A catalogue record for this book is available from the British Library.

Library of Congress Cataloging-in-Publication Data
A catalog record for this book is availabe from the Library of Congress

ISBN: 978-0-12-397012-1

Dedication

To my wife, Susan

Contents

Preface

I wrote this book largely as a means of interesting young people in the field of electronics, computers and, of course, robots. Back in the 1930s, 1940s, and 1950s, young people (mostly boys then) were attracted to electronics engineering largely as the result of radio kits and radio design. Although radios are taken for granted today, back then a radio was an exotic piece of hardware. A radio cost more than a week's salary for an average worker back in 1930, so it was hardly the commodity that we consider it today. A young person building a tube radio back then was working with something that was both cutting-edge technology and that was of real value when completed. If you talk to an electrical engineer who grew up back in the mid-20th century, chances are their interest in electronics was initially sparked by an interest in radio.

After the radio became commonplace and cheap, interest in engineering was sparked by such things as the space program in the 1960s and later by the emergence of personal computers. But today the national space program seems adrift in both purpose and achievement. Personal computers, while shrinking in size and growing in power, have become architecturally so complex and the functions so integrated, that a young hobbyist will find little opportunity to learn and explore, particularly with respect to the computer hardware.

To fill this void, a movement has sprung up in the last couple of decades to interest young people in building robots. There are now robot clubs at many of the middle schools and high schools in this country and there is also a network of performance events in which these clubs can compete. This is potentially an important movement.

However, just as radio clubs sprang up at high schools in the 1950s but individual students still wanted to build their own radios at home, it would seem to be the case with robots that individual students will want to enhance their abilities by building their own personal robots rather than just participating in a robot-building group at school.

The problem with building a robot at home is that the inexpensive kits or inexpensive approaches often result in robots that are slow, have limited ability to do anything, and, basically, just have a very low "pizazz coefficient", as a friend used to say. Many of them don't provide any insight into how the electronics work, with the result that the only skill learned is perhaps soldering. On the other hand, there are robots with decent performance available, but the price tag for the kits is typically in the hundreds of dollars – enough to discourage all but a few of the potential builders.

The intent of this book, then, is to provide the reader with designs for a robot that will be fast, interesting, and expandable, but one that can be built for a modest price (the choice of the MSP430 as the microcontroller was a direct result of this focus on cost, as the entire development system is sold for $4.30). If the book is successful, it will inspire young people to pursue electronics design, a career path that I have found to be very satisfying.

This book is written at a transitional point in my career. I have worked almost my entire adult life for McDonnell Douglas Corp., first as a college student in their cooperative education program and then, after graduation from college, as an engineer. After their acquisition by Boeing Corporation in the late 90s, I continued working there, retiring in 2011. Both companies treated me very well and I was blessed to have worked for them. McDonnell Douglas was willing to take a young, immature, but fairly bright high school student and give him the chance to develop the technical skills needed to be a productive engineer. My experience with them and with engineering convinces me that this can be a very rewarding vocation for young people with an interest in and aptitude for mathematics and science.

I was fortunate, during the creation of this book, to have a very able young engineer to assist me, my son Max. Max did the layout for the several versions of the robot electronics, built the boards, ordered the parts, wrote the software, and did most of the troubleshooting. In short, he took care of the actual creation of the robot, leaving me free to concentrate on writing the book about it. I doubt that I would have taken on this project without his assistance.

Finally, I was fortunate to have a wife, Susan, who was understanding of both my decision to retire early and my decision to pursue this project. I owe her a great deal for her willingness to let me pursue a new outlet for my creative side and for her support during the transition.

Dan Harres

CHAPTER 1

Introduction

Chapter Outline

This is a book intended for individuals who are just getting started in electronics and microcontrollers, as well as individuals who have some experience and simply want to learn how to make an exciting robot out of a modified radio-controlled car and a low-power, but powerful, microcontroller. For those with experience, the first several chapters, which are a review of theory, can be skipped. For those just starting out, the best idea is to read the book through.

Expected reader background

To make good use of this material, the reader should have, at a minimum, knowledge of the following:

1. Mathematics up through algebra, preferably trigonometry
2. Some programming experience
3. Some soldering experience
4. A high-school physics course covering basic electromagnetics.

Individuals with this much experience should be able to follow the material in the book and should be able to successfully build the robots detailed in the book, particularly if following closely the example circuits and software. Naturally, additional background, particularly in circuit theory, will be helpful.

Before diving into how to build a robot, it's important to define what we mean by the term. For example, is a radio-controlled (RC) car a robot? What about the mechanical arms that you see pictured in automobile factories, painting car panels or spot welding parts on the car?

MSP430-based Robot Applications.
DOI: http://dx.doi.org/10.1016/B978-0-12-397012-1.00001-1

For the purposes of this book, we're going to narrow the scope of what people generally mean by the term robot. Here a robot is a *mobile* device, capable of moving *autonomously* but, perhaps, augmented with external commands. So an unmodified remote-controlled car that you can buy at the toy store is not a robot, since it can't move autonomously. An automobile factory spot welder, although a robot in a more general context, is not what we're talking about either, since it is mounted in place and is therefore not mobile.

The parts of a robot

The overall functions of a robot can be broken down into roughly four parts:

1. Locomotion
2. Control
3. Power source
4. Platform

Locomotion consists of the motors and wheels. Control consists of the electronics and software for sensing as well as the electronics and software for steering, motor control, etc. The power source is typically just the batteries or other source of energy. And the platform is the frame and other parts needed to hold the whole thing together.

Most books on robot building spend considerable time on the platform (building the frame, attaching motor and wheels, etc.) and on locomotion (sizing the motors, driving the wheels, mounting wheels to axles, etc.).

Now, consider how to build a robot inexpensively. The way robot building is often approached in books on the subject is to choose materials from which to build the frame, then build that frame, choose and purchase motors that are then mounted to this frame, etc., etc. Not only is this a long, drawn-out process, but it's expensive, too. And the resulting robot may be slow-moving and not impressive.

In this book, we'll take a different approach. We can purchase inexpensive, high-performance, already-built platforms, in the form of radio-controlled (RC) cars. These cars are available at lots of stores, they generally sell for around $20 to $30, and they often have rechargeable batteries. Thus, three of the four functions in the above list are already completed by taking this route. By substituting our own electronics (in the form of a microcontroller and associated hardware and software) we can transform this vehicle into a high-performance autonomous robot, capable of operating at high speeds, avoiding objects, tracking lines, etc. The end product is not only cheaper and easier to build, but it's impressive as well.

So that leaves us with just the electronics and associated software to deal with. What are the functions that these electronics will have to perform to make this robot work? They include:

1. Motor drive – this includes the power transistors that route the battery current to the motor.
2. Steering drive – this includes the transistors that transfer the microcontroller's commands to turn left or right.
3. Situational awareness – this is the collection of sensors and the sensor driver and receiver electronics that give the robot information about the environment in which it is operating.

Associated with each of these functions are also microcontroller operations. For example, the motor drive must be able to control not only whether current is being supplied to the motor but how much current is supplied, so that the speed of the robot is properly controlled. The steering drive command must be timed properly by the microcontroller so that, for example, the robot executes a desired 90 degree turn, not a 180 degree turn. For the situational awareness, there is considerable timing, data acquisition, and logic required of the microcontroller.

Along with the microcontroller software associated with each of these functions, there is a need for overall processing, coordination, and scheduling software. This is the software that determines when each of the functions are performed and how often they are repeated. It processes information from the sensors and, based on that information, determines what the next actions are that need to be taken by the motor and steering drives. The block diagram of Figure 1.1 illustrates this interaction between the different hardware and software functions.

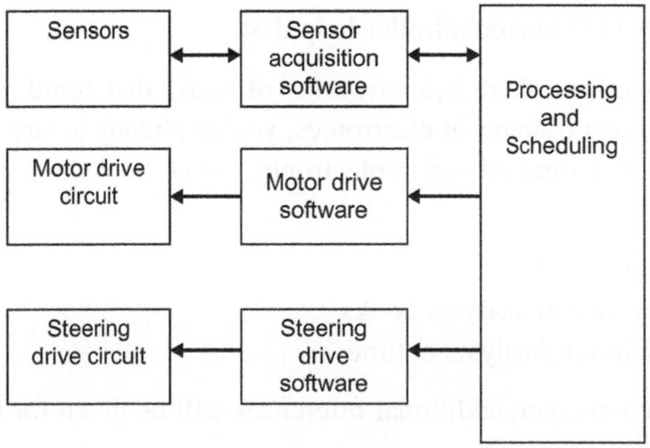

Figure 1.1
Robot hardware and software functions.

The design of these functions is what we'll be looking at in this book. Along the way, you'll learn how bipolar and MOSFET transistors work, what an analog filter does and how to design it, how to design a feedback amplifier and more. In addition, you'll learn about digital electronics. This includes computer logic, computer arithmetic, and computer programming. The digital portion of the book will be centered on a popular and very powerful microcontroller, the Texas Instruments MSP430.

The MSP430 has lots of features that make it a good choice for these projects. For example, it is actually a family of 16-bit microcontrollers that includes hundreds of different variations on the same basic processor. As a result, you can tailor the microcontroller to your application, choosing the types of data acquisition units included, the number of timers, the amount of memory, the number of input/output pins, and much more. The microcontroller is available as an inexpensive ($4.30) evaluation board, called the LaunchPad, that allows you to get started programming immediately.

Where to get help

For those who are just beginning their exploration of electronics, computers, and robots, this book may be a little daunting. Electronics is not an easy subject to master. But, like a lot of things that are difficult, perseverance pays off. It will also pay off to consult others for help. This includes, for the microcontroller:

1. Example programs in this book and from TI
2. The TI publication SLAU144H, "MSP430x2xx Family User's Guide"
3. TI MSP430 Users' Forum
4. TI Application Notes
5. Texas Instruments (TI) microcontroller help desk

For the electronics functions, there is a vast array of books that could help. If you have some formal college-level training in electronics, you're already aware of what books are available. If you're just getting started in electronics, some books that might be of interest to you are:

1. Electrical Engineering 101
2. Other introductory circuit analysis books
3. Schaum's Basic Circuit Analysis outline

As we go through each chapter, additional references will be given for the material in that particular chapter.

The internet is also a great source of information. There is, of course, always the caveat that anyone can post anything on the internet and that what you see in a particular posting may or may not be correct. Nevertheless, the internet is, in general, a great resource. In particular, Wikipedia is a good source of information.

Tools you will need

There are a number of tools that you will absolutely need, all at low cost, and one optional tool that's somewhat expensive. The essential tools are:

1. Diagonal cutters
2. Needle-nose pliers
3. Magnifying glass or loupe or magnifying visor
4. Soldering iron with fine tip
5. Solder
6. Wet sponge
7. Wire stripper
8. Multimeter
9. Electric drill
10. Screwdrivers
11. Wrench

Keep in mind that the components that you will be working with are small and the tools will need to accommodate these small components. So, for example, the diagonal cutters and needle-nose pliers that you use for electrical repair work around the house are not going to be suitable for use in these projects. Likewise, the soldering iron needs to have a fairly small tip and a small-diameter solder (of the order of 0.03" will work best). Figures 1.2 through 1.4 show typical tools for working with the small electronic

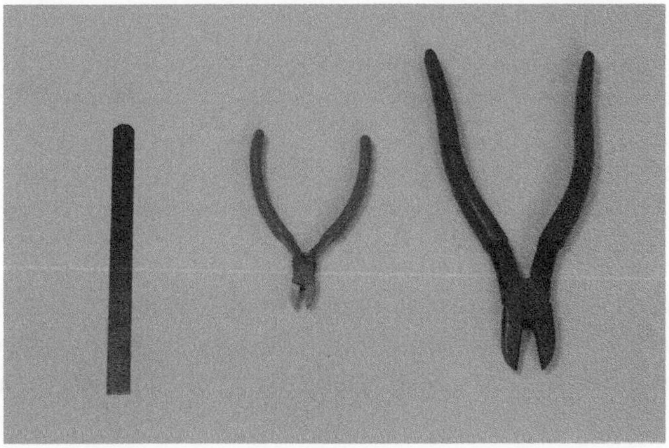

Figure 1.2
Diagonal cutters (the pair on the left is the correct size for this work).

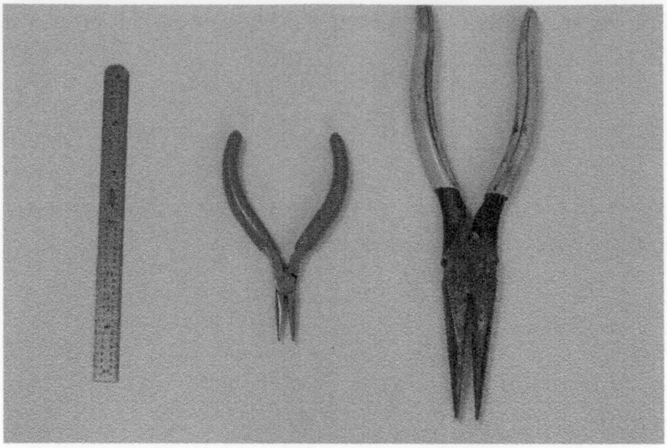

Figure 1.3
Needle-nose pliers (the pair on the left is the correct size for this work).

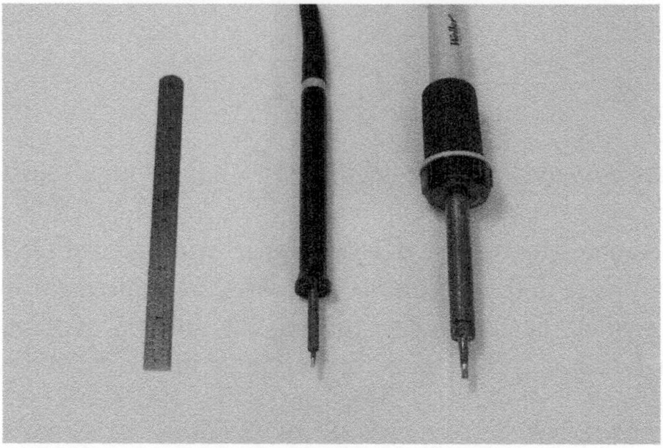

Figure 1.4
Soldering irons and tips (iron on the left is the appropriate one).

components in robotic projects. Alongside each tool is shown the larger version of the tool that is intended for home repair and is not appropriate for such small-component work.

There is one other tool that can be extremely useful, and that is the oscilloscope. No other diagnostic tool provides the insight into a circuit's behavior that the oscilloscope provides. Unfortunately, this tool is not cheap. The good news is that the cost of these instruments is coming down. A new, high-quality, two-channel scope can be purchased for less than $500.

Of course, that's still outside most hobbyists' price range. But there are alternatives. If you are a student, you may have access to oscilloscopes in your school's lab. There are also

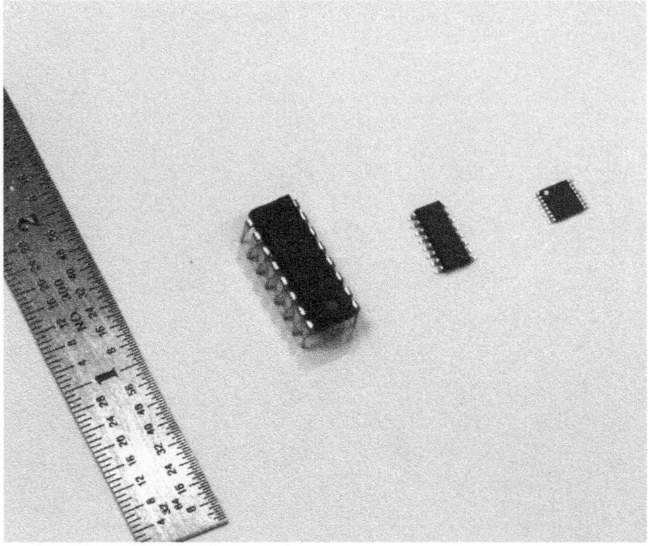

Figure 1.5
Comparison of package sizes — a DIP, an SOIC, and a TSSOP.

used scopes available, either locally or from market websites such as eBay. Finally, you can probably forego an oscilloscope altogether if you decide to stick strictly to the examples and designs presented in the text, particularly if you build the circuits from the printed circuit boards that are available for this purpose.

Components

Since this book is aimed at an audience with a wide range of electronic construction skills, the use of the larger dual-in-line packages (DIPs) and leaded passive components, rather than the smaller surface-mount technology (SMT) parts, is the goal in all designs. The industry's trend to smaller and smaller parts makes this approach a little challenging but, except in a few instances, it is still feasible and is used throughout the examples in this book.

For comparison purposes, Figure 1.5 shows three 16-pin packages, the dual inline package (DIP), a surface-mount small-outline integrated circuit (SOIC), and a surface-mount thin shrink small outline package (TSSOP).

CHAPTER 2

Mechanical and Electrical Disassembly of the RC Car

Chapter Outline

In order to modify the RC car that we've chosen, we'll first have to do a little disassembly and then, once that's done, examine the wiring and other electrical functions. We can do this just by inspection and by "ringing out" connections – using the multimeter in its "ohm" setting – to determine what's connected to what. This really doesn't require much in the way of electronics knowledge, so it's okay to consider this now, even before the chapters that deal with electronic operations.

One of the things that makes this a little tricky is that each different RC car is mechanically and electrically a little different from any other car. There is no standard as to what the wire color is for a particular function, or how the frame is attached to the body, or anything else. So the disassembly shown in the following paragraphs should be taken as simply an example.

Figure 2.1 shows one of the cars that we'll use as an example in this book. It is a Ford Mustang coupe 1:12 scale model purchased from Radio Shack. Its antenna is just a wire that is held rigid by passing it through a plastic "straw".

Mechanical disassembly

The first thing to figure out is how to get the car body off. Typically there are four screws that provide the attachment of the body to the frame. A picture of the car's underside (Figure 2.2) shows there are lots more than four screws, but most of these are for holding

MSP430-based Robot Applications.
DOI: http://dx.doi.org/10.1016/B978-0-12-397012-1.00002-3

Figure 2.1
A typical RC car.

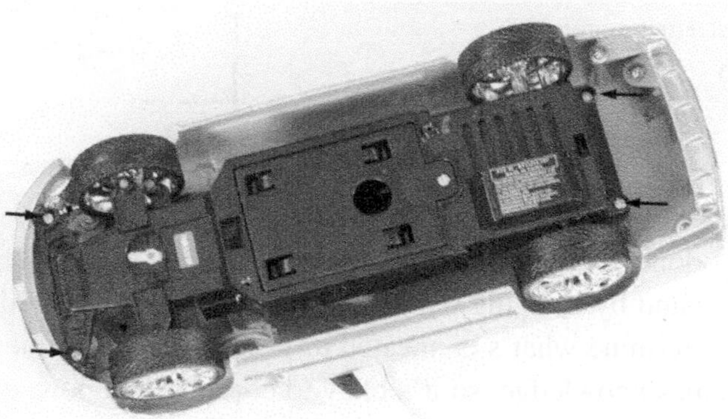

Figure 2.2
The body was attached to the frame on this particular RC car by four screws.

different parts of the car's body together. You can ignore those other screws. Set the car body aside, as we will not use that for the robot.

In some cases, the car will have LED lights embedded in the body, with wires running between the frame and body for this purpose. Just clip these wires, as they are not needed.

After removing the body, the car looks like Figure 2.3. In some cases the drive motor, steering motor, and electronics will be exposed at this point. In the case of this particular car, there was a plastic cover over the drive motor and electronics, which could be removed easily by prying the two tabs on each side of the frame.

The exposed parts will look like the Figure 2.4 picture. At this point you have completed the mechanical disassembly and it's time to begin the determination of which wire is

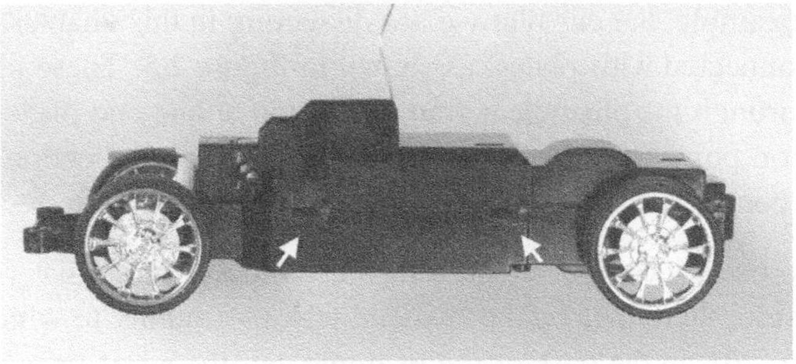

Figure 2.3
Two tabs on each side had to be pried out to remove the top cover.

Figure 2.4
Once the cover is removed on this car, the electronics and drive motor are exposed.

associated with which function. Like the car body, the plastic cover, if it's included, can be set aside as we will have no further use for it.

Electrical inspection

The first thing to determine is where the + and − connections to the battery are made. With some cars, particularly with rechargeable battery packs, there will just be two wires sticking up through a hole in the frame and soldered or otherwise attached to the main circuit board, so it's easy to see what's what.

In other cases, for example, the car that we are dissecting in this chapter, there are multiple AA batteries interconnected with plates, as shown in Figure 2.5. These plates have little tabs that protrude through the plastic car frame. The tab at one end of the battery collection is then the connection point for the positive side of the battery collection, and the tab at the opposite end is the connection point for the negative side of the battery collection.

After examining the battery compartment, turn the vehicle back so that you can again see the top. On the topside, there will be the two end tabs protruding, to which wires are attached (see Figure 2.6). In the case of the example car, the wires were red and black, with red the positive wire and black the negative. Those are traditionally the colors that are used in electronic design for those wires, but RC cars in general seem not to follow any specific color coding, so you'll need to check.

To check which wire is positive and which is negative, you can simply install the batteries and take a voltage reading. Put your multimeter on the "DC volts" setting, put your red multimeter probe on the connection point that you think is the positive, and your black multimeter probe on the connection point that you think is negative. If the voltage you read is positive, you have guessed correctly about which is positive and which is negative. If the voltage you read is negative, then the polarity is opposite from your guess.

On/Off switch

Next, let's figure out how the battery power is turned on and off. The switch will be in series with either the positive or the negative wire. Moving the switch to Off simply opens the connection for the particular side of the battery that it's in series with. On the example car in this chapter, the switch is in series with the negative side of the battery. We

Plate connects battery
to its neighbor

Figure 2.5
After removing the battery cover, the battery compartment shows the way the batteries are series connected.

Figure 2.6
Connection points to the battery.

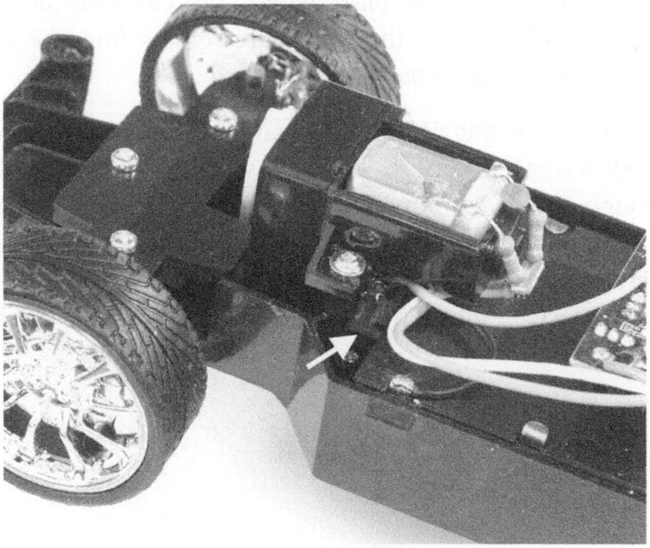

Figure 2.7
In this car, the On/Off switch on the car's underside had its electrical connections on a small circuit board under the steering motor.

determine this by observing that the black wire from the battery is routed to the tiny switch circuit board under the steering motor, which contains the switch's connection points (see Figure 2.7).

Note in Figure 2.6 that the other wire coming from the switch is green, and it is routed to the main circuit board. When the switch is On, this green wire completes the circuit path for the RC car. We'll leave that switch in and use it for the robot electronics in the same way.

Note that the green wire connection is the "switched ground" for the RC car's main electronics board. It will serve this function for the robot of this project as well. The term "ground" in this application is a little misleading. There is, in fact, no connection to earth ground for any part on the board. What is really meant by this term is just a reference. By choosing this negative battery terminal voltage as the reference, all other voltages can be simply stated as a single, unqualified number, such as "6 volts", rather than saying "6 volts with respect to such and such point", which is cumbersome and confusing.

Connections to the drive motor

There should be just two wires to the drive motor. In the case of the example car, they are the blue and black wires of Figure 2.6. Here is an example of where color coding by the RC car manufacturer is misleading. The black wire in this case does not mean "ground" or negative battery polarity, as it did with the black wire coming from the battery. Here, it is simply one of the two wires going to the motor.

To verify that the black wire is not simply a ground, the voltages were measured using an oscilloscope. Although we haven't talked yet about voltage and current, the reader is assumed to have enough of a background to understand that this is the drive signal to the motor. The RC car was driven first forward (Figure 2.8a) and then in reverse (Figure 2.8b). As can be seen in those photos, the motor responds to a +7.5 V and 0 V (referenced to the "ground" or negative battery terminal) for forward and to a 0 V and +7.5 V for reverse. In

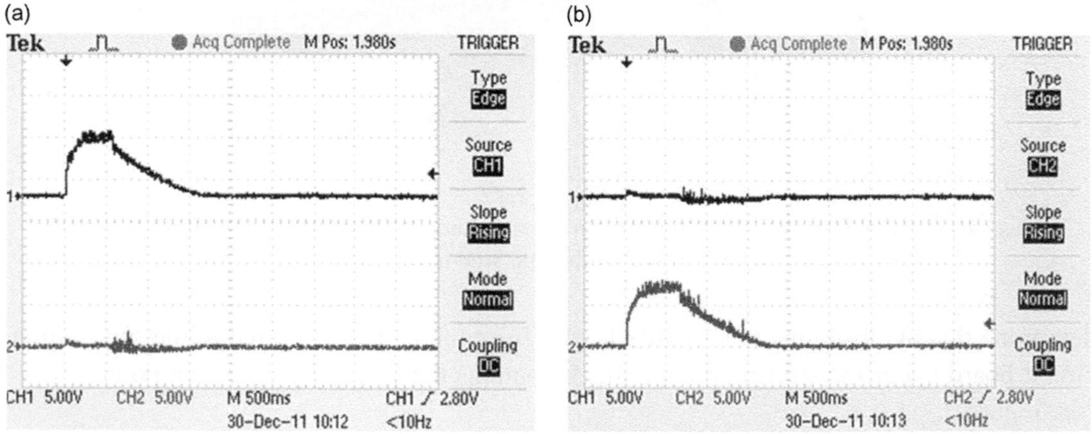

Figure 2.8
(a) Motor drive waveforms – forward drive. (b) Motor drive waveforms – reverse drive.

a subsequent chapter we'll see how this is achieved using a so-called "H-bridge". The waveform colors correspond to the wire colors.

Connections to the steering motor

The steering motor is intended to just operate forward or reverse. On most RC cars this turns a plastic arm that moves a plastic "tie rod", causing the wheels to turn right or left. On the example car, the motor has just two wires, like the drive motor, and on this car they are colored white and orange. Note that they are attached to what appear to be green resistors (see Figure 2.7). In fact, these are inductors that, along with a capacitor, are intended to smooth the steering drive signal that is coming from the RC car's electronics.

Electrical disassembly

Before taking things apart, it's a good idea to make a sketch or at least some notes as to what wire goes where. Most of this will be obvious, but it might help later.

The wires can be clipped near the electronics board or the connections at the board can just be heated with a soldering iron and the wires lifted from the board. Remove the RC car's electronics assembly and set it aside (you may or may not have a use for this later, depending on whether you want to use the car's RF system for sending commands to the robot).

That completes the disassembly. The next step is to start building, but first, let's review the basics of electronics, software, and the MSP430 microcontroller.

CHAPTER 3

Beginning Electronics — Resistors, Capacitors, and Inductors

Chapter Outline

A popular (and overused) phrase is "to think outside the box". By this is meant, I think, that the person is supposed to "be creative", that is, do things in a way that is outside the conventional approach. For young circuit designers, this type of advice often leads to their trying approaches that, had they known more about circuits, they would have realized defied some basic principle. Because the individual is on shaky theoretical ground, mistakes of the past are simply repeated. Far too often, the result of this "thinking outside the box" stuff is failure.

A much better approach, at least for those just starting their engineering career, is to think "inside the box", but make sure that that box is a very, very large one. In other words,

MSP430-based Robot Applications.
DOI: http://dx.doi.org/10.1016/B978-0-12-397012-1.00003-5

learn as much as you can about the subject and make full use of all of the understanding that's been passed down through time. Once you've mastered these fundamental concepts, you're in a much stronger position to try novel approaches to problems.

Electronic circuit design is a topic where that box is extremely large. For electronics engineering students, the curriculum typically includes several semester-long courses in circuit design, in addition to lots of other important topics, like control systems, communications, digital signal processing, computer programming and more. For such a vast topic as circuit theory, we shouldn't expect to cover everything in one book, much less just a few chapters of a book.

In this book, the basics of electronic design are covered to the extent needed to create the specific robot projects discussed. Since, with even that narrowed scope, there will be deficiencies in the explanations given, a list of references is added at the end of this chapter. Note that the reader is assumed to have some rudimentary knowledge of what is meant by terms such as resistance, capacitance, inductance, etc. – that is, the topics related to electricity and magnetism that would be covered in a high-school physics class. For readers not familiar with these terms, please consult that list of references.

The topics covered in this chapter include:

- Ohm's law
- Kirchoff's laws
- Passive components – resistors, capacitors, inductors
- Circuits that eliminate low-frequency signals
- Circuits that eliminate high-frequency signals

Some basic laws

It's assumed the reader is familiar with basic electromagnetic concepts presented in high-school physics classes. The reader should already know what is meant by voltage and current – if not, do some background reading in one of the books listed at the end of the chapter. So we'll get started with some of the fundamentals of electronics design and analysis. First one up is Ohm's law.

Ohm's law

Ohm's law is one of those simple ideas that you will use over and over in electronics design. It is simply the linear relationship:

$$V = IR. \tag{3.1}$$

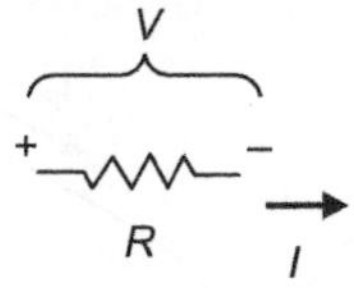

Figure 3.1
Ohm's law.

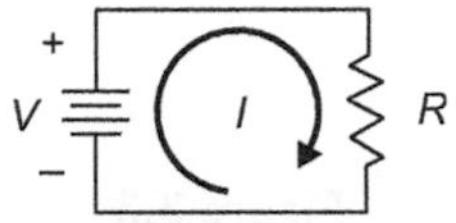

Figure 3.2
Applying Ohm's law to determine current drawn from battery.

That is, the voltage across a resistor is equal to the product of the current through the resistor and the resistance of the resistor. The idea is illustrated in Figure 3.1.

Using Ohm's law, we can immediately determine the current flowing in the circuit of Figure 3.2. If the battery (the voltage source, *V*) is 10 V and the resistor, *R*, is 100 ohms, then, by rearranging Eq. 3.1, the current is $I = V/R = 0.1$ amperes.

Kirchoff's laws

In the mid-nineteenth century, Gustav Kirchoff developed two useful laws, one regarding voltages in a circuit and the other regarding currents. The first of these, Kirchoff's current law, says that the net current flowing into a circuit node is zero. A node is just a common connection point. All this is saying is that the sum of the currents flowing into the node depicted in Figure 3.3 must be zero:

$$\sum I_n = 0 \qquad (3.2)$$

Note that sometimes it will be more convenient to think of one or more of the currents as flowing in the opposite direction from the others, and that this is easily accommodated by changing the sign in Eq. 3.2. For example, if we choose to think of I_B as flowing away from the node in Figure 3.3 (see Figure 3.4) then we would write:

$$I_A - I_B + I_C + I_D = 0$$

or equivalently,

$$I_B = I_A + I_C + I_D.$$

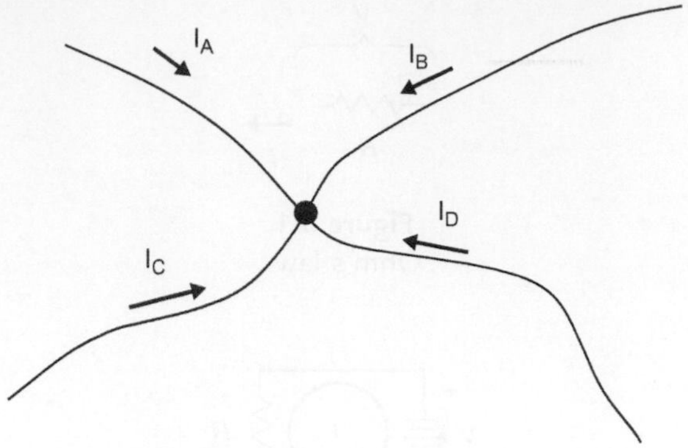

Figure 3.3
Kirchoff's current law states that the sum of the currents into a node is 0.

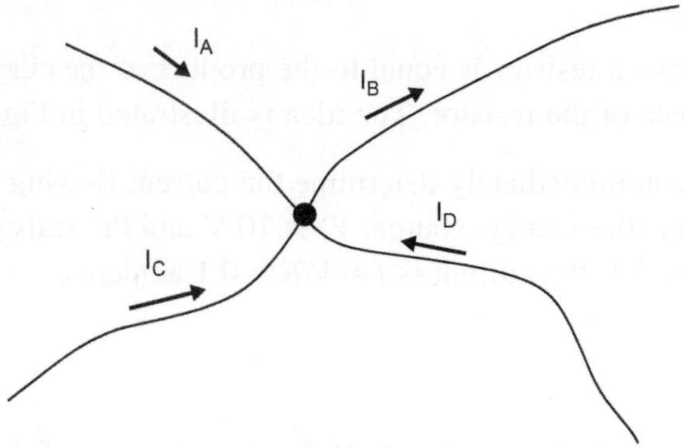

Figure 3.4
Here I_B is referenced in the direction opposite to other currents and $I_B = I_A + I_C + I_D$.

Kirchoff's *voltage law* is similar. It states that the sum of the voltages around any closed path of a circuit is zero:

$$\sum v_n = 0. \tag{3.3}$$

Using Kirchoff's voltage law for the circuit of Figure 3.5, we can quickly write the equation for the circuit voltages:

$$V_{B1} + V_{B2} + V_{R1} + V_{R2} = 0 \tag{3.4}$$

As with Kirchoff's current law, we can change the reference direction of one or more of the components and accommodate this change in direction by reversing the sign for that particular component. For example, we could redraw the Figure 3.5 circuit diagram so that the R_1 and R_2 polarities are referenced with positive polarity at the top (Figure 3.6).

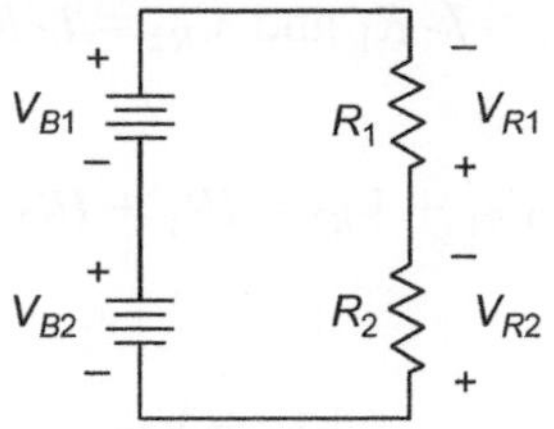

Figure 3.5
Example circuit to demonstrate Kirchoff's voltage law.

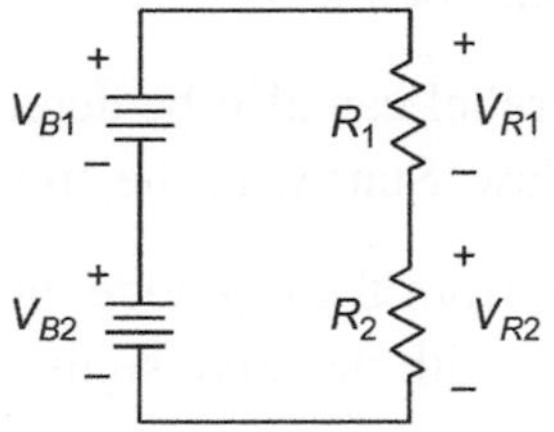

Figure 3.6
Same as Figure 3.5 but with the polarity reference on the resistors reversed.

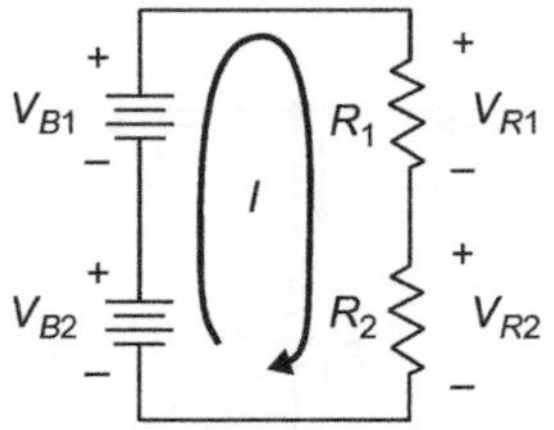

Figure 3.7
Circuit example.

In this case, the sum of voltages is written as:

$$V_{B1} + V_{B2} - V_{R1} - V_{R2} = 0. \tag{3.5}$$

(using the Figure 3.6 circuit diagram) which can be rearranged as:

$$V_{B1} + V_{B2} = V_{R1} + V_{R2} \tag{3.6}$$

Now, let's say that we are asked to determine the current through this circuit. The circuit is the same as in Figure 3.6 and we assume that the current is the same anywhere within the loop (Figure 3.7).

From Ohm's law we know that $V_{R1} = I \cdot R_1$ and $V_{R2} = I \cdot R_2$ and substituting this into Eq. 3.6 gives:

$$V_{B1} + V_{B2} = IR_1 + IR_2 \tag{3.7}$$

or

$$I = \frac{V_{B1} + V_{B2}}{R_1 + R_2} \tag{3.8}$$

Equation 3.8 demonstrates yet another important principle – that resistors in series add. We could simply replace the two resistors in Figure 3.7 with a single resistor equal to the sum of the two, and the current would be the same.

The behavior of two resistors in parallel can also be determined quite easily, using Kirchoff's current law and Ohm's law. Start with the simple circuit of Figure 3.8.

Since I_1 and I_2 are referenced away from the top node, but I_3 is referenced into that node, the first two currents will be written with negative signs in the node equation:

$$-I_1 - I_2 + I_3 = 0 \tag{3.9}$$

or

$$I_3 = I_1 + I_2. \tag{3.10}$$

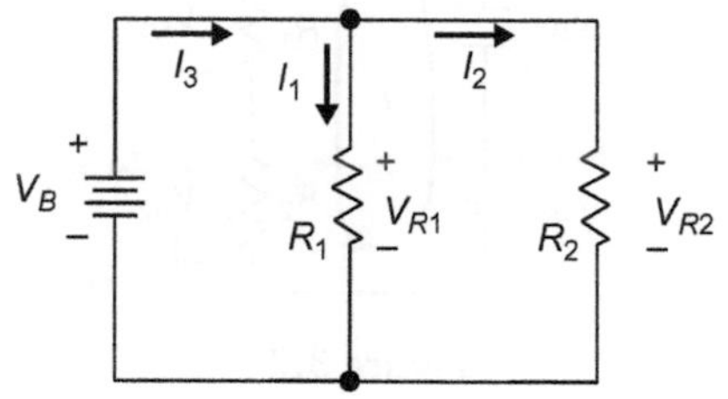

Figure 3.8
Two resistors in parallel.

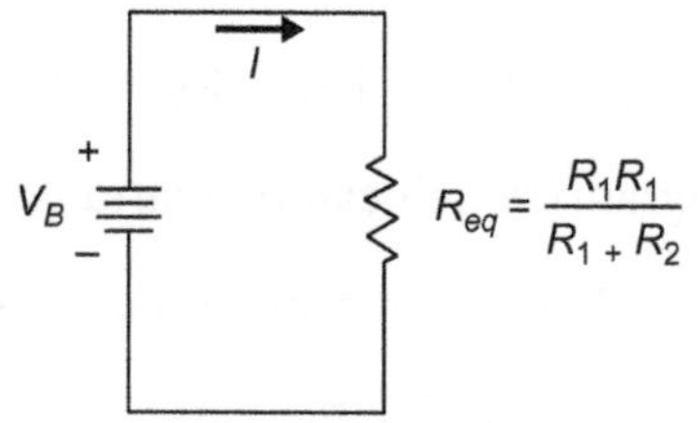

Figure 3.9
Simplified circuit substituting single resistor for two parallel resistors.

In this case, both V_{R1} and V_{R2} must be equal to V_B. Making use of Ohm's law, currents I_1 and I_2 are:

$$I_3 = \frac{V_B}{R_1} + \frac{V_B}{R_2} = V_B\left(\frac{1}{R_1} + \frac{1}{R_2}\right) = V_B\left(\frac{R_1 R_2}{R_1 + R_2}\right) \tag{3.11}$$

If we wished to simplify the circuit and substitute a single equivalent resistance, R_{eq}, for the two parallel resistors, we would draw it as in Figure 3.9.

Note that we've discussed only resistors in illustrating Kirchoff's laws. However, the laws apply to any of the components that we will introduce in this chapter.

There are a number of additional theorems and laws, such as superposition, Thevenin's equivalent, Norton's equivalent, and many others, that can be extremely useful in circuit analysis and design. However, armed with just Ohm's law and Kirchoff's laws, plus the additional component behavior described in the next sections, the designer of the robotics projects described in this book should be prepared to tackle the design and analysis of passive component circuits.

Resistors, capacitors, and inductors

Resistors

We've already seen how resistors behave in circuits in the previous section. They follow the linear relationship of Ohm's law (Eq. 3.1). Before going on to the next topic, there's a resistor circuit called the resistor divider that is used quite often in electronics design. It is simply two resistors in series (Figure 3.10), in which our interest is the voltage across the bottom resistor.

Of course, we've already encountered a circuit similar to this in Figure 3.7. We can again use Ohm's law and Kirchoff's voltage law to write down the equations:

$$I = \frac{V_B}{R_1 + R_2} \tag{3.12}$$

$$V_{R2} = I \cdot R_2 = \frac{V_B}{R_1 + R_2} R_2 = V_B \frac{R_2}{R_1 + R_2} \tag{3.13}$$

This divider circuit can be used any time a smaller voltage needs to be created from a larger voltage.

DC and AC voltages

A constant voltage is called a DC voltage (DC stands for direct current, although it is almost always referred to as DC). A voltage which varies with time is referred to as an AC

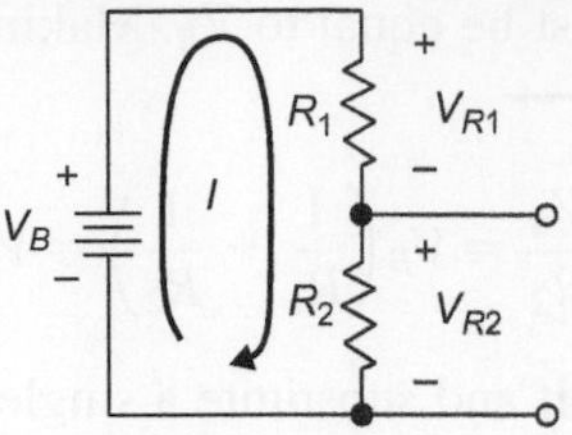

Figure 3.10
A voltage divider.

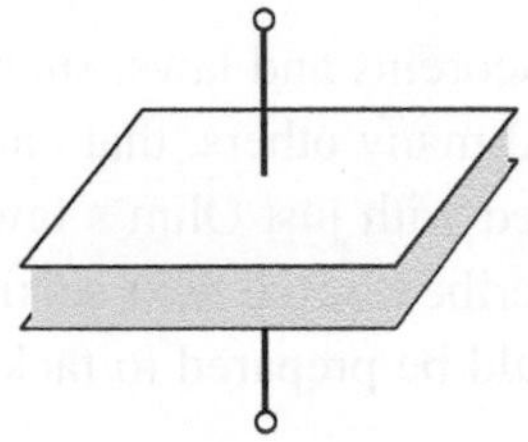

Figure 3.11
Parallel-plate capacitor.

voltage (AC standing for alternating current, although, again, almost everyone uses just the acronym AC). All of the voltage sources shown in the previous figures were DC voltages. This distinction between DC and AC voltages will become important when we discuss capacitors and inductors in the next sections.

Capacitors

A capacitor consists of two conductors separated by an insulator. The parallel-plate capacitor of Figure 3.11 is an example, although higher-valued capacitors often use more complex structures.

When a constant (DC) voltage is applied across this device, no current flows but the two conducting plates become charged. On the other hand, if the voltage changes, current flows through the device (Figure 3.12).

The capacitor of Figure 3.12 responds to current flowing into it by increasing its voltage (with respect to the negative terminal). Its approximate behavior is given by:

$$\frac{I}{C} \approx \frac{\Delta V}{\Delta T}. \tag{3.14}$$

Given this equation, it's easy to see one of the unique characteristics of the capacitor – a large change in voltage over a short period of time requires a huge amount of current. *To state it another way, a capacitor resists changes in voltage.*

To see how Eq. 3.14 works in a real circuit, let's combine a resistor and a capacitor, as shown in Figure 3.13. At first glance, this circuit might appear to have a constant voltage source, V_B, and therefore no current should be flowing through the capacitor. However, note that there is a switch in the circuit, which is closed at time $t = 0$. Closing that switch causes the voltage across the resistor/capacitor branch to jump from 0 V to V_B (assuming the initial capacitor voltage was zero). Even though V_B is constant, the voltage across the resistor and capacitor changes, and therefore current will flow through the capacitor. So, let's analyze the circuit.

To avoid the confusion of what happens at *exactly* $t = 0$, we'll talk about the current, I, at some moment just after 0. We'll call this time $t = 0^+$. We'll also, at times, refer to the circuit current and voltages as functions of time, for example, $V_C\ (t = 0^+)$, or just $V_C\ (0^+)$.

Continuing on, apply Kirchoff's voltage law clockwise around the circuit and use the convention that, if we encounter the negative terminal of the voltage first, it is treated as a negative voltage. The equation becomes:

$$-V_B + V_R + V_C = 0 \tag{3.15}$$

or

$$V_B = V_R + V_C. \tag{3.16}$$

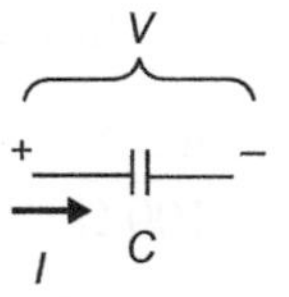

Figure 3.12
Charging a capacitor.

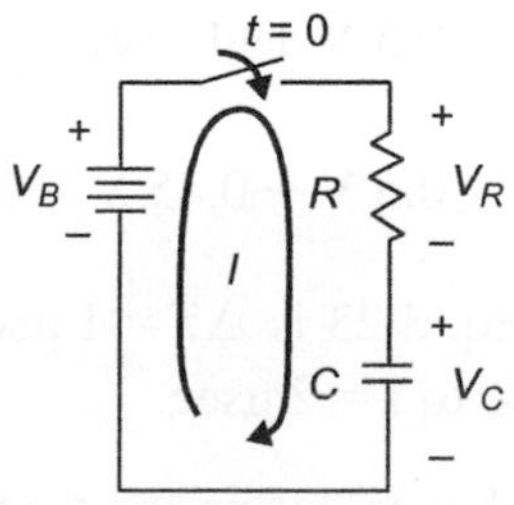

Figure 3.13
A simple RC filter.

The initial capacitor voltage, $V_C(0^+)$, is zero. Therefore, at time, $t = 0^+$, Eq. 3.16 becomes just:

$$V_B = V_R. \tag{3.17}$$

So we can immediately determine the current through the resistor using Ohm's law:

$$I(t = 0^+) = \frac{V_R}{R} = \frac{V_B}{R}. \tag{3.18}$$

Let's put some values to the circuit parameters. Make $V_B = 5$ V, $R = 100\ \Omega$, and $C = 0.1\ \mu$F. Then, just after closing the switch, $I = 5/100 = 50$ mA. Now we can determine the rise in capacitor voltage. Rearranging Eq. 3.14 gives:

$$\Delta V_C \approx \Delta T \frac{I}{C} = \Delta T \frac{V_R}{RC}. \tag{3.19}$$

So, after 1 μsec, the capacitor voltage is:

$$V_C(t = 1\ \mu\text{sec}) \approx (1\ \mu\text{sec}) \frac{5\ \text{V}}{(100\ \Omega)(0.1\ \mu\text{F})} = 0.5\ \text{V} \tag{3.20}$$

How about after 2 μsec? Well, the capacitor voltage at the beginning of the 1 μsec-to-2 μsec interval is now 0.5 V. Applying Kirchoff's voltage law around the circuit loop and rearranging the terms gives the new resistor voltage:

$$V_R = V_B - V_C = 5\ \text{V} - 0.5\ \text{V} = 4.5\ \text{V}. \tag{3.21}$$

The current through the resistor (and therefore through the capacitor) is given by:

$$I = \frac{V_R}{R} = \frac{4.5\ \text{V}}{100\ \Omega} = 45\ \text{mA}. \tag{3.22}$$

The rise in capacitor voltage, during the interval from $t = 1$ μsec to $t = 2$ μsec is added to the existing capacitor voltage at $t = 1$ μsec to find the voltage at $t = 2$ μsec:

$$\begin{aligned} V_C(t = 2\ \mu\text{sec}) &\approx V_C(t = 1\ \mu\text{sec}) + \Delta V_C \\ &= 0.5\ \text{V} + (1\ \mu\text{sec}) \frac{4.5\ \text{V}}{(100\ \Omega)(0.1\ \mu\text{F})}. \\ &= 0.5\ \text{V} + 0.45\ \text{V} = 0.95\ \text{V} \end{aligned} \tag{3.23}$$

Note that the time interval used in Eq. 3.23 is $\Delta T = 1$ μsec, since Eq. 3.23 determines the circuit behavior from time $t = 1$ μsec to $t = 2$ μsec.

So now we know how the circuit behaves during the first 2 microseconds after the switch of Figure 3.13 is closed. We could continue making these computations by hand but it

Microsoft Excel - numerical example of filter.xls

C3 =C2+E2*deltaT/Capacitor

	A	B	C	D	E	F	G	H	I
1	Time	time (usec)	Capacitor Voltage	Resistor Voltage	Current	Exact			
2	0.0000E+00	0	0.00	5.00	5.000E-02	0			
3	1.0000E-06	1	0.50	4.50	4.500E-02	0.475813			
4	2.0000E-06	2	0.95	4.05	4.050E-02	0.906346			
5	3.0000E-06	3	1.36	3.65	3.645E-02	1.295909		deltaT=	1.00E-06
6	4.0000E-06	4	1.72	3.28	3.281E-02	1.6484		Capacitor=	1.00E-07
7	5.0000E-06	5	2.05	2.95	2.952E-02	1.967347		Resistor=	1.00E+02
8	6.0000E-06	6	2.34	2.66	2.657E-02	2.255942		VoltageSource=	5
9	7.0000E-06	7	2.61	2.39	2.391E-02	2.517073			
10	8.0000E-06	8	2.85	2.15	2.152E-02	2.753355			
11	9.0000E-06	9	3.06	1.94	1.937E-02	2.967152			
12	1.0000E-05	10	3.26	1.74	1.743E-02	3.160603			
13	1.1000E-05	11	3.43	1.57	1.569E-02	3.335645			
14	1.2000E-05	12	3.59	1.41	1.412E-02	3.494029			

Figure 3.14

Excerpt from spreadsheet used to compute response of RC combination.

would really be tedious! Fortunately, there's an easy way to do this and that's to use Excel or Calc (Open Office) or some other spreadsheet program. Figure 3.14 shows an excerpt from a spreadsheet that computes the capacitor voltage iteratively. It steps through at 1 microsecond intervals (note that the constant, deltaT, is 1.00E-6) and, at each new step, uses the current found in the previous step to compute the new estimate of the capacitor voltage. Setting things up like this should allow you to enter an equation in just one row of spreadsheet cells, then copy and paste for as many iterations as you wish to examine.

The response of the capacitor voltage is plotted against the time, in microseconds, in the graph of Figure 3.15. Also plotted in that figure is the function:

$$V = 5\left(1 - e^{-\frac{t}{RC}}\right). \tag{3.24}$$

It is no coincidence that this function produces nearly the same plot as the numerical approximation. Equation 3.24 is, in fact, the exact solution for Eq. 3.19 as ΔT becomes very small. Deriving this relationship between capacitor voltage and time requires calculus, but making use of the function is straightforward, as all scientific calculators provide the exponential function.

Figure 3.15 demonstrates one of the reasons that capacitors are useful. The capacitor "remembers" past voltages and hangs on to those old values, rather than jumping immediately to the final value. As a result, it is often used to "average" or "smooth" signals that otherwise vary too much.

Before proceeding to the next section, it is important to mention how capacitors combine. Unlike resistors, two capacitors in *series* create an equivalent capacitance that is *smaller* than either of the two capacitors (Figure 3.16).

On the other hand, two capacitors in parallel have an equivalent capacitance that is the sum of the two capacitors (Figure 3.17).

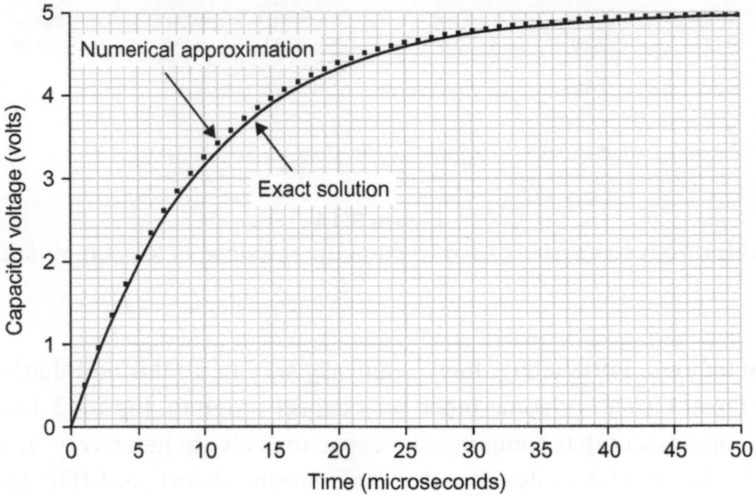

Figure 3.15
Capacitor voltage for the RC circuit of Figure 3.13.

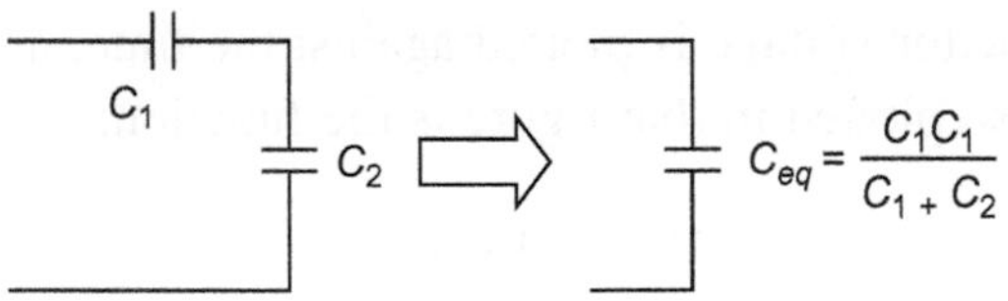

Figure 3.16
Capacitors in series.

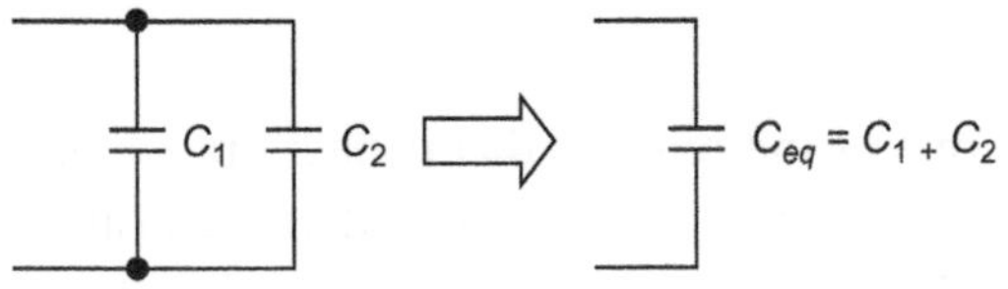

Figure 3.17
Capacitors in parallel add.

Inductors

Inductors are less commonly used than capacitors but are still important to understand. Their use is particularly important in the switching power supplies that are common today.

The inductor, shown in Figure 3.18, has the relationship of voltage to current given by:

$$\frac{V_L}{L} \approx \frac{\Delta I}{\Delta T}. \tag{3.25}$$

Note the similarity of Eqs. 3.14 and 3.25. By swapping V and I in Eq. 3.14 and by replacing C with L, we arrive at Eq. 3.14's *dual* in Eq. 3.25.

The inductor/resistor circuit of Figure 3.19 will produce an exponential response just like the capacitor/resistor circuit of Figure 3.13.

Let's analyze this circuit in a manner similar to the resistor/capacitor circuit. We'll use values for the components of $R = 100\ \Omega$, $L = 1$ mH, and $V_B = 5$ V. Assume no current flows through the inductor prior to the switch closing. In that case, there is no current flowing through the resistor as well, so the voltage across the resistor is just 0 V. Therefore the entire voltage, V_B, is across the inductor ($V_L(0^+) = V_B$).

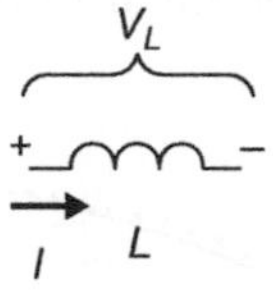

Figure 3.18
The inductor.

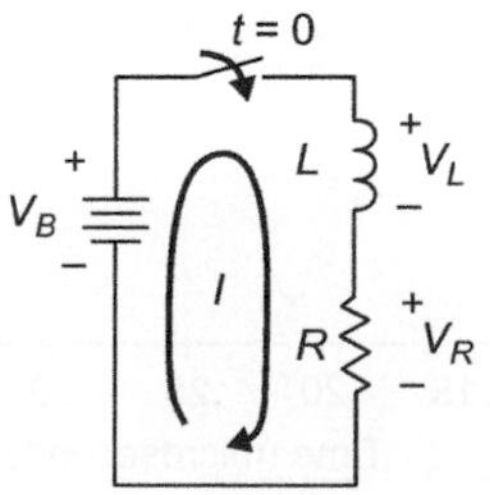

Figure 3.19
Inductor/resistor circuit.

What is the resistor's approximate voltage at 1 microsecond? To answer that, we'll need to calculate the current through the resistor. From Eq. 3.25, the current rises during this interval from 0 to:

$$\Delta I \approx \Delta T \frac{V_L}{L} = (1\ \mu\text{sec}) \frac{5}{1\ \text{mH}} = 5\ \text{mA}. \tag{3.26}$$

Since $I(t = 1\ \mu\text{sec}) \approx 5$ mA, $V_R(t = 1\ \mu\text{sec}) \approx I \cdot R = (5\ \text{mA})(100\ \Omega) = 0.5$ V and, by Kirchoff's voltage law, $V_L\ (t = 1\ \mu\text{sec}) \approx V_B - V_R = 4.5$ V.

The entire process can be iterated to find the change in current for the interval from 1 μsec to 2 μsec:

$$\Delta I(1\ \mu\text{sec} < t < 2\ \mu\text{sec}) \approx \Delta T \frac{V_L}{L} = (1\ \mu\text{sec}) \frac{4.5}{1\ \text{mH}} = 4.5\ \text{mA}. \tag{3.27}$$

Since this is just the *increase* in current, ΔI, the result at 1 μsec needs to be added to determine the current at $t = 2$ μsec, which is $I = 5\ \text{mA} + 4.5\ \text{mA} = 9.5$ mA. As with the resistor/capacitor combination, this process could be continued by hand for as many microseconds as we might be interested in computing, but the process would be very tedious and is in need of automation, something that spreadsheet programs are well-suited to doing. Once again, the calculations are entered into a spreadsheet and plotted (Figure 3.20).

Comparing Figures 3.15 and 3.20 shows that, for these particular values of resistor and inductor, the response of this circuit is identical to the resistor/capacitor circuit analyzed

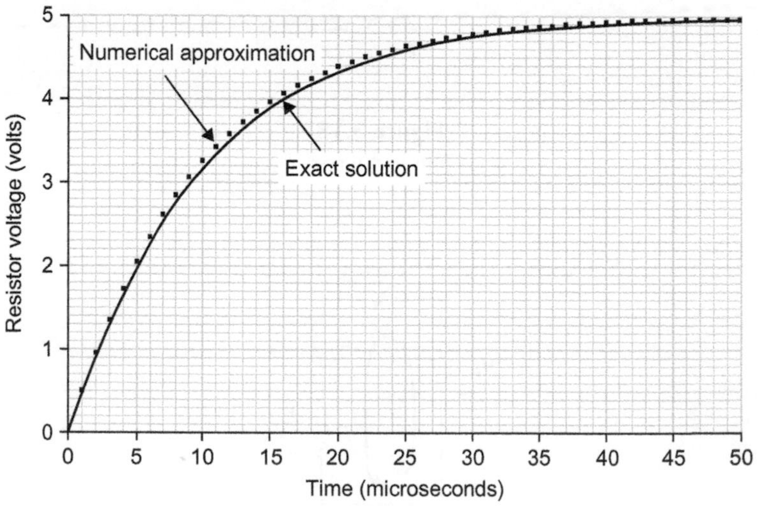

Figure 3.20
Resistor voltage for the example resistor/inductor circuit.

earlier. As with the resistor/capacitor circuit, this circuit has an exact response that can be derived using calculus and is given by:

$$V = 5\left(1 - e^{-\frac{tR}{L}}\right). \tag{3.28}$$

In both the resistor/inductor circuit and the resistor/inductor circuit, the steady-state value is 5 V.

Inductors in series and parallel

Series and parallel combinations of inductors behave in the same way as series and parallel combinations of resistors. That is, series inductors add and parallel inductors are mathematically combined in the same manner as was shown for resistors in Figure 3.9.

Electric and magnetic fields

The subject of electromagnetic fields is an important one in electrical engineering, but it's one that we can almost get by without talking about in this book. However, when we discuss motors in the next chapter, some basic understanding of magnetic fields will be needed.

There is, of course, an electric field as well as a magnetic field. For example, a battery connected to the parallel-plate capacitor of Figure 3.11 creates charge on each of the plates, which, in turn, induces an electric field, *E*. Figure 3.21 shows such a situation.

The electric field can easily be calculated for this. However, an understanding of electric fields beyond what you've encountered in high-school physics isn't required to build the robot project of this book, so we'll leave the topic of electric fields and venture on to magnetic fields.

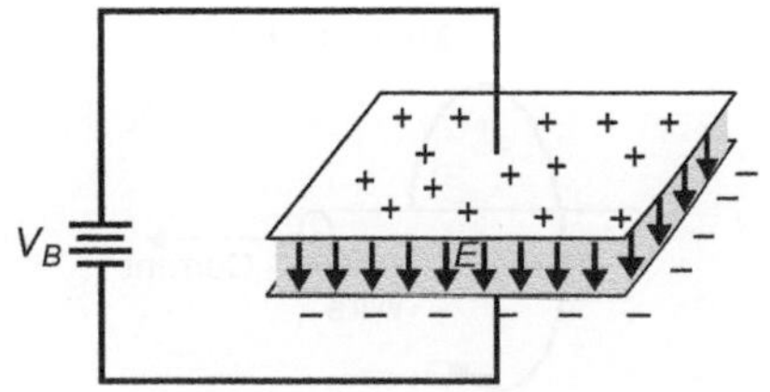

Figure 3.21
Parallel-plate capacitor charged by battery.

Magnetic fields

We have all had experience with magnetic fields as they arise from permanent magnets. When permanent magnets' like poles are placed near one another they repel. Permanent magnets with opposite poles facing one another attract. Without really thinking much about it, we are witnessing the interaction of the magnets' magnetic fields.

However, magnetic fields can also be produced from *electromagnets*. In general, a current flowing through a conductor produces a magnetic field. An important association to make is:

$$\text{Current} \rightarrow \text{Magnetic Field}$$

$$\text{Voltage} \rightarrow \text{Electric Field}$$

A current flowing through a wire creates a magnetic field oriented as shown in Figure 3.22. The magnetic field rotates perpendicular to the wire. By coiling the wire, the magnetic field lines align and reinforce one another, creating the situation depicted in Figure 3.23. Note that the magnetic field now looks much like the magnetic field depicted for permanent bar magnets.

The wire loop in Figure 3.23 is actually just an inductor. In most instances the wire will be wound around some type of ferromagnetic material, such as iron but there are also inductors made that have no material in the center. These inductors are referred to as "air-core inductors".

Storing energy in capacitors and inductors

The electric field built up in a capacitor and the magnetic field built up with an inductor can be very useful. Consider the capacitor of Figure 3.21. If we remove the battery connection to that capacitor, it continues to stay charged and maintains the voltage across its plates until such a time as the capacitor is connected to a circuit that draws current from

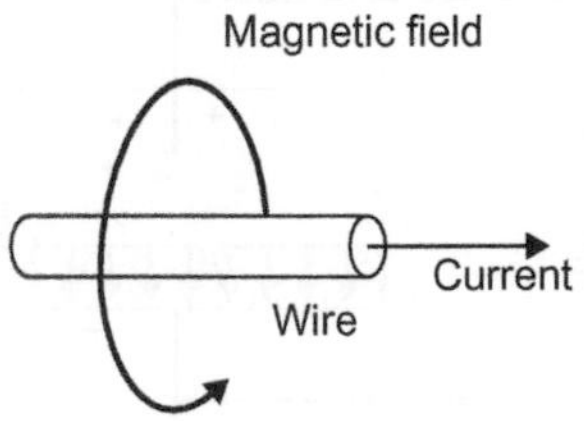

Figure 3.22
Magnetic field created by current flowing through wire.

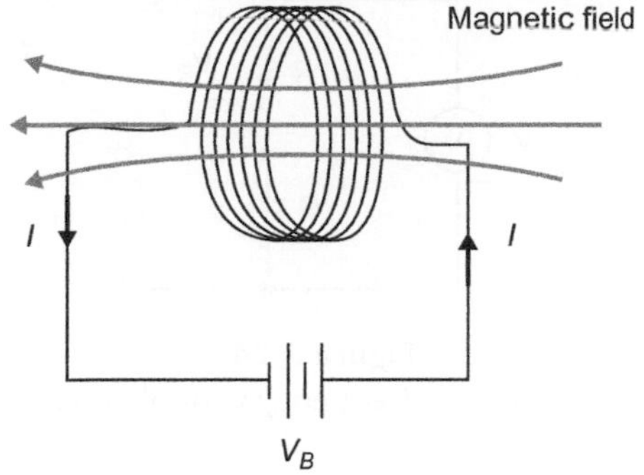

Figure 3.23
Coiling wire increases magnetic field.

the capacitor, at which point its voltage will begin to drop in accordance with Eq. 3.14. Large-value capacitors are sometimes used in this way as a replacement for batteries.

We will also see in a later section of this chapter that the capacitor's ability to store charge and its corresponding ability to resist rapid changes in voltage makes it very useful in holding up voltages in circuits that may experience fluctuations in supply voltage.

Making analog filters out of resistors, capacitors and inductors

An analog filter is a circuit that operates on AC signals. Remember from the beginning of this chapter that an AC signal is one which is changing. As we've seen earlier, capacitors impede changes in voltage and inductors impede changes in current. Equations 3.14 and 3.25 give us a clue that the amount of this impedance is time-dependent which, in turn, tells us something about the device's response to changing signal frequency.

Resistor/capacitor filters

Frequency is the number of cycles of change per second. It is given the unit hertz (abbreviated Hz), where 1 Hz = 1 cycle per second. The higher the frequency of a signal, the faster it changes with time. Reviewing Eq. 3.14, this means that the current through the capacitor increases with increasing frequency. Another way of saying that is that the capacitor's impedance to current decreases with increasing frequency. In fact, that's exactly what happens, and we can talk about a capacitor's impedance as follows:

$$X_C = \frac{1}{2\pi * f * C}. \tag{3.29}$$

The impedance identified in Eq. 3.29 has units of ohms, the same as resistance. What this is saying is that, for a circuit like that of Figure 3.24, we can think of the capacitor as behaving much like a resistor.

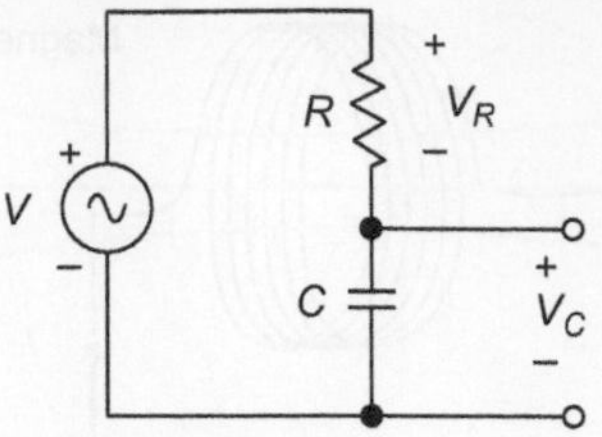

Figure 3.24
The capacitor as a frequency-variable impedance.

(The reason we can't say that the capacitor, at some given frequency, behaves exactly like a resistor is that the capacitor produces a phase shift in its voltage. Explaining exactly what is meant by a phase shift is beyond the scope of this book and isn't really necessary for what we'll be doing with robot projects but it will be important if you pursue electronics in more depth.)

Okay, so back to our impedance divider resistor/capacitor circuit, a.k.a. the filter.

For values of $C = 0.1\ \mu$F, and frequency, $f = 5000$ Hz, the capacitor has, according to Eq. 3.29, an impedance of 318 Ω. If the resistor, R, is 100 Ω then, using the same approach as was used to create the resistor divider equation (Eq. 3.13), the voltage across the capacitor, V_C, is:

$$V_C = \frac{X_C}{X_C + R} V_S = \frac{318}{318 + 100} V_S = 0.76 V_S. \tag{3.30}$$

Therefore, for an alternating voltage, V_S, of 1 V zero-to-peak (see Figure 3.25) and with frequency 5000 Hz, the capacitor voltage is 0.76 V zero-to-peak.

What if the frequency of V_S was 50,000 Hz? In that case, $X_C = 31.8\Omega$ and the capacitor voltage is only

$$V_C = \frac{X_C}{X_C + R} V_S = \frac{31.8}{31.8 + 100} V_S = 0.24 V_S. \tag{3.31}$$

As expected, the capacitor impedance has gone down with increased frequency (it impedes less at this frequency). When we perform the X_C calculations and the impedance divider calculations at each of several frequencies between 1000 Hz and 100,000 Hz, the graph of Figure 3.26 is obtained.

Actually, a more useful way of plotting this, and the way it is normally presented in textbooks, is to plot both axes logarithmically to obtain the plot of Figure 3.27.

Figure 3.27 makes clear what it is that is useful about this simple circuit and why it goes by the name of *lowpass filter*. It gets that name from the fact that it passes

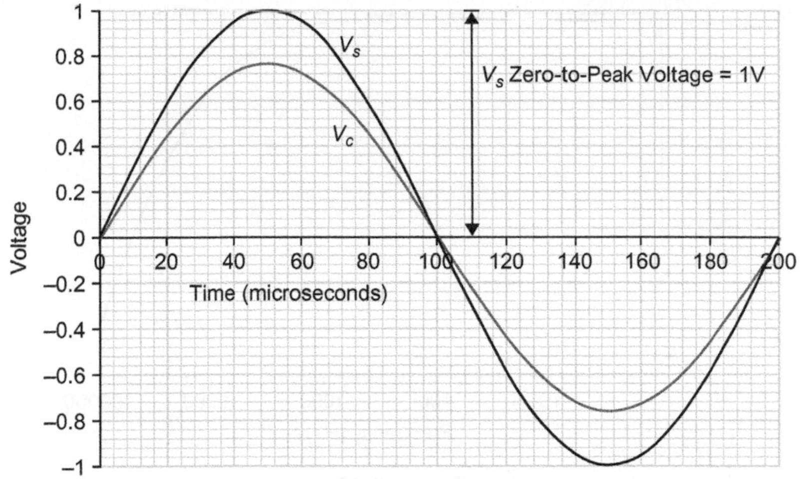

Figure 3.25
Capacitor voltage at 5000 Hz.

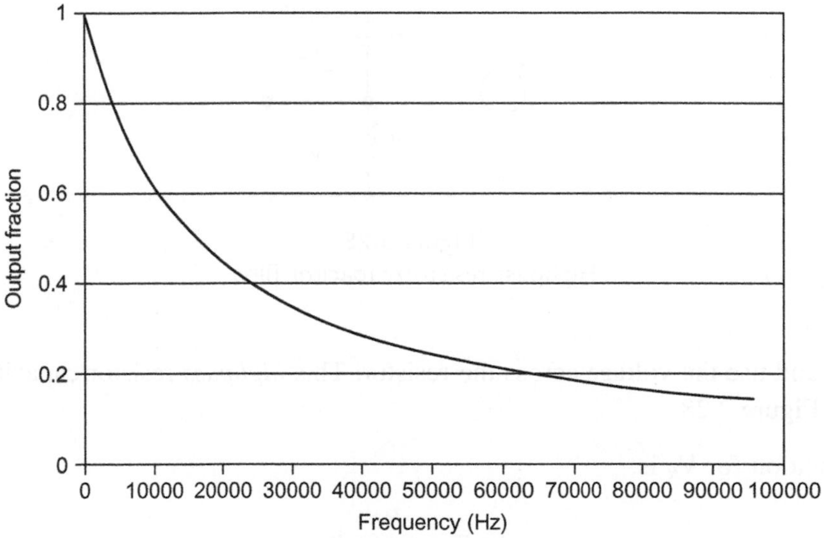

Figure 3.26
Plot of capacitor voltage as a fraction of the supply voltage.

low-frequency signals with little attenuation but, at higher frequencies, significantly attenuates the signal.

What if we wanted to do the opposite? What if we wanted to just pass high-frequency signals? That's easy. We just reverse the position of the resistor and capacitor in

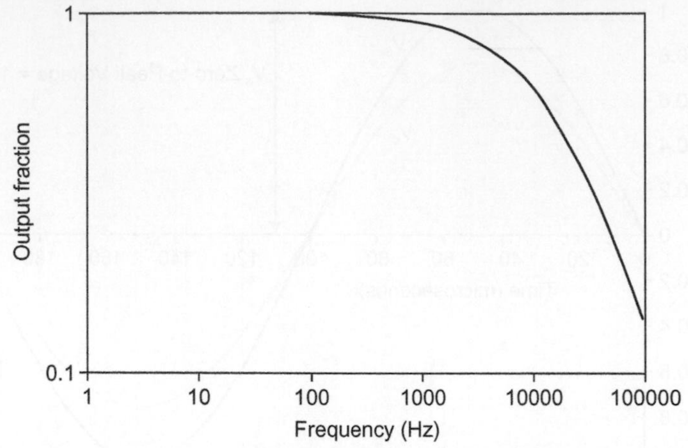

Figure 3.27
Logarithmic/logarithmic plot of the resistor/capacitor filter circuit.

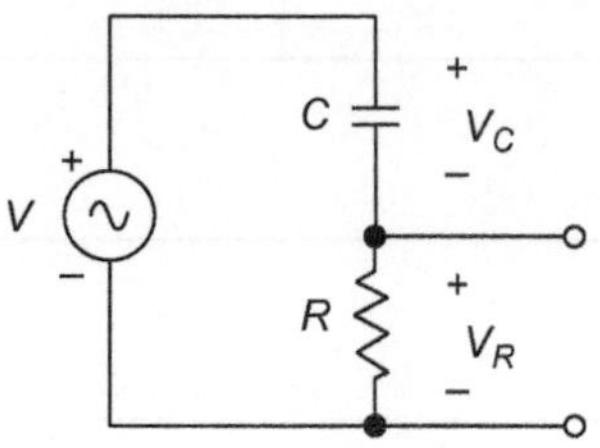

Figure 3.28
Highpass resistor/capacitor filter.

Figure 3.24 and use the voltage across the resistor. This *highpass* resistor/capacitor filter is shown in Figure 3.28.

Here the equation for V_R is:

$$V_R = \frac{R}{X_C + R} V_S. \tag{3.32}$$

Computing the value of X_C at each frequency, using the previous component values of $C = 0.1\ \mu\text{F}$ and $R = 100\ \Omega$, and plotting the response, V_R, obtained from this equation results in the graph of Figure 3.29.

Resistor/inductor filters

Like the capacitor, the inductor's impedance changes with frequency. From Eq. 3.25 we can see that, for a given voltage and inductance, decreasing ΔT produces decreasing current,

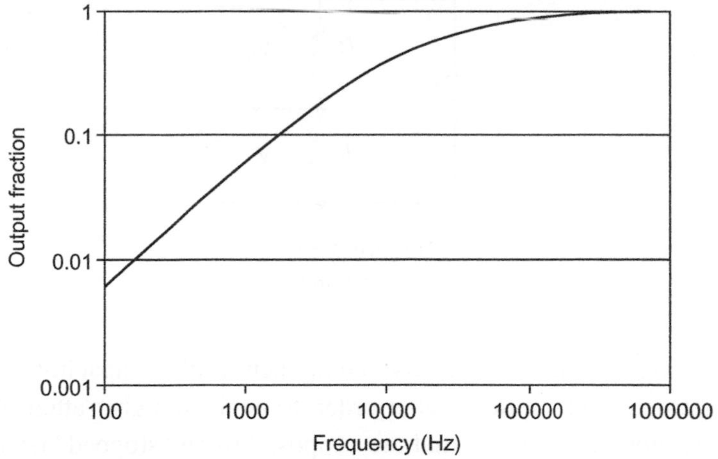

Figure 3.29
Highpass response of the circuit of Figure 3.28.

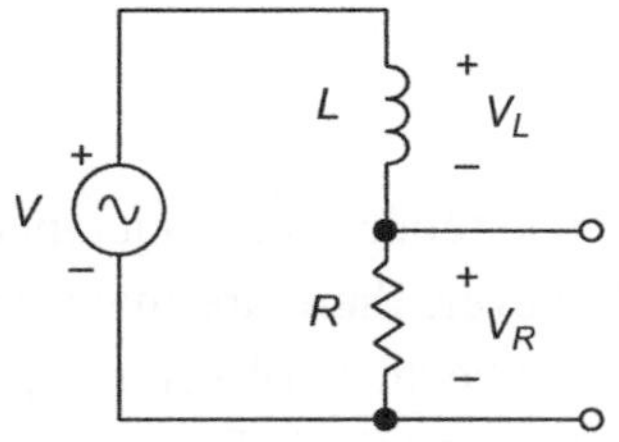

Figure 3.30
Resistor/inductor lowpass filter.

which implies that its impedance *increases* with increasing frequency. Indeed, this is what happens. The equation for inductor impedance is:

$$X_L = 2\pi * f * L. \tag{3.33}$$

Just as with resistors and capacitors, lowpass and highpass filters can be constructed with resistors and inductors. The only difference between the two sets of filters is that the position of the resistor and inductor is reversed (Figures 3.30 and 3.31), with the resistor being in parallel with the output in the lowpass filter and the inductor being parallel with the output in the highpass filter.

A final word about filters

The topic of filters is actually quite vast and we've only scratched the surface with the filters presented here. Many universities offer semester-long courses on just this topic. The filters that we talked about in this chapter are known as first-order filters because they

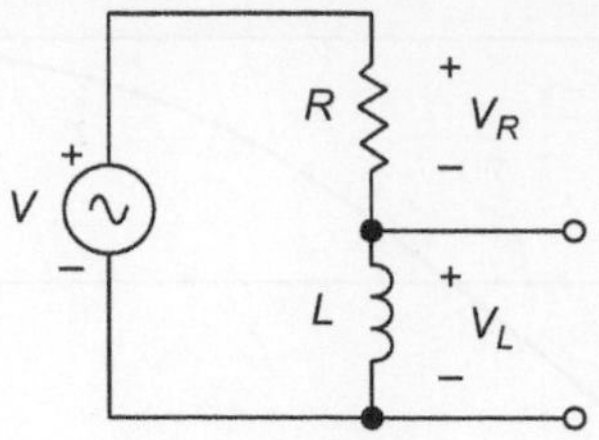

Figure 3.31
Resistor/inductor highpass filter.

incorporate only a single frequency-sensitive component (either capacitor or inductor). Filters can be constructed with much higher order, to achieve higher attenuation in the "stopband" (that is, those frequencies that are supposed to be "stopped" or attenuated) or to achieve particular "passband" shapes ("passband" being those frequencies that the filter is supposed to "pass"). The filters presented in this chapter, although simple, first-order versions, will suffice for the projects encountered in this book.

Stray capacitance and inductance

So far we've been talking about capacitance and inductance as things that we add to circuits to produce some desired result. However, there are many instances where the capacitance or inductance is unwanted and occurs because of some aspect of circuit layout. Such "parasitic capacitance" and "parasitic inductance" can be minimized or their effects can be minimized, but only if the circuit designer is aware of their presence and is concerned about their effects on circuit behavior.

How do we wind up with stray inductance and stray capacitance on a board? Let's look at inductance first. Remember that it was mentioned that the coil of Figure 3.23 is an inductor. It creates a magnetic field, as that diagram depicts. But even the wire of Figure 3.22 creates a magnetic field and is an inductor, albeit one with a very small value of inductance. *Any conductor has some inductance.* Because it's a very small amount of inductance, we can usually ignore it.

But what happens when a large amount of current starts rushing through this inductor in a small amount of time? Equation 3.25 tells us that, in that case, even with L small, the effect on V_L may be significant.

Here's an example of how this stray inductance can cause unwanted (and possibly unexpected) circuit behavior. The circuit of Figure 3.32 has two resistors of equal value, 10 Ω. One of these is always connected to the battery, V_B; the other is connected through the switch. The wire from the + side of the battery to the rest of the circuit is several inches long and has an inductance of 200 nH.

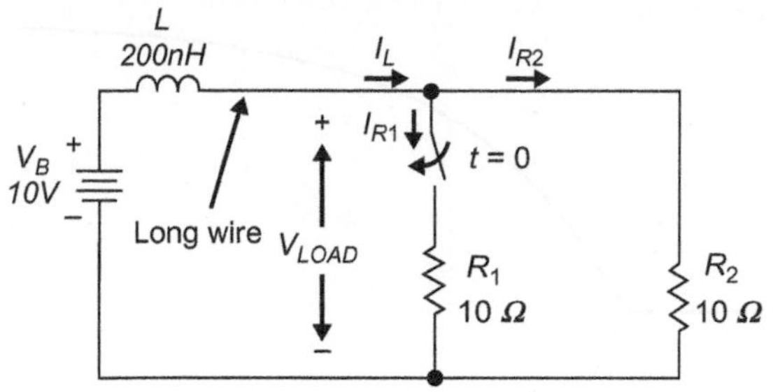

Figure 3.32
Circuit demonstrating effect of parasitic inductance.

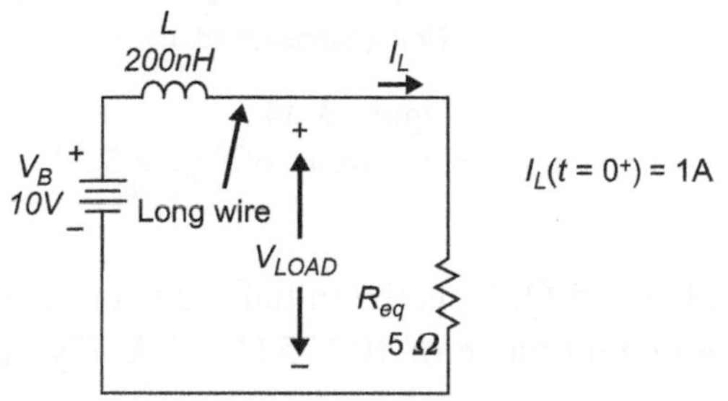

Figure 3.33
Circuit after switch closure.

Assume that the circuit has been connected for a long time prior to $t = 0$. In that case, and with the switch open that entire time, we know from the earlier discussion of the resistor/inductor circuit that the steady-state voltage across R_2 is the battery voltage (that is, $V_{LOAD} = 10$ V) and that the inductor current is 10 V/10 Ω = 1 A.

At $t = 0$, the switch closes. Now, remembering that the inductor current won't change instantaneously, we know that the inductor current at $t = 0^+$ is still $I_L = 1$ A. However, at this point, there are two 10 Ω resistors in parallel. They divide the current I_L evenly (since they are of equal resistance). Viewed a little differently, we could simply say that after $t = 0$, we have the equivalent of a single resistor, 5 Ω, with an initial inductor current of 1 A. Figure 3.33 is the equivalent circuit for this situation.

This circuit is a little more involved to analyze than the previous resistor/inductor case, due to the non-zero initial condition on the inductor current. The solution for the current is:

$$i_L(t) = [i_{Li} - i_{Lf}]e^{-tR/L} + i_{Lf}. \tag{3.34}$$

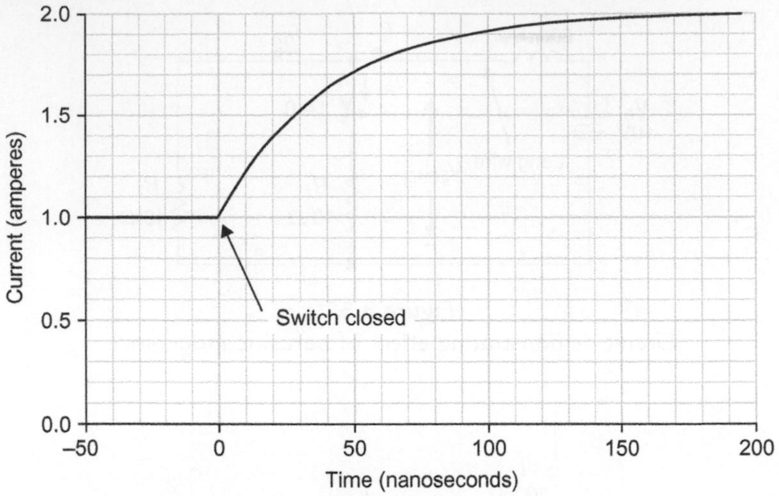

Figure 3.34
Supply current to circuit of Figure 3.33.

R in this case is the R_{eq} value of 5 Ω, i_{Li} is the initial inductor current value of 1 A, and i_{Lf} is the final inductor current value of 10 V/5 Ω = 2 A. Figure 3.34 shows the current for this circuit.

Given this current waveform, we can calculate the output voltage, V_{LOAD}, during this interval. We know that prior to $t = 0$, only resistor R_2 is connected, so the voltage up to this time is just the product of 1 A and 10 Ω, or $V_{LOAD} = 10$ V. Following switch closure at $t = 0$, the load changes to 5 Ω and the current follows Eq. 3.34, so the voltage V_{LOAD} is given by:

$$V_{LOAD} = \begin{cases} 10, & t < 0 \\ 10 - 5e^{-25 \times 10^{-6} t}, & t \geq 0 \end{cases} \tag{3.35}$$

The output is plotted in Figure 3.35.

Will this 50-nanosecond "glitch" in the voltage affect your circuit? Maybe, maybe not. A low-frequency circuit, for example, an audio amplifier, might not respond to such a change at all, but as you build faster, more sophisticated circuits, such anomalies can create unexpected results.

So what's the remedy for this problem of parasitic inductance? One thing you can do is to use short leads wherever possible. Another solution, and one that is a characteristic of almost every good circuit design, is to use "bypass" capacitors. These are simply capacitors, often with values of the order of 0.01 μF to 0.1 μF, connected across the + and − power supply lines as a way to "hold up" the supply voltage when glitches like this occur.

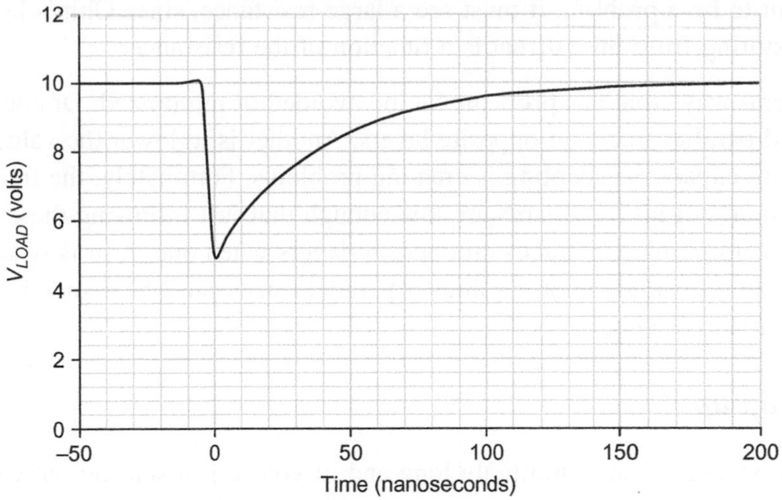

Figure 3.35
Voltage of Figure 3.33 circuit.

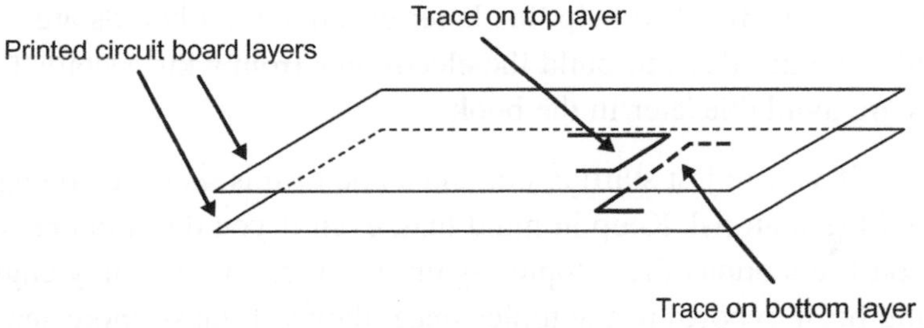

Figure 3.36
Parallel traces on top and bottom layers of a two-sided printed circuit board.

The capacitors are most effective when placed near the IC or other component whose voltage needs to be maintained.

Okay, so that's parasitic inductance. What about parasitic capacitance? One common source of parasitic capacitance on printed circuit boards is the result of two circuit board traces being routed one above the other on different layers of the board. An example is shown in Figure 3.36.

The result of these parallel traces is that they form a long, skinny parallel-plate capacitor. Equation 3.14 shows that such a capacitance will pose the most problem when there is a large-amplitude voltage changing very rapidly on one trace. Of course, that's exactly what a digital signal does. It switches several volts in just a few nanoseconds. In that case, even a small capacitance induces a significant current in the opposite trace. For this capacitively coupled current to be a problem, it must see a large resistance, since Ohm's law tells us that the voltage resulting from this current is a function of the resistance.

There are several ways that this problem can be avoided or minimized. Of course, one way is to avoid parallel traces on opposite layers. Another is to lower the values of resistors in circuits where capacitive coupling is creating problems. Fortunately, the frequencies of signals on the robot project boards are low enough that just following these two guidelines – avoiding parallel traces on opposite layers when one trace is susceptible to the other and using lowered resistor values if parallel traces can't be avoided – will be sufficient to guarantee success.

Chapter wrap-up

This and the next chapter are admittedly long and, if you're just starting out with electronics, may be daunting. Don't worry too much about that. We've set this book up so that individuals with some background can use the examples and information in the book as just a starting point, from which they can do their own experimentation with design. But we've also tried to keep in mind that this book and these projects should be for those just beginning to tackle electronics. To help this latter group, circuit boards are available at a nominal fee to allow individuals to build the electronics from a kit. Details for where to obtain these kits are available later in the book.

Nevertheless, even if you're just starting out and some of this seems challenging, do your best to understand the material. Keep in mind that, even if you don't understand everything you read here, you'll encounter these topics again if you choose to study engineering in the future and, having been exposed to the topics once, they will make more sense when you encounter them in the future.

Bibliography

[1] D. Ashby, Electrical Engineering 101, second ed., Newnes, 2009.
[2] W. Hayt Jr., J. Kemmerly, Engineering Circuit Analysis, second ed., McGraw-Hill, 1971.
[3] J. Mills, Electro-Magnetic Interference, Prentice Hall, 1993.
[4] J. O'Malley, Basic Circuit Analysis (Schaum's Outlines), second ed., McGraw-Hill, 1992.
[5] F. Ulaby, Fundamentals of Applied Electromagnetics, Prentice Hall, 2001.

CHAPTER 4

Basic Electronics — Semiconductors

Chapter Outline

So far we've talked only about resistors, capacitors, and inductors as elements of electronic circuits. But as the reader is no doubt aware, the device that has made possible the electronics revolution of the last 50 years is the transistor. In this chapter we'll talk exclusively about the transistor and its cousin, the diode. As in the previous chapter, we'll

MSP430-based Robot Applications.
DOI: http://dx.doi.org/10.1016/B978-0-12-397012-1.00004-7

only cover those topics needed to build the robot projects described in this book. Even with that limited scope, there'll be plenty to cover.

Specifically, we'll cover:

1. Diodes
2. Bipolar transistors
3. Metal oxide semiconductor field-effect transistors, which are almost always referred to by their acronym, MOSFET (pronounced *Moss-fet*)
4. Operational amplifiers, referred to as op amps
5. Power drivers

Operational amplifiers aren't single transistors but, rather, collections of transistors (and sometimes diodes) that create a very high-gain voltage translator. A collection of transistors like this is referred to as an "integrated circuit", often called an "IC" (pronounced "*Eye-see*"). Such an op amp IC can have either bipolar transistors or MOSFETs as its basic elements.

"Power drivers" is a somewhat nebulous term that is included here because there are several functions on a robot that need such a device. The drive motor, the steering motor, and certain actuators all need circuits that can supply large amounts of both current and voltage simultaneously.

The above list is missing several important items. If we're going to include op amp and power driver ICs, then we should also include logic ICs, microcontrollers, and data acquisition ICs. It isn't that we've forgotten them, it's just that each of these is such a large topic that they receive one or more chapters devoted just to them. For now, just looking at the items in the list will take some time.

P–n junctions

A p–n junction is formed in a semiconductor material, such as silicon, by adding impurities in one end of the material to create a region of negative charge carriers (the n-type material) and a region of positive charge carriers (the p-region). A semiconductor is a material with electrical resistivity intermediate in magnitude between that of a conductor and an insulator. The boundary between these two oppositely charged regions is called a p–n junction. The semiconductor allows current to flow from the p-type side (called the anode) to the n-type side (called the cathode), but not in the opposite direction. All semiconductor transistors and diodes are made from one or more p-type and one or more n-type regions.

Admittedly, this is a really, really brief description of what a p–n junction is (when I was an undergraduate, I had a semester-long course on the subject). But as with other topics in this book, we are looking for the depth needed to successfully build a robot, with the idea that, should you desire to study these topics in the future, you'll pick up the detail then.

The diode

A diode is a quite useful and common p–n junction device in electronics. Unlike a resistor, it is nonlinear – its resistance changes dramatically according to the voltage across its terminal. In fact, it is sometimes modeled as a device with zero resistance when current flows from anode to cathode, and infinite resistance when current flows, or attempts to flow, in the opposite direction. That's actually very close to the way a real diode operates.

Figure 4.2 shows the actual current-voltage characteristics of a 1N4148 diode shown in Figure 4.1. At any voltage below about 0.5 V, the current is virtually zero. Above 0.5 V, the current rises rapidly. For most applications, the diode can be modeled as a device that looks like an open for negative current and looks like a 0.7 V battery for positive current.

The 0.7 V (or thereabouts) voltage is referred to as the forward voltage of the diode.

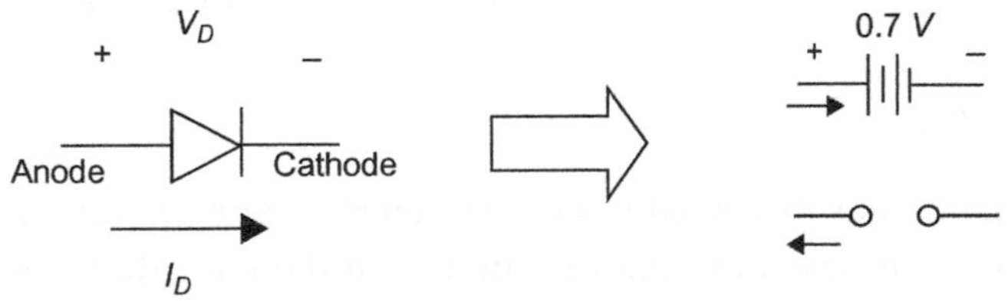

Figure 4.1
The diode and the model for forward and reverse current.

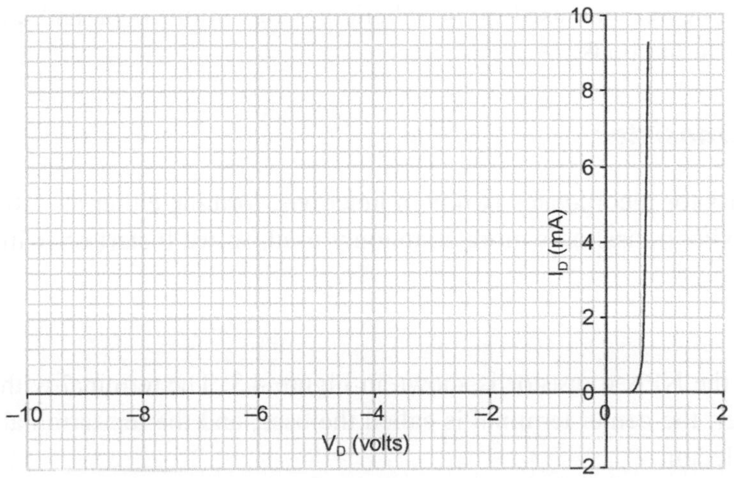

Figure 4.2
Voltage-current characteristics of a 1N4148 diode.

The diode's reverse breakdown voltage

The diode tested for the Figure 4.2 voltage-current graph will block current when a negative voltage is applied. However, there is a limit to how large we can make that negative voltage and this voltage is referred to as the reverse breakdown voltage. For the 1N4148 diode, that limit is around 100 V. Beyond −100 V (referenced as in Figure 4.1) current will flow in the reverse direction. Since the applications for diodes in this book are for battery-operated robots with voltages well below this limit, there should be no instances where reverse breakdown comes in to play. We can use the straightforward model of Figure 4.1 without qualification – positive current from anode to cathode creates a 0.7 V forward voltage across the diode and has virtually zero resistance, while current flowing from cathode to anode is blocked.

There is one type of diode in which the diode is manufactured in such a way that it has a fairly precise reverse voltage. This diode is specifically intended to be operated in this reverse breakdown mode. Such a diode is called a Zener diode, and we will have occasion to use such a device in a later chapter when we develop a high voltage supply.

Special diodes

There are two special diodes that need to be mentioned. These are the light-emitting diode (LED), which converts electrical current into light, and the photodiode, which converts light into electrical current. LEDs are not usually made from silicon (they are generally made from gallium arsenide, abbreviated GaAs) and their forward voltage is generally higher than the 0.7 V for a conventional silicon diode.

Photodiodes are typically operated with zero bias or with reverse bias. Both of these types of *optoelectronic* devices are used in the robotic projects in this book and are discussed in more detail later in the book.

The bipolar transistor

The bipolar transistor incorporates two p–n junctions, forming either an "NPN" or a "PNP" transistor. The NPN transistor is more commonly used, so we'll start off with that one.

NPN transistor

The symbol for the NPN transistor is shown in Figure 4.3. The terminal with the arrow is called the emitter, the center terminal is called the base, and the top terminal in the diagram is called the collector.

Like the diode, current generally flows in only one direction at each terminal – in the case of the NPN transistor, current flows *into* the base and the collector and *out of* the emitter.

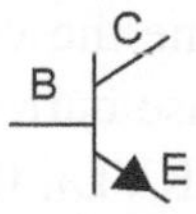

Figure 4.3
The NPN transistor.

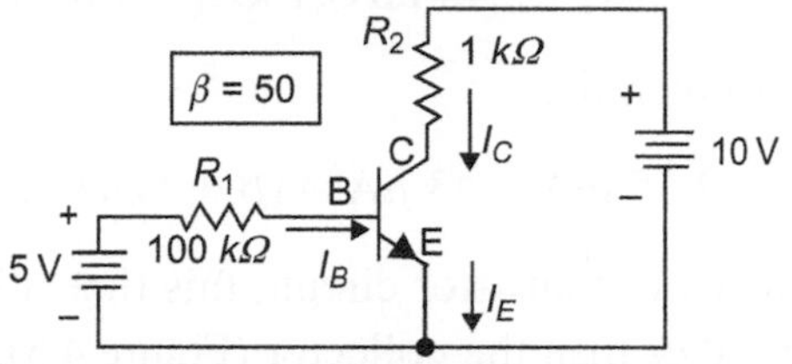

Figure 4.4
A single NPN transistor circuit.

And in a manner similar to the diode, the base-to-emitter voltage is at approximately 0.7 V when the transistor is conducting current.

So what is a transistor good for and how do we use it? To answer that, we need to know a little more about the transistor. First, when the transistor is operating in its "linear" region, its collector current is linearly related to its base current. This relationship is written as:

$$I_C = \beta \times I_B. \tag{4.1}$$

So, if we control the base current, we control the collector current. β is called the current gain of the transistor and most NPN transistors have β values in the range 30 to 100.

The emitter current is simply the sum of the base and collector currents:

$$I_E = I_C + I_B = (\beta + 1)I_B. \tag{4.2}$$

Since $\beta >> 1$, a valid approximation in most applications is:

$$I_E \approx I_C. \tag{4.3}$$

Armed with only this much information about the NPN transistor, we can analyze the circuit of Figure 4.4.

Before analyzing this circuit, let's write down the things that we know so far about NPN transistors that operate in their linear region:

1. The base-to-emitter voltage (V_{BE}) is approximately 0.7 V
2. $I_E = I_C + I_B = (\beta + 1)I_B \approx I_C$
3. $I_C = \beta \cdot I_B$
4. The collector-to-emitter voltage is positive

With only these key facts, we can determine the voltage at and the current into (or out of) each of the terminals. We start with the base current, I_B. Since V_{BE} is approximately 0.7 V, then, by applying Kirchoff's voltage law (KVL), the current into the base is [(5 V − 0.7 V)/100 kΩ] = 43 μA. Since $I_C = \beta \cdot I_B$, the collector current is $I_C = 50 \cdot 43$ μA = 2.15 mA. By KVL, the collector voltage, with respect to the emitter, is given by:

$$V_{CE} = 10 - (2.15\ \text{mA})(1\ \text{k}\Omega) = 7.85\ \text{V} \tag{4.4}$$

Finally, the emitter current is given by:

$$I_E = I_C + I_B = 2.15\ \text{mA} + 43\ \mu\text{A} = (\beta + 1)\ 43\ \mu\text{A} = 2.2\ \text{mA}. \tag{4.5}$$

Before going on, let's try one more transistor circuit, this time a little trickier. This one has R2 connected to the emitter rather than the collector (Figure 4.5).

Here we'll use KVL on the left hand loop to give:

$$5 = I_B R_1 + V_{BE} + I_E R_2 = I_B(100\ \text{k}\Omega) + 0.7 + (\beta + 1) I_B (1\ \text{k}\Omega) \tag{4.6}$$

The only unknown in this equation is I_B, which can easily be determined to be:

$$I_B = \frac{4.3\ \text{V}}{151\ \text{k}\Omega} = 28.5\ \mu\text{A} \tag{4.7}$$

The resulting currents and voltages are given in Table 4.1.

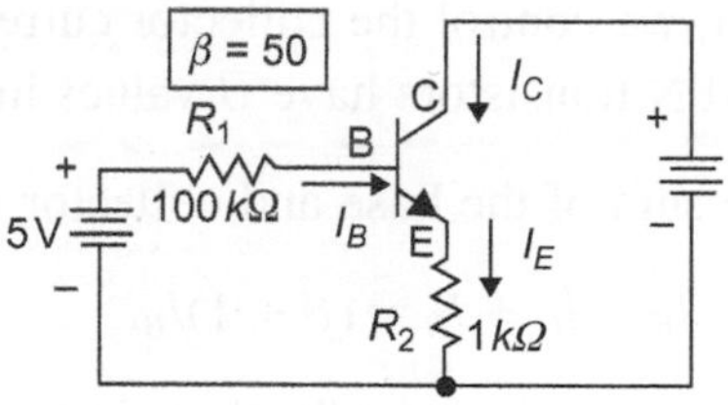

Figure 4.5
Single transistor with emitter load resistor.

Table 4.1: Transistor currents and voltages for the example of Figure 4.5.

Transistor Parameter	Value
Base current, I_B	28.5 μA
Collector current, I_C	1.43 mA
Emitter current, I_E	1.45 mA
Base-to-emitter voltage, V_{BE}	0.7 V
Emitter voltage, V_E (with respect to lower rail)	1.45 V
Collector voltage, V_C (with respect to lower rail)	10 V

A side note — ground conventions in electronics schematics

Before continuing on, we'll talk about a convention that is often used in electronic design regarding a "ground". In the previous examples, mention was made of measuring the voltage with respect to the "lower rail", meaning the bottom line in the circuit drawing. In everyday electronics terminology, this lower rail is often referred to as "ground". Quite often this point is actually connected to earth ground, through a wall outlet grounding plug or some other means of connecting to earth ground. But in battery-operated devices there is usually no connection of any circuit point to true ground. Nevertheless, electronics schematics are usually drawn with this "ground" point depicted using the triangle symbol of Figure 4.6. It is then understood that, if no reference is made to the second of two points in a voltage measurement, the second point is this "ground". For example, in Table 4.1, using this convention we could have simply said "Emitter voltage, V_E" and left off the parenthetical reference to the lower rail. We will follow this convention throughout the remainder of the book.

NPN transistor saturation

Let's return to the circuit of Figure 4.4, but this time, let's change the value of R_2 to 10 kΩ. The circuit is shown in Figure 4.7.

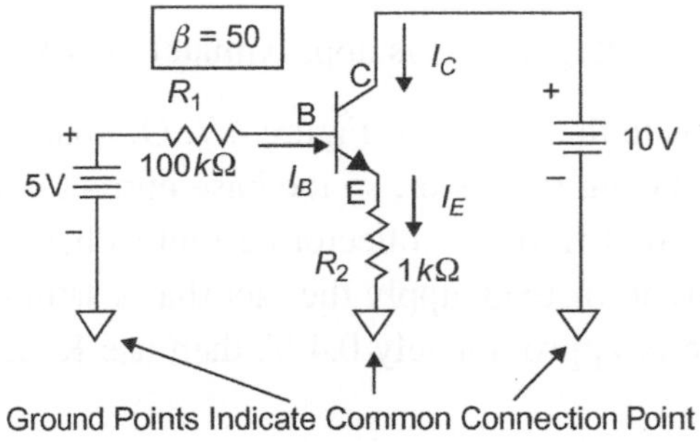

Figure 4.6
The circuit of Figure 4.5 drawn using ground points for the lower rail.

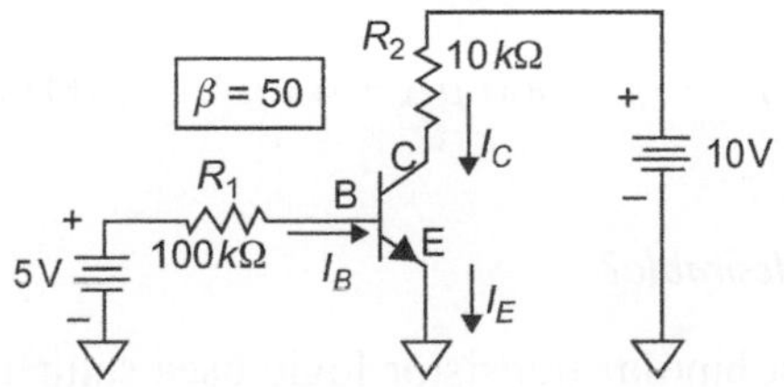

Figure 4.7
The circuit of Figure 4.4 with R_2 at 10 kΩ.

Let's analyze this circuit exactly as before. The base current, I_B, is 43 μA which should then give a collector current of 2.15 mA, right? Well, using that value for current and applying Ohm's law and KVL, we find that:

$$V_{CE} = 10 - I_C R_2 = 10 - (2.15\text{ mA})(10\text{ k}\Omega) = -11.5\ V(?). \tag{4.8}$$

Okay, hold everything! Our list of conditions for an NPN linearly operating transistor stated that the collector-to-emitter voltage should always be positive, so how did we manage to get a negative voltage? The answer is: we didn't. *Equation* 4.8 *is bogus.* We employed the relationship $I_C = \beta \cdot I_B$, which holds for a *linearly operating* transistor. The fact that we arrived at the nonsensical result of Eq. 4.8 is a tip-off that the transistor is no longer operating in such a linear fashion. *The transistor is saturated.*

A saturated transistor is one in which there is not enough collector current available to maintain the linear relationship of Eq. 4.1. In the circuit of Figure 4.7, the 10 kΩ resistance of R_2 restricts the current to such an extent that it cannot satisfy Eq. 4.1, no matter what positive voltage between 0 and +10 V V_C might be. Just as we wrote down the conditions for a linearly operating transistor, we can write down the conditions for a saturated NPN transistor:

1. The base-to-emitter voltage (V_{BE}) is approximately 0.7 V
2. $I_E = I_C + I_B$
3. $I_C < \beta \cdot I_B$
4. The collector-to-emitter voltage, V_{CE}, is approximately 0.4 V

Let's apply these conditions to the circuit of Figure 4.7. Of course, item 1 on the list is the same as for the linearly behaving transistor, so the base current is found to be the same, $I_B = 43$ μA. However, we won't find the collector current as before, since it is no longer a linear function of base current. Instead, apply the fact that a saturated transistor has a collector-to-emitter voltage of approximately 0.4 V, then use KVL around the right hand loop to obtain:

$$V_{CE} + I_C R_2 - 10\text{ V} = 0 \Rightarrow I_C = \frac{10\text{ V} - 0.4\text{ V}}{10\text{ k}\Omega} = 960\ \mu\text{A}. \tag{4.9}$$

We can also write down:

$$I_E = I_C + I_B = 960\ \mu\text{A} + 43\ \mu\text{A} = 1.003\text{ mA}. \tag{4.10}$$

Is a saturated transistor ever desirable?

The answer is: you bet. Most bipolar transistor logic uses saturated transistors. Think about the circuit of Figure 4.7, redrawn in Figure 4.8. When V_{IN} is low (that is, ~0 V) then the transistor is off, since V_{BE} is less than 0.7 V. Since the base current is zero, the collector

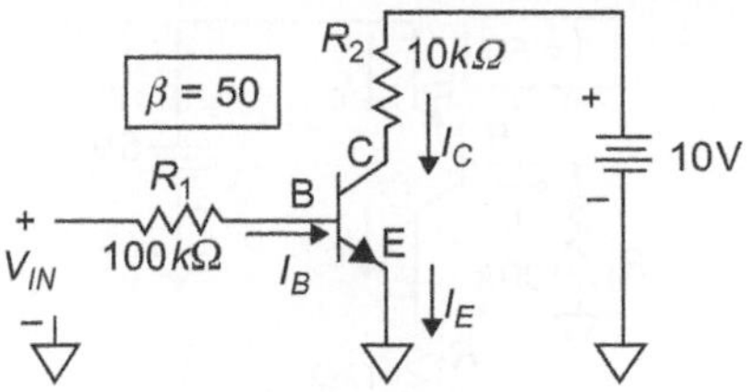

Figure 4.8
The collector-coupled transistor as a logic element.

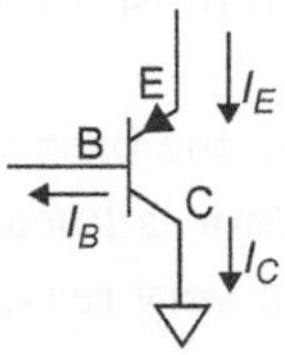

Figure 4.9
The PNP transistor.

and emitter currents are also zero. In that case, $V_C = 10$ V. On the other hand, when V_{IN} is high (say, 5 V) then we have the situation just analyzed, and $V_C = 0.4$ V. Subsequent logic circuits can easily distinguish between these two very distinct voltages, so V_C makes an excellent logic output.

When is it linear and when is it saturated?

When we look at a transistor in a circuit schematic, it's often not obvious whether it will operate in a linear or saturated fashion. So, how can we know? The answer is that it's okay to make a guess about the transistor's state. If you guess wrong you'll know from the nonsensical answers that you get. For example, for the circuit in Figure 4.7, we initially guessed that the circuit operated as a linear device, but we quickly learned that that gave us results which violated our understanding of circuits and transistors.

The PNP transistor

The PNP transistor symbol is shown in Figure 4.9. Unlike the NPN transistor, it is the PNP transistor's *emitter-to-base* that is at 0.7 V (rather than base-to-emitter) when it is on. And unlike the NPN transistor, the PNP transistor's emitter current flows *into* the transistor and its base and collector currents flow *out of* the transistor. Also, the PNP transistor's *emitter-to-collector* voltage is positive (whereas the NPN transistor's *collector-to-emitter* voltage is positive).

Let's look at an example. Figure 4.10 shows a single PNP transistor circuit. The only independent voltage source in this circuit is the 5 V battery. Regardless of whether the

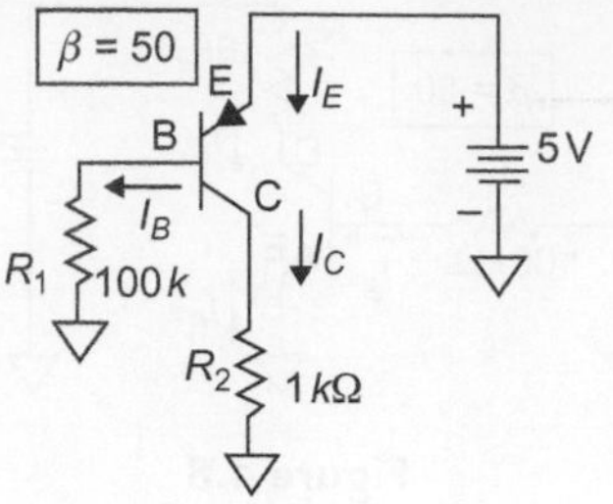

Figure 4.10
PNP transistor circuit.

transistor is in its linear or saturated state, the base voltage is 4.3 V (5 V − 0.7 V). So the base current is 43 μA. Assume the transistor is linear. Then $I_C = \beta \cdot I_B = 2.15$ mA and $V_C = 2.15$ V. This makes $V_{EC} = 2.85$ V, a very reasonable voltage for a linearly operating PNP transistor.

Metal oxide semiconductor field-effect transistor (MOSFET)

Another type of transistor is the field-effect transistor (FET). Whereas the bipolar transistor can be thought of as a current-controlled device (the base current being the control parameter), the FET is a voltage-controlled device.

There are two general categories of FETs, junction field-effect transistors (JFETs) and metal oxide semiconductor field-effect transistors (MOSFETs). The specific type of FETs used in these robot projects is the MOSFET. The MOSFET can be further broken down into the categories of enhancement-mode or depletion-mode. The type of MOSFETs that we'll be using and talking about is the enhancement-mode devices. Okay, that's a lot of categorization. Let's see if a little roadmap can help (Figure 4.11).

Like the bipolar transistor, the FET comes in two varieties, N-channel and P-channel, similar to the way bipolar transistors can be either NPN or PNP. And like the bipolar transistor, one of these two types, the N-channel, is more commonly used (just as the NPN bipolar transistor is more common than the PNP). So, we'll start off looking at the N-channel enhancement MOSFET.

The enhancement N-channel MOSFET

Figure 4.12 shows the schematic symbol for this device. It is a three-terminal device, like the bipolar transistor, with a gate (G), drain (D), and a source (S). The way enhancement N-channel MOSFETs work is that the gate-to-source voltage, V_{GS}, changes the conductivity between the drain and the source.

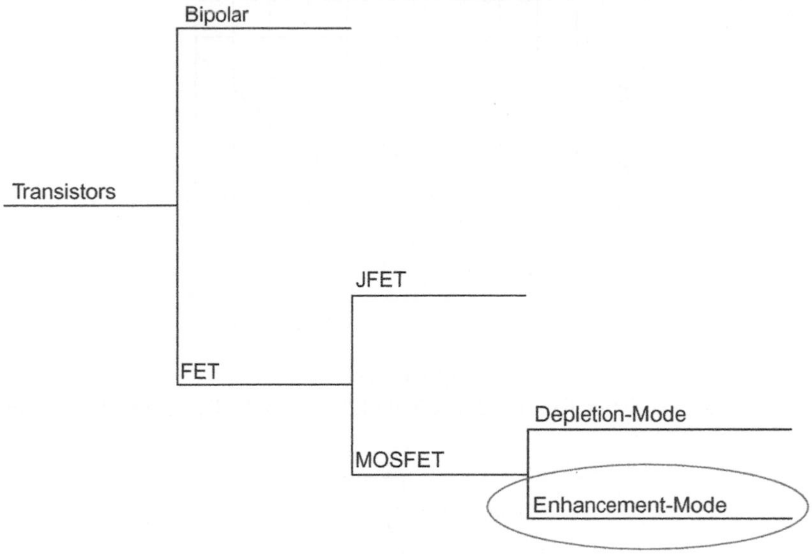

Figure 4.11
Transistor categories.

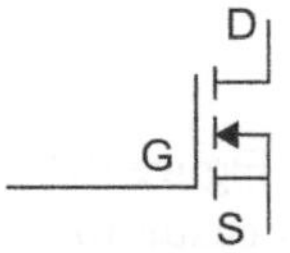

Figure 4.12
Enhancement N-channel MOSFET schematic symbol.

Although MOSFETs can and are used to create analog circuits, our interest in this discussion will be in using them as power switches, which is a very common application of them. They are well-suited to this job because, when their gate-to-source voltage (V_{GS}) exceeds some threshold voltage (V_T), the drain-to-source behaves very much like a low-resistance resistor. Unlike the bipolar transistor, which has a saturation voltage of around 0.4 V, the MOSFET has no such saturation voltage and only a low resistance (typically less than an ohm).

Figure 4.13 shows the transfer characteristics for this device. A test circuit to produce this graph is shown in Figure 4.14. The supply voltage between the gate and the source is variable (indicated by the arrow through the supply). The ammeter could be a standalone ammeter or, more likely, the ammeter setting that exists on your multimeter. Ammeters ideally have a resistance of zero, so it should have no effect on the operation of the circuit.

What the Figure 4.13 graph tells us is that, above some threshold voltage, in this case about 3 V, the device begins to conduct current. However, as you might have gathered from looking at that graph, this threshold voltage can be a little misleading. Let's say that we

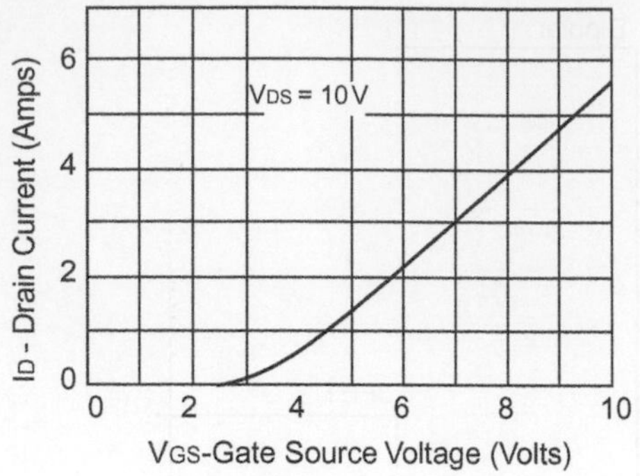

Figure 4.13
Transfer characteristics for the Zetex ZVN4206 MOSFET *(courtesy Diodes, Inc.)*.

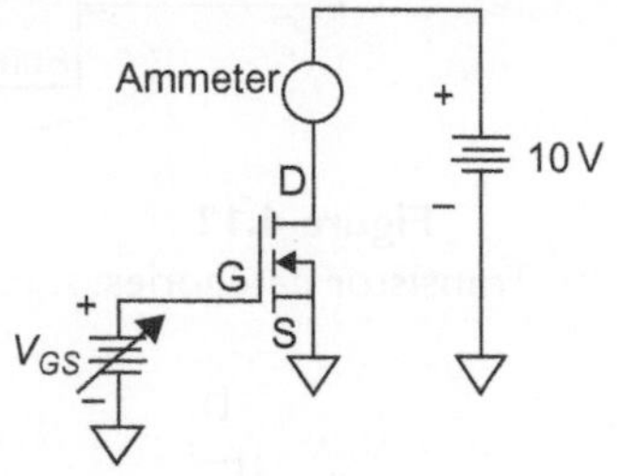

Figure 4.14
Test circuit for measurements of Figure 4.13.

want to use this transistor to switch on the motor for our robot, which might draw of the order of 500 mA. If we just provide a gate-to-source voltage of 3 V (a typical output voltage from a microcontroller) Figure 4.13 shows that there will be a problem. Even though the manufacturer's data sheet says that the threshold voltage is 3 V and even though the graph says that current is flowing at 3 V, it's not enough current for our application.

A better graph for that purpose is the one given in Figure 4.15. The first thing to note is that, for a V_{GS} of 3.5 V or less, 500 mA is simply not achievable. Of course, we already knew that from Figure 4.13. But Figure 4.15 also gives us insight about what the minimum V_{GS} will need to be. For example, with a V_{GS} of 4.5 V, we should be able to sink 500 mA, provided we can live with a little over 1 Ω resistance. At 500 mA, that means that the drain will be about 500 mV (1 Ω × 500 mA) above the source, which is connected to ground.

Using the enhancement N-channel MOSFET as a saturating switch is just about that simple. You drive the gate high with respect to the source and the drain looks like it's shorted to the source (okay, with a very small resistance). Drive the gate so that it is at about the same voltage as the source (that is, $V_{GS} \approx 0$) and the drain looks like an open with respect to the source. This simplified model is shown in Figure 4.16.

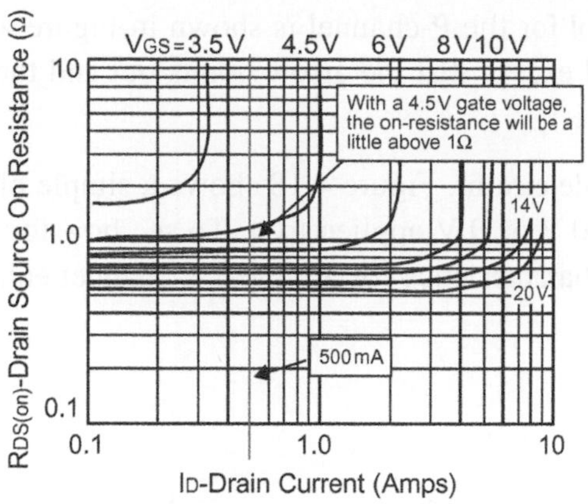

Figure 4.15
Voltage saturation characteristics for the ZVN4206 *(courtesy Diodes, Inc.)*.

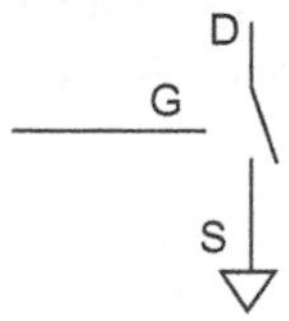

Figure 4.16
Simplified model for the enhancement N-channel saturating MOSFET switch.

Actually, things aren't quite that simple, at least if the intent is to turn the switch on and off very rapidly. For example, the ZVN4206 data sheet indicates that the transistor has a rise time less than 15 nanoseconds – a very impressive number given that this device is switching on or off relatively high current. But that's in addition to the time it takes to charge the gate capacitance. The MOSFET has significant capacitance between its gate and source and between its gate and drain, and the actual charge time is considerably more than the data sheet capacitance would lead one to believe, due to something called the *Miller effect*. The Miller effect is an advanced topic that we will not explore in detail in this book, but suffice it to say that the result of the Miller effect is that the MOSFET gate capacitance will typically be more than an order-of-magnitude greater than the MOSFET's gate-to-drain capacitance. The upshot of this is that a MOSFET can often require a fair amount of current to turn on quickly, although once it reaches its final state, the current drive required at the input is near zero.

The enhancement P-channel MOSFET

P-channel MOSFETs work very similarly to N-channel MOSFETs except that their gate-to-source voltage must be made more *negative* than the source voltage in order to conduct current from drain to source. Let's take a look at an example, to see how this works.

First, the schematic symbol for the P-channel is shown in Figure 4.17. Note that it looks like the N-channel symbol except that the arrow points out and the positions of the source and drain are swapped.

Let's use this in an example circuit. Figure 4.18 shows a simple circuit in which a load resistor is to have either 10 V or 0 V applied to it. To see how this works, let's take a look at the voltage saturation characteristics for a typical P-channel enhancement MOSFET, the ZVN3306A (Figure 4.19).

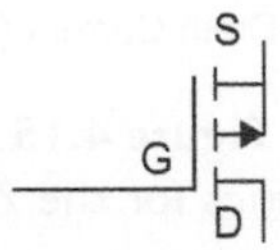

Figure 4.17
The P-channel MOSFET schematic symbol.

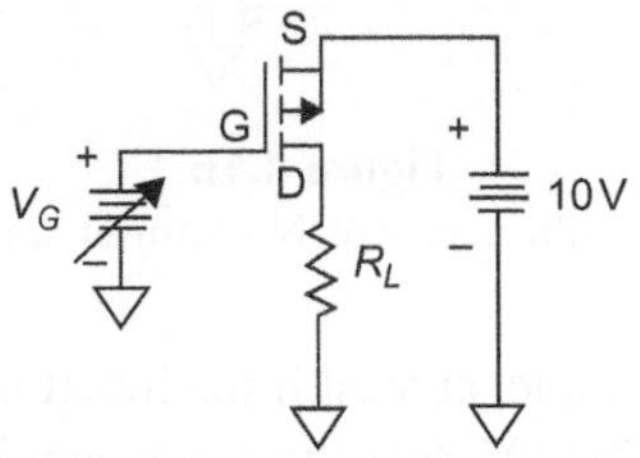

Figure 4.18
Example circuit for P-channel MOSFET.

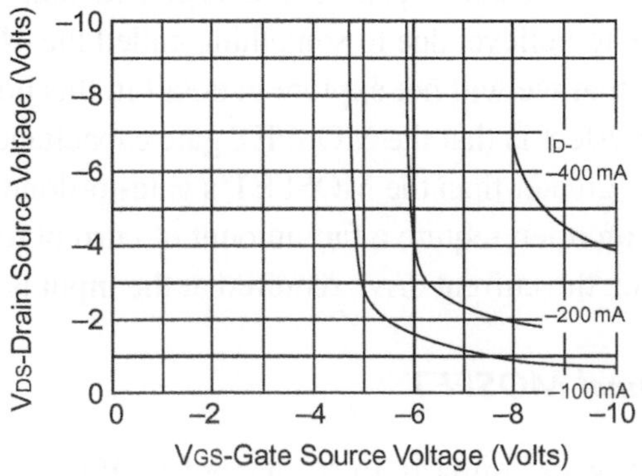

Figure 4.19
Voltage saturation characteristics for the ZVN3306A *(courtesy of Diodes, Inc.)*.

P-channel characteristics like those of Figure 4.19 look a little weird because the gate-to-source voltage is shown as a negative number and the drain current values are shown as negative numbers. That might seem confusing, given that the only independent voltage sources in the Figure 4.18 example are positive voltages. However, keep in mind that V_{GS} is the voltage measured with respect to the source terminal:

$$V_{GS} = V_G - V_S. \tag{4.11}$$

Think about that. To turn this MOSFET off we would need to drive the gate voltage, V_G, to about +10 V. Then $V_{GS} = 0$ V. On the other hand, by making V_G, say, +5 V produces a V_{GS} that is −5 V, a value that guarantees, according to the Figure 4.19 graph, saturation for loads requiring 100 mA or less.

The negative drain current is due to the fact that MOSFET drain currents are referenced positive when they flow from drain to source. Since, in this circuit, the current flows from source to drain, it is a negative value.

The operational amplifier

The operational amplifier (op amp) is a versatile and cost-effective analog building block that is used over and over again in analog hardware designs. The operational amplifier is made from a collection of transistors (either MOSFET or bipolar) and resistors. Schematically, it is drawn as in Figure 4.20.

V_1 is usually referred to as the *inverting input voltage* and V_2 as the *non-inverting input voltage*. V_O is the *output voltage*. Note that when the voltages are drawn as in Figure 4.20, it is implied that these voltages are with respect to ground.

The Figure 4.20 symbol gives us no insight as to how this device works, but the block diagram of Figure 4.21 should help.

In an ideal op amp, $R_i = \infty$ and $R_o = 0$. In that case, we just have:

$$V_o = A_V(V_2 - V_1). \tag{4.12}$$

A_V is called the op amp's *open-loop gain*. It is generally quite large, anywhere from 1000 to 1,000,000.

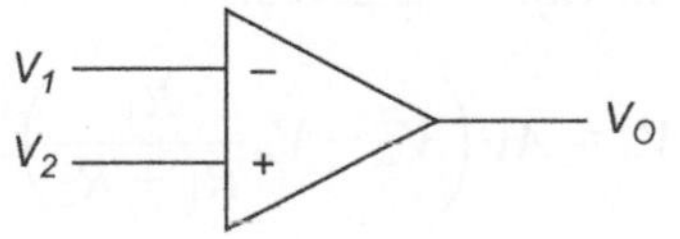

Figure 4.20
Basic operational amplifier.

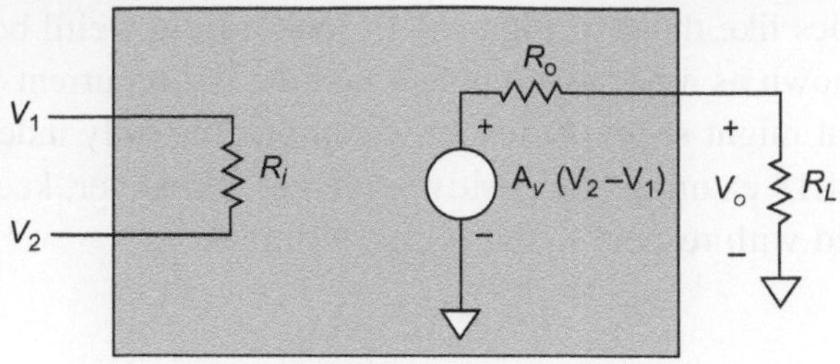

Figure 4.21
Circuit model of an operational amplifier.

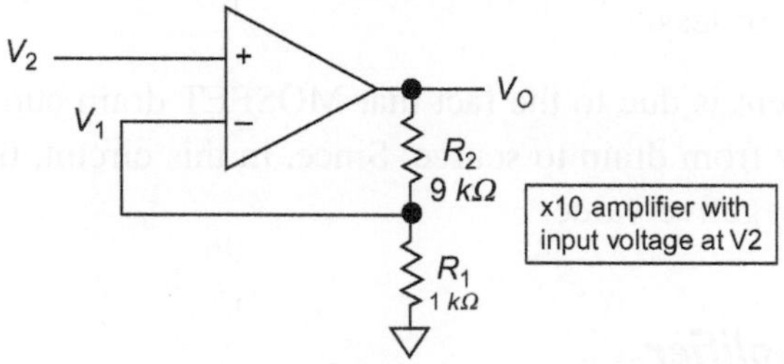

Figure 4.22
Gain-of-10 amplifier with V_2 as the input.

The non-inverting amplifier

Okay, so what's this thing good for? Well, let's add some negative feedback, as in Figure 4.22. Here we've taken the output signal, V_o, sent it into a resistor divider composed of R_1 and R_2, and fed the divided-down signal back to the inverting input, V_1.

Now, we'll write down what we know from the previous discussion. First, we've assumed that the input resistance between V_2 and V_1 is infinite, so no current can flow into either of the inputs. That means that V_1 is just:

$$V_1 = V_o \frac{R_1}{R_1 + R_2}. \tag{4.13}$$

Substituting this for the V_1 term in Eq. 4.12 gives:

$$V_o = A_V\left(V_2 - V_o \frac{R_1}{R_1 + R_2}\right). \tag{4.14}$$

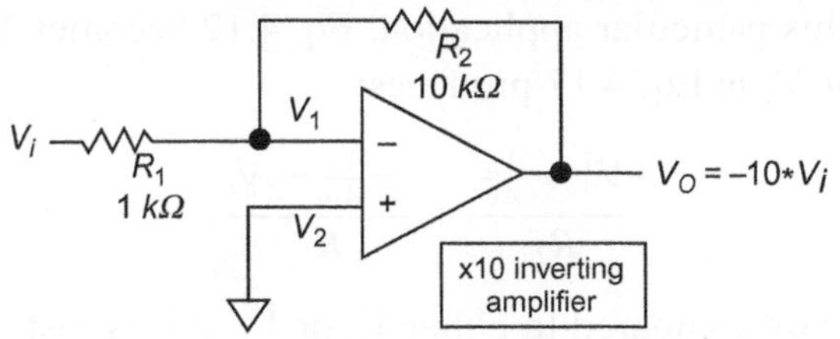

Figure 4.23
An inverting amplifier with gain of −10.

After a little algebraic reduction, we arrive at the intermediate result of:

$$V_o\left(\frac{R_1}{R_1+R_2}+\frac{1}{A_V}\right)=V_2. \tag{4.15}$$

At this point, consider that A_V is typically 100,000 for the op amps that we'll be using for robotics. So $1/A_V$ is of the order of 10^{-5}. The first term in the parentheses of Eq. 4.15, involving the resistor divider, has a value of 10^{-1}. So the resistor term is much, much larger than the open-loop gain inverse term, and we are justified in simplifying Eq. 4.15:

$$V_o\left(\frac{R_1}{R_1+R_2}\right)\approx V_2 \Rightarrow V_o \approx V_2\left(\frac{R_1+R_2}{R_1}\right)=10V_2 \tag{4.16}$$

In general, such non-inverting amplifiers have a gain of $(1+R_2/R_1)$.

Note from Eq. 4.13 that $V_2 \approx V_1$. This is true of any op amp that is operating in its linear range (that is, anywhere that Eq. 4.12 holds true).

Inverting amplifiers

Quite often we want an amplifier that produces an output that is a linear function of the negative of the input voltage. Figure 4.23 shows an example. This can also be readily analyzed using the conditions discussed earlier.

As before, we assume that no current flows into either the V_1 or V_2 terminals of the op amp, since we assumed R_i in Figure 4.21 is infinite. We also know, from Kirchoff's current law, that:

$$\frac{V_o-V_1}{R_2}=\frac{V_1-V_i}{R_1} \tag{4.17}$$

Now, since $V_2 = 0$ for this particular application, Eq. 4.12 becomes $V_o = -A_V V_1$. Substituting $-V_o/A_V$ for V_1 in Eq. 4.17 produces:

$$\frac{V_o + \frac{V_o}{A_V}}{R_2} = \frac{-\frac{V_o}{A_V} - V_i}{R_1}. \tag{4.18}$$

Since V_o/A_V is insignificant compared to either V_o or V_1, this is just:

$$\frac{V_o}{R_2} = \frac{-V_i}{R_1} \Rightarrow V_o = \frac{-V_i R_2}{R_1} \tag{4.19}$$

Note that we could have also arrived at this answer by using the fact that $V_1 \approx V_2 = 0$.

Powering the op amp

So far we've pretended that the op amp was some sort of magical element that required no power and produced the linear results that we needed. Of course, the op amp is really a collection of transistors that requires one or more DC voltages to make those transistors work. Typically, op amps are powered from dual DC voltage sources in the range ±1.5 V up to ±15 V or from a single DC supply from +3 V up to +30 V.

Single-supply op amp power

An op amp circuit intended for operation from a ±15 V DC supply can be converted to one that works from a single +30 V supply without much in the way of changes. The way this is usually approached is to create a reference for the single-supply op amp circuit that is at half the supply voltage (+15 V if the supply voltage is +30 V). An example is shown in Figure 4.24.

Note that the input is assumed to be on the half-voltage (15 V) pedestal and that the output produced is on a 15 V pedestal. Think of this half-supply-voltage as a "pseudo-ground". It takes a little getting used to, to design these single-supply circuits, but this is an important skill for anyone designing a battery-operated device, such as a robot.

Input and output range of the op amp

One thing to be aware of is that many op amps, especially those designed years ago, have a limited input range. For example, the TL061 op amp, when powered from a dual-voltage ±5 V power supply, won't behave properly if the input voltages go past ±2 V.

On the other hand, the TLV2731, which is a much more recent design, can handle inputs anywhere in the entire range of −5 V up to nearly +5 V when powered from a ±5 V

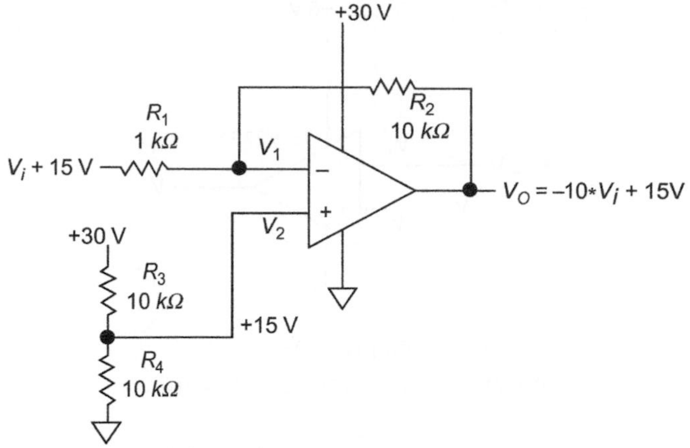

Figure 4.24
The single-supply inverting amplifier with 15 V reference.

supply. This range of allowable input voltages is known as the *common-mode input voltage range.*

In addition, some op amps have a limited range of voltages over which their outputs can go. Some op amps can only go to within a few volts of the supply voltages, before no longer behaving linearly.

Op amps that can operate with their outputs anywhere within the range of the power supply voltages are referred to as having *rail-to-rail outputs*. As noted, some op amps also have *rail-to-rail inputs*, meaning that those op amps' common-mode input voltage extends to the entire power supply voltage range.

Op amps as comparators

Just as transistors can be operated as saturated devices, rather than operating in their linear region, so op amps can be operated as saturated devices. For example, we might have a microcontroller that operates from +3 V but need to turn some device on or off that needs a 0 V to +6 V digital signal. What can we do?

The Figure 4.25 circuit is a very simple way to achieve this type of *level translation.* This circuit uses a TLV2731 op amp, which has rail-to-rail outputs and which can respond properly to inputs all the way down to its negative supply voltage, which in this case is ground. The two input voltages are +1.5 V, which can easily be generated with a resistor divider from +3 V, and the microcontroller voltage, V_i, which is 0 V when logic low and +3 V when logic high. When V_i is 0 V, V_o attempts to go to $A_V(V_2 - 1.5)$, which would be

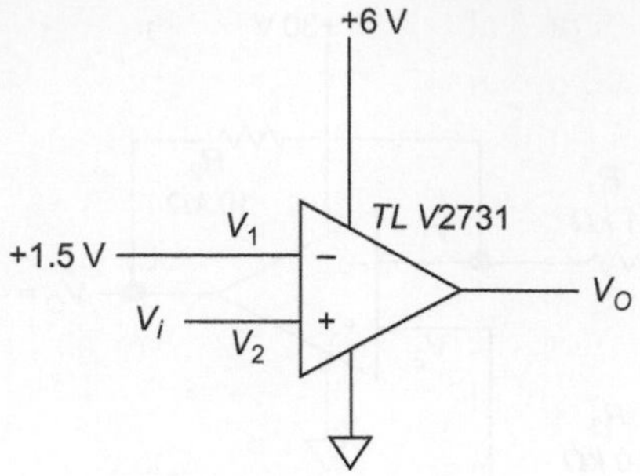

Figure 4.25
An op amp as a voltage translator.

some huge negative number (since A_V is a huge number). Of course, it can't go below the negative rail, which in this case is ground, so it saturates at $V_o = 0$ V.

When $V_i = 3$ V, the output voltage again tries to go to $A_V(V_2 - 1.5)$, which in this case would be a huge positive number. But the output can't go above the positive supply voltage, +6 V, so it saturates at that voltage.

Recapping, the output for this voltage translator circuit is:

$$V_O = 0 \text{ V if } V_i = 0 \text{ V}$$

$$V_O = +6 \text{ V if } V_i = +3 \text{ V}$$

Of course, we could have achieved the same results with a ×2 linear amplifier (assuming we were using a rail-to-rail output type of op amp). However, digital signals, like the one from the microcontroller, aren't generally very precise. A microcontroller operating on +3 V could have a low voltage anywhere from 0 V to as much as 1 V and a high voltage anywhere from 2 V to 3 V. The uncertainty in the exact voltage from the microcontroller gets passed along if we use the linear amplifier but is eliminated by using a saturated device.

This type of op amp is called a voltage comparator. In fact, there are integrated circuits made for just this purpose. Such comparator ICs don't make good op amps but they are optimized for these type of saturated-output applications.

Bandwidth and slew rate

So far we've ignored the speed of these op amps. However, every transistor has, associated with it, capacitance and other speed-limiting characteristics that constrain the device's ability to change. Since an op amp is a collection of many transistors, it will definitely have some limit as to how fast it can change its output.

We talked briefly in the last chapter about frequency and that's an important concept when we talk about the speed of these op amps. As the input signal's frequency increases, the op amp has an increasingly difficult time keeping up with the signal. Generally, for a given amplifier, there is some frequency beyond which the amplifier's performance is deemed too slow to be effective. This cut-off frequency is referred to as the amplifier's *bandwidth*.

In applications like the comparator, the *slew rate* is also important. It's one thing for the op amp to be able to handle high-bandwidth, small signals, but when the output is required to change, or *slew*, quickly over several volts, it may still have a problem. For applications like that, it's also important to make sure that the op amp will have sufficient slew performance.

Other constraints on op amp performance

As mentioned, an op amp will have its inverting and non-inverting inputs nearly equal in voltage when operating linearly. In reality, an op amp will typically have an offset between these two voltages. One or the other inputs will have to be driven anywhere from a few microvolts to a few millivolts higher than the other input to produce the expected output voltage. This can limit performance in some instances. Other constraints on performance include input bias currents and output current limit.

While these are important to understand for advanced applications, they are outside the scope of this book. The reader is referred to the list of references at the end of the chapter to find out more about these op amp performance constraints.

The H-bridge

Like the op amp, the H-bridge is a collection of transistors, sometimes contained in a single integrated circuit, sometimes fashioned out of discrete transistors. Bridge circuits are used mostly to drive motors and transducers. It is simply a method of driving the device from both sides of the device rather than just one.

For example, let's say that we have a DC motor to which we'd like to apply either +6 V (let's call this the "forward" direction for the motor) or −6 V (to reverse the direction of the motor). One way to do this is to have both polarities of 6 V available and have a switch (mechanical or transistor) that applies one or the other voltage according to our desired direction (Figure 4.26).

As is usual, the +6 V and −6 V sources are assumed to have the other side tied to ground.

For battery-operated applications (like our robots) this is a problem. To generate both +6 V and −6 V we would need to carry two batteries, connected so that they generate the two polarities. A better way might be to just add a second switch as in Figure 4.27.

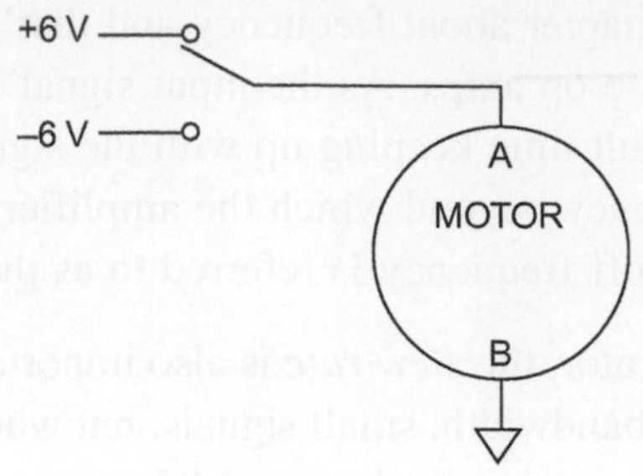

Figure 4.26
Motor with bipolar voltage drive.

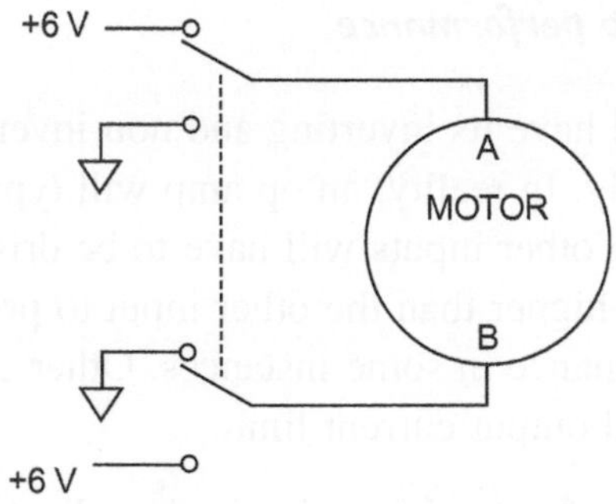

Figure 4.27
Motor with H-bridge drive and single supply.

In this case, the motor receives the +6 V at terminal A, ground at terminal B when the switches are in the up position, and it receives ground at terminal A and +6 V at terminal B when the switches are in the down position. The current flows from terminal A to terminal B in the up position and in the opposite direction when the switches are in the down position. Thus, we achieve both forward and reverse directions while using only a single supply voltage.

We will use this idea for several functions: it is used with the drive motor, to produce forward and reverse drive; it is used for the steering motor, to produce the left and right motion on the front wheels; and it is used for the ultrasonic transmitter, to produce the bipolar drive voltage for this transducer. The bridge switching circuit for the motors is a special integrated circuit capable of handling the hundreds of milliamperes that the motors require.

Bipolar transistor implementation of an H-bridge

Figure 4.28 shows an example implementation of half of the H-bridge, made with a rail-to-rail-output op amp, an NPN transistor and a PNP transistor. The output, V_O, is high (near 6 V) when V_{SW} is greater than 1.5 V, and is low (near 0 V) when V_{SW} is less than 1.5 V.

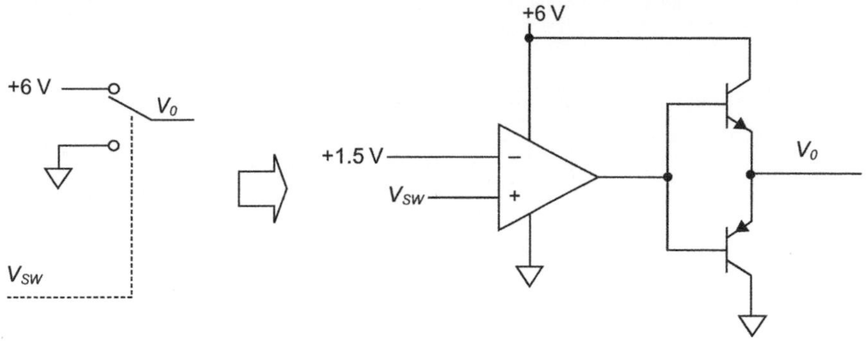

Figure 4.28
Bipolar transistor/op amp implementation of half of an H-bridge.

Semiconductor wrap-up

Okay, this and the last chapter were a lot to hit you with, particularly if you haven't seen this material before. If you're still with us at this point, the rest should get a little easier. Hang in there!

Bibliography

[1] Application Note AN-558. Introduction to Power MOSFETs and Their Applications. Texas Instruments, Literature Number SNVA008.
[2] D. Ashby, Electrical Engineering 101, Newnes, 2009.
[3] J. Millman, Micro-Electronics – Digital and Analog Circuits and Systems, McGraw-Hill, 1979.

CHAPTER 5

DC Motors

Chapter Outline

The reader is, no doubt, familiar with the fact that two magnets repel one another when poles of the same type face each other and attract one another when opposite poles are near one another. In addition, we've seen in one of the previous chapters that a loop of wire creates a magnetic field that points along the axis of the loop; that creating multiple loops intensifies the magnetic field; and that inserting an iron rod inside the loops increases the field still further. With these few facts, the concept of a motor can be understood.

MSP430-based Robot Applications.
DOI: http://dx.doi.org/10.1016/B978-0-12-397012-1.00005-9

Learning by doing

The next section is about a device, Beakman's motor, that isn't used in these robot designs at all. So why include it? Well, understanding motors is important for robot design as well as lots of other types of machine design. One of the best ways to learn how to do something is to actually do it. For example, you could study music and guitar theory for years, but if you really want to excel at playing guitar, the best way is to get a guitar and start playing it. The theory is great but actually doing it is really important.

Similarly, if you've ever read a book about computer programming, you may have had the experience of finishing the book, putting it down, yet not knowing even how to start on your first program. What you've probably found is that, to really learn programming, you need to program. The book can be a valuable resource, but mainly as a reference to be consulted, rather than a sole source of learning.

So while the next section is optional, it is highly recommended. If you can make Beakman's motor and understand how it works, you're well on your way to understanding how DC motors, in general, work.

Beakman's motor

This DC motor is from the TV show *Beakman's World*. A version of this motor is shown in Figure 5.1. If you want to see a video of one operating, just search on the Internet for "Beakman's Motor". There are lots of examples out there. The motor consists of just a long piece of wire, coiled several times around a form (the coil in Figure 5.1 has 10 turns).

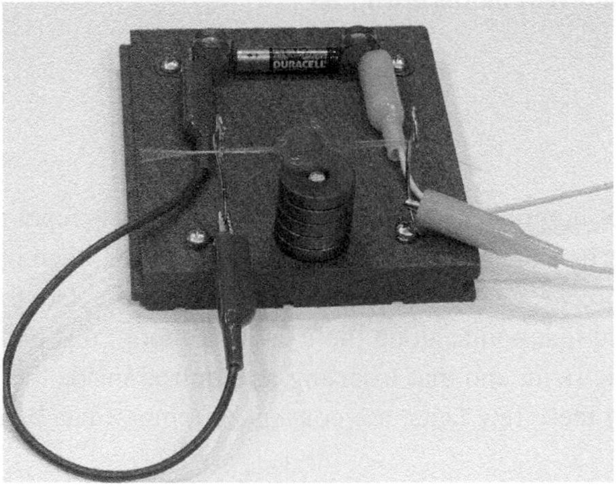

Figure 5.1
Beakman's motor.

A 5/8" dowel rod is what I wound my wire around, but anything round with a diameter of that order will work as a form. You could wind it around a spare AA battery, if nothing else.

When I made my motor, I used "magnet wire" from Radio Shack. Magnet wire is just wire that has the insulation "painted on", using a lacquer, rather than the usual plastic insulation. Anyway, on one "tail" of the coil, all the insulation is stripped, to about a ¼" of the coil. This can be done easily by scraping repeatedly with a knife edge. On the other end, the insulation is only stripped on half the wire (see Figure 5.2). *However, the orientation of this scraped-off half of the insulation with respect to the wire coil is important.* Figure 5.3 shows the orientation, with the insulated half of the wire on top when the coil is held vertically.

To make the supports, I used some 14-gauge wire, the kind that is used for home wiring. It's nice and stiff and can be bent with pliers, as shown in Figure 5.1, and held in place on a block of wood with a couple of screws. These stiff wires serve as both the electrical "brushes" and the mechanical supports for the loop. You'll need one or more permanent magnets – the ones in Figure 5.1 were purchased at Radio Shack and there's a long screw in the middle that holds them in place. A battery, some springy copper contacts for the battery (I used a couple of copper plumbing brackets from the hardware store) and some little jumper cables (or you can just use wire) and that's it. When you complete the circuit with the battery, you should find that, with a push from your finger to get the thing started, it will rotate. Congratulations! You've just constructed your first motor.

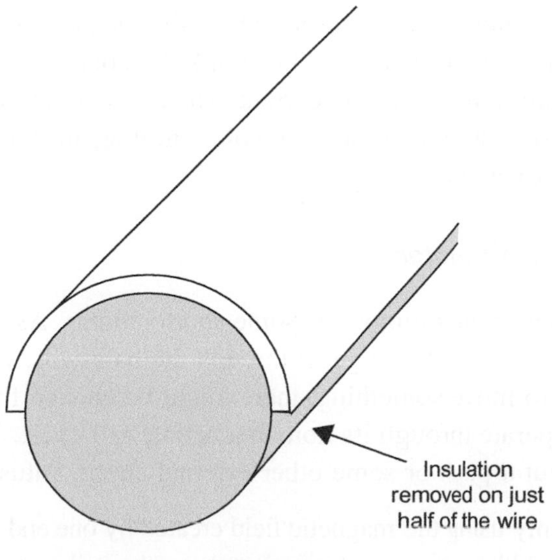
Insulation
removed on just
half of the wire

Figure 5.2
Insulation is stripped on just half the wire on one end (the other end is stripped completely).

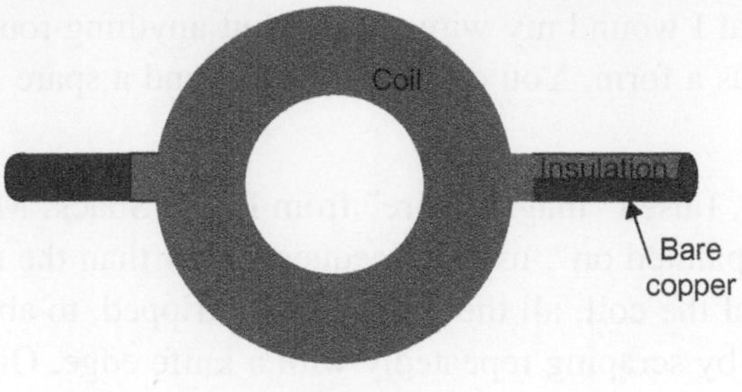

Figure 5.3
Orientation of the stripped part of the wire with respect to the coil.

How does Beakman's motor work?

Recall from Chapter 3 how a coil of wire produces a magnetic field that is oriented predominantly in the axial direction. That is, if we imagine the coil oriented around an axle, the magnetic field is in the direction of that axle. In one direction is the "South" end of the coil's magnetic field and in the other direction is the "North" end. The determination as to which end is which depends on which way the current runs through the coil. Since we just want our motor to spin and aren't concerned about which direction it spins, this isn't important right now.

To see what's going on with Beakman's motor, take a look at the sequence in Figure 5.4.

Assume the coil starts off with its insulated side of the wire up. With current running through it, the coil produces a magnetic field. One end of the magnetic field is attracted to the magnet and one end is repelled by it, so the coil starts moving.

Just as the coil reaches its horizontal position (where the magnetic force is now at a maximum, due to the magnetic fields being lined up), the connection to the battery is broken, due to the insulation blocking the current. The coil continues spinning due to inertia and, after spinning without current for half a revolution, it again starts conducting current and the cycle begins all over again.

Shortcomings of Beakman's motor

As you might expect, Beakman's motor has some shortcomings. As you're no doubt aware, the motor relies on the coil's inertia to get it through the half cycle when no current flows. If it were actually used to move something, there might be enough force acting against the motor that it couldn't operate through its non-conducting half cycle. If that's the case, then it's stuck, at least without a push or some other external energy infusion.

Also notice that we are only using the magnetic field created by one end of the electromagnet (a.k.a. the coil). There could be a permanent magnet above the coil, one with the opposite polarity face from the bottom magnet pointed toward the coil. This would improve motor output.

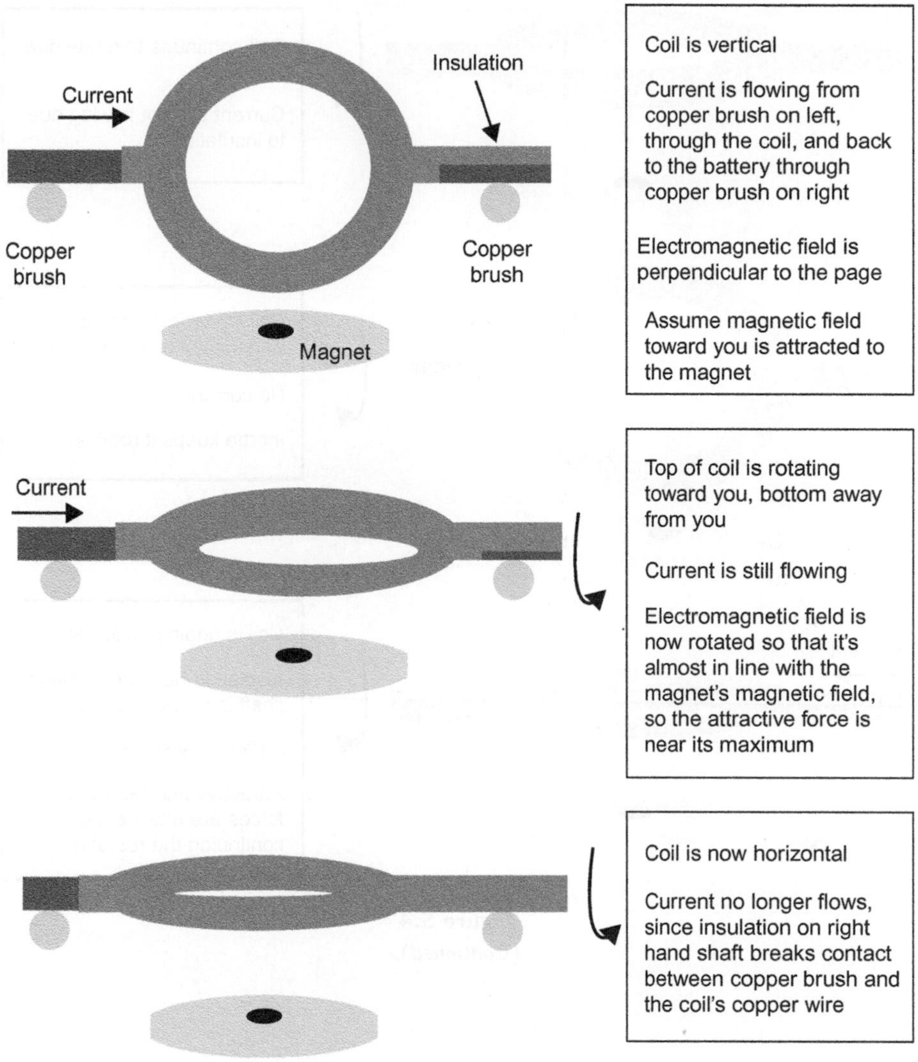

Figure 5.4
Sequence of Beakman's motor states.

The "phase" of the half-circumference stripping, illustrated in Figures 5.2 and 5.3 (that is, its orientation on the shaft with respect to the coil), cannot be achieved with much precision with just the crude method of scraping the insulation with a knife.

In addition, it's obvious, as you watch the coil wobble around in its "cradle", that this motor is not a precision-built piece of equipment. The spacing between coil and magnet is considerable, when compared to a commercial motor. The wobble of the coil and the coil

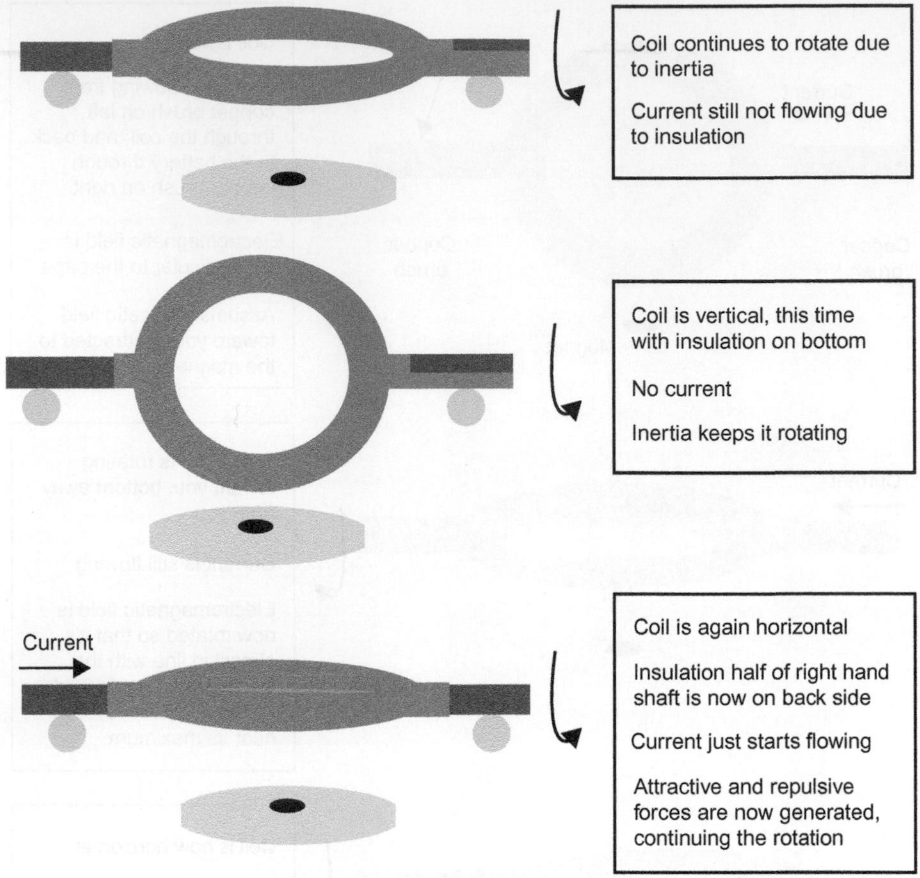

Figure 5.4
(*Continued*).

shaft friction consumes energy without contributing to desired propulsion. Finally, the number of coil turns and the DC voltage are not optimized.

None of this is a slam against Beakman's motor as a great educational tool. But it gives us good insight as to why we will want to purchase off-the-shelf motors rather than attempt to build our own.

Improving on Beakman's motor

A number of the problems with Beakman's motor have to do with mechanical precision. However, even if the motor is well-constructed, using precise, modern manufacturing methods, problems remain. The motor must "coast" every half cycle. And there's no magnet above the coil to take advantage of the coil's magnetic field above.

Putting a magnet on top of the coil, in addition to the one on the bottom, is no big deal, at least in theory. Getting the motor to be propelled by magnetic force for the entire rotation cycle, or nearly the entire rotation cycle, will be a little trickier.

Improved commutation

So, let's address the coasting problem. The way motor manufacturers solved this problem is by coming up with a slightly more complicated *commutator* than the one on Beakman's motor. On Beakman's motor, the half-insulated/half-stripped shaft of the coil is really a rotating switch. It turns battery power on for a particular half cycle of rotation and turns it off for the other half cycle. The term for this in motor jargon is *commutation.*

Motor manufacturers take care of this problem of absent magnetic field during half the cycle by changing the magnetic field every half cycle. Figure 5.5 shows how this can be done. The shaft is now a separate metal rod, rather than just an extension of the magnet wire. Imagine that a layer of insulation surrounds the shaft and over this insulation are two metallic half-arcs. These two metallic surfaces make up the commutator. One of these is connected to one end of the motor coil (which we will call "Side A") and the other is connected to the other end of the motor coil ("Side B").

When the shaft is oriented as in Figure 5.5, Side A of the coil gets the "+" side of the battery through the top brush and Side B of the coil gets the "−" side of the battery through the bottom brush. When the shaft rotates far enough, the coil's voltage is reversed, with Side B positive and Side A negative. In this way, there is a magnetic field propelling the shaft rotation nearly 100% of the shaft rotation.

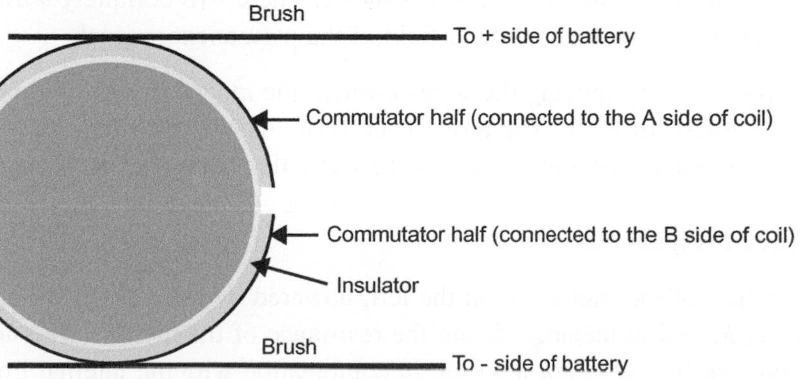

Figure 5.5
Conventional brushed DC motor commutator.

Improving the DC motor still further

Well, the theoretical improvements to the DC motor are almost complete. But there's still a problem. What happens when the motor shaft is rotated 90° from its Figure 5.5 orientation? In that case, each of the two brushes will be in contact with *both* metallic commutator halves. The battery will be shorted.

We could reduce the circumferential coverage of the two arcs, so that at the 90° rotation the brushes contact *neither* of the commutator halves. But then what happens if the motor stops in exactly that orientation? The next time we try to start up the motor, it won't get any current. This was one of the problems with Beakman's motor, although the probability of the problem arising there was much higher.

To fix this problem, motor manufacturers generally build a motor with more than just two poles. Each pole has its own winding and at least one of the poles is energized at any given time.

One final note about modern DC motors: whereas Beakman's motor consisted solely of a wire coil, with no material inside the coil, modern DC motors have the wire wound around iron. The iron not only provides the support for each of the motor's poles but also improves the motor performance.

Is a DC motor also a DC generator?

Perhaps you've heard someone say that a motor isn't much different from a generator. Let's modify that statement a little. A DC motor *is* a DC generator. We already know that a current moving through a wire creates a magnetic field, a necessary phenomenon to create a motor. But it's also true that a wire moving through a magnetic field creates a current. That's a generator! Think about Beakman's motor. There was definitely wire and it was definitely moving through the magnetic field of the permanent magnet.

The voltage produced by moving the wires through the magnetic field is referred to as the "electromotive force" or EMF. The term "back EMF" is also used for this voltage. Although it's a voltage, not a force, this is the name that was given to it way back when and it's the name that sticks to this day. At this point we can draw a pretty good model of the motor circuit (Figure 5.6).

The two terminals of the motor are on the left, powered by an external voltage source, *V*. The resistance, *R*, and inductance, *L*, are the resistance of the motor's wire and the inductance created by the motor's wiring in combination with the internal iron. The EMF is labeled *E* and, as Figure 5.6 shows, it creates a DC voltage that is said to *oppose V*.

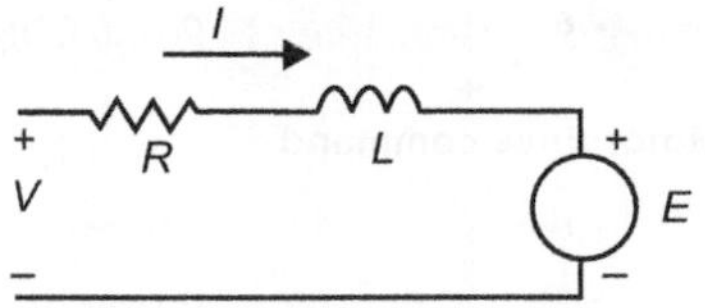

Figure 5.6
Equivalent circuit for the DC motor.

That might sound like a bad thing, but it's actually very good. Let's sum the voltages around the loop to see what's going on:

$$V \approx IR + L\frac{\Delta I}{\Delta t} + E. \tag{5.1}$$

The reason that the "≈" is used rather than the "=" is that the expression for the inductor voltage is an approximate expression, as discussed in the chapter on inductors. To write the exact expression requires knowledge of calculus, which isn't assumed for the reader. But it really doesn't matter in this discussion – when the motor is operating under steady-state conditions, that is, constant voltage, V, and constant motor speed, the current will be constant, so that the inductor voltage will be zero.

In fact, for now, let's just simplify Eq. 5.1 to:

$$V \approx IR + E. \tag{5.2}$$

Now, let's re-write this equation to find the current through the motor:

$$I \approx \frac{V - E}{R}. \tag{5.3}$$

Now we can see why the EMF, E, is such a helpful thing. When the motor is spinning it's producing an EMF that subtracts from the input voltage, V. As a result, the current that the voltage source must supply is greatly reduced.

So when is it a motor and when is it a generator?

The answer is pretty simple – if you're externally supplying current to the machine, it's acting as a motor, and if the shaft is moving, it's acting as a generator. So what that means is that most of the time, it's both.

Figure 5.7 demonstrates the EMF voltage clearly. Here, the wheels of the RC car are driven by a transistor drive circuit turned on and off by the input drive waveform at top. When the input waveform is high, the motor receives approximately 6 V, and when the input is low, the transistor is off and presents a high impedance to the motor.

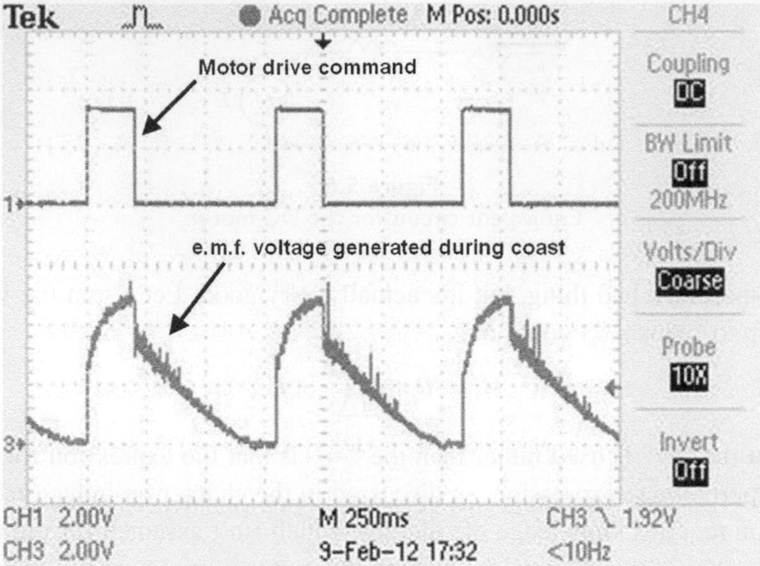

Figure 5.7
Robot motor on and coasting, showing EMF generated during coast.

Because the transistor looks like an open circuit when the drive voltage is low, the motor coasts during that interval. During that time, the motor is behaving strictly as a generator. Since the wheels are still spinning, the voltage across the motor during this period does not immediately go to zero. Instead it ramps down at a fairly linear rate and the EMF voltage reduces accordingly, as the wheels slow down.

Note in Figure 5.7 the "spikiness" of the motor drive waveform. This is due to the commutation in the motor – the commutator plates of the motor making and breaking connection, causing the relatively large voltage spikes. These spikes are easily transferred to the supply voltage of our robot and, unless proper precautions are made, can spoil our sensor measurements and other sensitive circuits. The cure for this is referred to as isolation and will be designed into the robot circuits in this book.

Torque, force, and current

When current flows through a motor it produces a force. Because the force is rotational, it is usually referred to in terms of torque. Torque is just force acting at some radius.

The torque from a motor is proportional to the net current flowing into a motor. But what about when the motor is rotating but no input voltage is applied? In that case it's a generator. Is there a torque produced then? The answer is that there is and that it is a *negative* torque. Instead of the torque acting to propel the car/robot, it resists movement. In fact, we can use this negative torque as a way to "brake" the vehicle.

Dynamic braking

If we wish to slow the robot very rapidly, we can simply ground the input. Or we could connect both sides of the motor to the +6 V or whatever DC voltage we're using to power the motor. In either case, both sides of the motor are at the same potential – the equivalent of shorting the motor. The only resistance that this voltage sees is the resistance of the motor winding. The current will be quite large and therefore the negative torque will be large. This is an effective way of stopping our robot fast.

Powering the motor

Okay, so we've got the basics of how a motor works. Now let's figure out how to power it.

An RC electric car comes with a motor already installed, so we really don't need to do anything extra to use it except create the drive voltage, *V*. We already know what the voltage is that will propel the RC car at maximum speed – it's whatever the car's battery rating is. So, for example, if our RC car comes with a 6 V battery (four AA batteries, for example, or a 6 V rechargeable battery) then we will want to deliver around that voltage to achieve maximum speed.

But what if we want to power the robot/car at a lower speed? Our robot may be executing some difficult maneuvers that require lower speed – maybe much lower speed. To achieve that reduced speed will require a lower drive voltage, *V*, to the motor. What's the best way to generate that?

Digital and analog – definitions

Before talking about how to drive the motor, let's define some terms. You've probably heard the term "analog" many times. Lots of people use the term, including lots of people with only a vague notion of what they're talking about. The term "analog", in an electrical engineering context, means a continuously-valued signal.

Okay, maybe that definition didn't help much. An example will illustrate the idea. If someone asks us to build a resistor divider that produces a 2 V signal from a 5 V signal, we can do it. If they tell us they want the voltage to actually be 1.97 V, we can do that. We might have to connect some extra resistors in parallel or series with the original resistors in the divider circuit to get just the right voltage, but we can do it. If they tell us they want 1.968 V, well, given enough time and sufficiently accurate measuring equipment, we could do that too. The divider circuit is capable of taking on a continuous range of outputs, which is what allows us to adjust the output. This is an analog signal.

A digital signal typically has only two values, for example, 1 and 0. We might claim that +3 V represents a "logic 1" and 0 V represents a "logic 0". But actually, any voltage above, say, 1.5 V, represents a logic 1 and any voltage below that threshold, a logic 0. An entire range of voltages is mapped to a single value, 1, and another entire range of voltages, those below 1.5 V, are mapped to 0. This is a digital signal. We can combine several of these individual two-level signals to form larger numbers, but there is still a countable number of discrete values that can be represented. In a later chapter we'll talk more about digital logic and digital arithmetic.

The analog DC motor drive

A straightforward way to set the speed of the motor is to have our robot electronics generate the analog voltage, *V*, needed to attain the desired speed. This could be done, for example, by having our microcontroller (which we'll be talking about a little later in the book) send a number to a circuit called a digital-to-analog (D/A) converter (another topic which we'll discuss later in the book). If we wanted to drive the motor with a voltage of, say, 2 V, we would have the microcontroller generate the special number that causes the D/A converter circuit to create the 2 V output.

What's a load and how do we drive a heavy one?

One problem with the idea just described, of using the output of the D/A to drive the motor, is that a circuit output voltage isn't really a pure voltage source. It's more accurately modeled as in Figure 5.8 – a pure voltage source with a series resistance. It's not that there's necessarily a physical resistor in the circuit that you can point to as that series resistor. But every transistor, wire, etc., has some resistance and in some circuits that resistance may be relatively high.

What if R_S in Figure 5.8 is, say, 10 Ω while the load resistor, R_L, is, say, 100 Ω? Let's say that the voltage source, *V*, is 2 V. What we have is a resistor divider. The output voltage, rather than being the hoped-for 2 V, will actually be 1.8 V. In engineering jargon, the output is "heavily loaded". The load circuit is the circuit that the output must drive. A

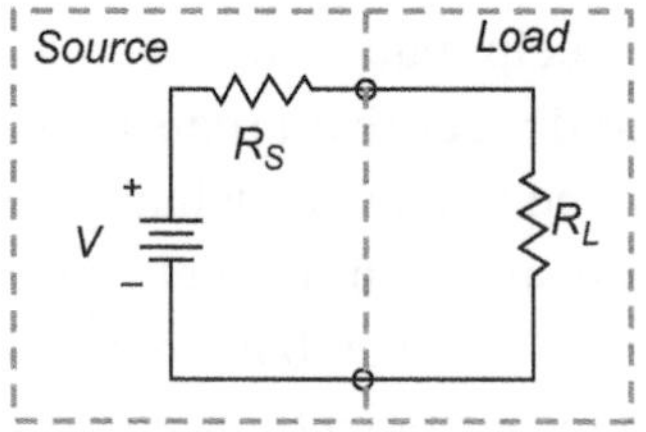

Figure 5.8
Model of an analog output voltage with load.

heavily loaded output is one which requires a sufficiently large amount of current that the output voltage begins to change significantly from the desired value.

Most D/A circuits can supply only a few milliamperes at their output. On the other hand, motors typically need *hundreds* of milliamperes to operate. To say that the motor would be a heavy load for the D/A circuit is quite an understatement.

Could we make this work? Yes, by simply adding a follower op amp circuit that can drive the motor. Now, most op amps can't supply more than a few tens of milliamperes, so it might seem like we're still stuck. However, we could combine an op amp with a transistor, as in Figure 5.9. Most transistors can supply the hundreds of milliamperes that we'll need.

To analyze this, let's assume that the motor initially has no current flowing through it, so that V_M is 0 V. Let's say that v_{IN} is +2 V. Initially, the op amp input difference is 2 V. Let's say the gain of the op amp is 100,000. The op amp tries to go to $2 \times 100{,}000$ V! Of course it doesn't make it. But the output definitely starts slewing up fast. Keep in mind that V_M is the transistor emitter voltage, which is equal to the op amp output voltage minus the transistor's v_{BE} voltage of 0.7 V.

During this time, the amplifier is not behaving in a linear manner because it does not satisfy:

$$v_O = A(v_{IN} - v_M). \tag{5.4}$$

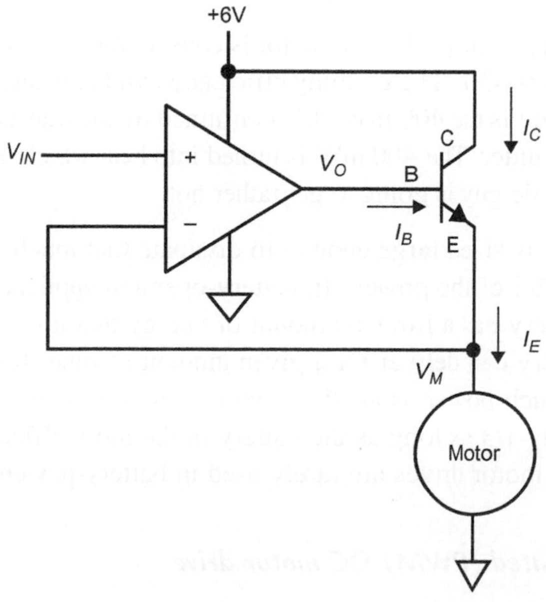

Figure 5.9
Power amplifier created with op amp and PNP transistor.

However, as V_M rises it will, within a few microseconds, reach the point where the amplifier behaves linearly. At that point, Eq. 5.4 is satisfied, so that:

$$v_{IN} - v_M = \frac{v_O}{A} \approx 0 \Rightarrow v_{IN} = v_M. \tag{5.5}$$

Thus, we've transferred the input voltage, v_{IN}, to the motor but, because the transistor is the final driving element, there is no issue about the heavy load of the motor.

Disadvantages of an analog DC motor drive

One disadvantage with the analog DC motor drive just described is the fact that the maximum voltage attainable is around 5.3 V rather than the full 6 V that our battery provided. However, an even bigger problem with the analog DC motor drive is its efficiency. Efficiency is another one of those words you hear spoken a lot, often by people with only a vague idea of what it means. *Electrical efficiency is the amount of power delivered to the load, divided by the total power consumed.* Confused? The analog DC motor drive will serve as an excellent example.

Let's say that we want the motor to operate at a certain speed that corresponds to a 2 V input and which will cause the motor to draw 100 mA. That means that the transistor emitter must produce 100 mA, which means that the transistor collector current is right around 100 mA. So we're drawing 100 mA from the +6 V supply, which means that total power *consumed* is 6 V × 100 mA = 600 mW.

But the motor is operating at just 2 V. The motor is consuming just 2 V × 100 mA = 200 mW. The efficiency is 0.2/0.6 = 33%. The crummy efficiency number is due to the fact that 400 mW is wasted. Where is the 400 mW? It's consumed by the transistor, which has 4 V between collector and emitter. The 400 mW is turned into heat which, if the transistor is small, means the poor little guy is going to get rather hot.

Assuming the transistor is sized large enough to dissipate that much power, is there any harm done by wasting 2/3 of the power? In battery-operated applications, the answer is: definitely. A given battery has a limited amount of energy that it can store. This energy is the power that the battery can deliver for a given amount of time. If one application consumes 3 times as much power as another application, the battery in the former application will last only 1/3 as long as the battery in the more efficient application. For this reason, analog DC motor drives are rarely used in battery-powered applications.

The pulsewidth-modulated (PWM) DC motor drive

If the analog DC motor drive isn't the way to go, then what is? The circuit of Figure 5.10 shows a very simple way of controlling the motor voltage using a switch that is opened and

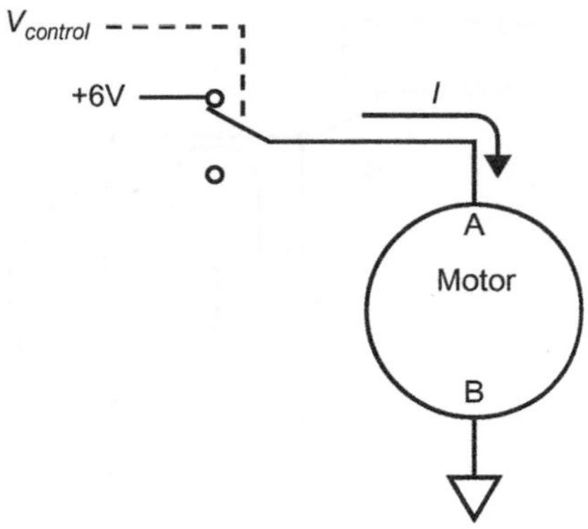

Figure 5.10
A pulse-width-modulated motor drive.

closed by the command of a signal labeled $v_{control}$. When the switch is closed, the entire 6 V is applied to the motor. When the switch is opened, the motor coasts.

Now, if this closed/open sequence is performed at a high enough rate, the vehicle will basically move at a near-constant speed. Why? Well, for one thing, the vehicle itself has a fair amount of mass and its inertia damps out any changes in speed. For another, the motor has inductance that impedes rapid changes in current.

The inductance mentioned above is something we need to keep in mind as we build this PWM drive. Recall that the current through an inductor cannot change instantaneously. So what happens when the switch in the Figure 5.10 circuit changes from its connection to +6 V to its open position? If we leave the circuit as shown, the answer is that the input lead will rise in voltage until the current can continue flowing. This will be a voltage high enough to create a spark, if that's what it takes to keep the current flowing. That could be a voltage high enough to do damage to the switch (if the switch is a solid-state device such as a transistor) or to other components in the circuit.

This problem can be avoided by simply adding a diode, as in Figure 5.11. These diodes go by various names such as flyback diode, snubber diode, freewheeling diode, suppressor diode, or catch diode. The idea is that, when the switch is connected to the +6 V, the diode is reverse-biased and is not involved in the circuit at all, but when the switch opens, the current can keep going through the motor by simply circulating through the diode. The current will eventually fall to zero, given time, but has a path when the switch is first opened.

Example of a PWM DC motor drive

The waveforms in Figure 5.12 show a motor that is being driven alternately between "on" and "coast", as was just discussed. The top waveform is the input to a driver circuit.

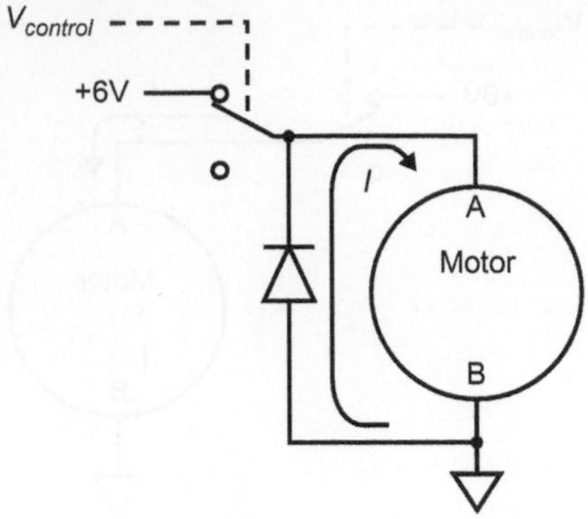

Figure 5.11
PWM drive with flyback diode.

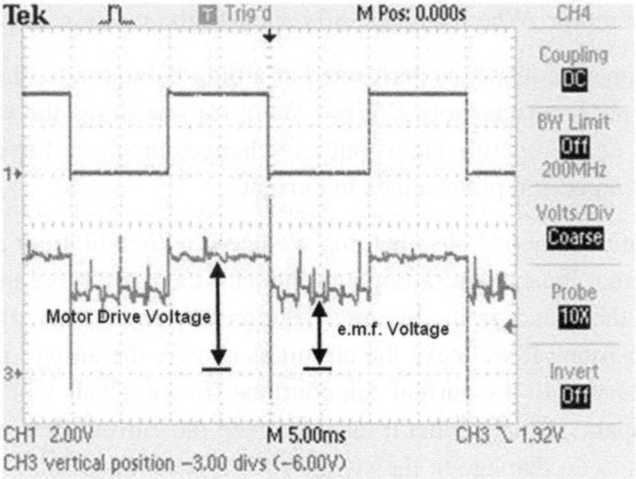

Figure 5.12
PWM motor drive waveform.

Note that when the driver circuit is off (input voltage is zero), the motor voltage does not go to zero but, rather, goes to EMF "coast" voltage. Note that, unlike Figure 5.7, the coast voltage shows no discernible drop, due to the short (10 msec) coast intervals.

PWM power advantages

Okay, so we can drive the motor with a fast series of pulses rather than a constant, analog voltage. But what's so great about that? The great thing about that is that at any given time, the motor is either connected directly to the battery voltage or it is open. In theory, that means that there is no power dissipated in the drive electronics. When the switch is closed, there may be lots of current but there's no voltage across the switch. An ideal switch has zero resistance, so the power consumed by the switch is $I^2R = 0$. When the switch is open, there is voltage across the switch, but no current. So in either state, the switch dissipates zero power. All of the power goes directly to the motor.

In practice, the PWM drive electronics do have some power dissipation. The drive transistor, for example, will have either a saturation voltage when it is on (if it is a bipolar transistor) or will have some resistance (if it is a MOSFET). Either way, the product of the motor current and the "on" voltage results in power dissipation. Nevertheless, PWM motor drives are generally much more efficient than an all-analog, linear motor drive.

Brakeable and reversible PWM motor drive

The motor drive described so far has only two modes – "on" (presumably *forward*) and "coast". We can easily add a third mode, "brake", by adding another position to the switch, as shown in Figure 5.13. In that case, the switch can be commanded to go to the position connected to ground and the motor, in a strictly generator role, will dump all of its current rapidly through the short created, resulting in the motor stopping very quickly.

Adding a reverse mode for this motor drive is pretty easy – we just duplicate the switch and diode of Figure 5.13 on the other side of the motor, as in Figure 5.14. Like the forward-only drive, this one can operate with pulse trains (PWM) to lower the reverse speed. Table 5.1 shows all nine possible states for the drive and what the effect is on the motor.

To summarize: any time either side is open, the motor coasts. When the A side is connected to +6 V and the B side to ground, the motor moves the platform forward. When the two

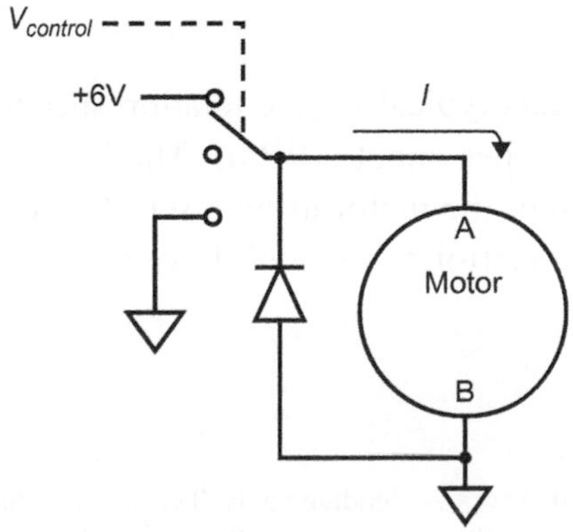

Figure 5.13
PWM motor drive with braking function added.

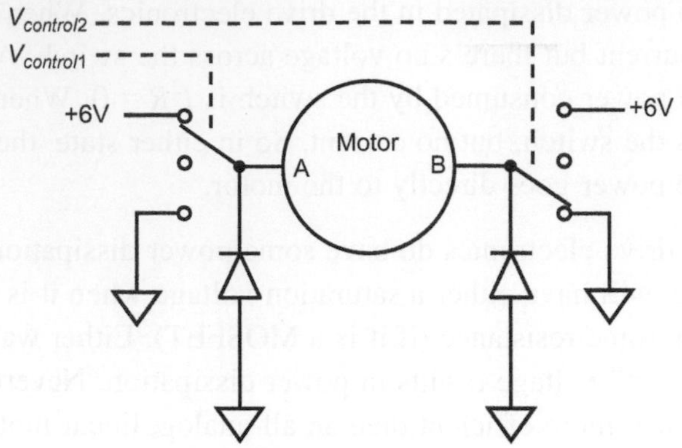

Figure 5.14
Reversible DC motor drive.

Table 5.1: Motor states for each switch state.

		B Side		
		+6V	Open	0V
A Side	+6V	Brake	Open	Forward
	Open	Open	Open	Open
	0V	Reverse	Open	Brake

sides of the motor have the opposite voltage arrangement, the platform moves backward. And when both sides are at the same potential, either ground or +6 V, the motor is braked.

Gears

The motors available for RC cars typically have rotation speeds of the order of thousands or tens of thousands of revolutions per minute (RPM). That's too high for our car/robot. The solution is to reduce the speed of the motor using a set of gears. Fortunately, by choosing to use an existing RC car for our platform, we don't have to actually construct this gear system.

Bibliography

[1] A. Hughes, Electric Motors and Drives – Fundamentals, Types and Applications, third ed., Newnes, 2006.
[2] F. Ulaby, Fundamentals of Applied Electromagnetics, Prentice Hall, 2001.

CHAPTER 6

Inexpensive Ways to Perform Circuit Simulation

Chapter Outline

We've covered a lot of territory so far. Hopefully you understand what was presented, but you may have questions. Or maybe you think you understood everything but find that when you try to apply the ideas for new circuit designs, you aren't quite sure about things.

The latter situation is actually quite common among new engineers (and even some experienced engineers). What if you go to the trouble of building the whole thing up, using one of the prototyping techniques that we'll talk about in a later chapter, and it turns out that the circuit doesn't work because of some aspect of electronics that you didn't understand? You'll have wasted a lot of time learning a painful lesson, right?

Fortunately, there's a way to get around some of this. It turns out that there exist simulation programs that will run on your computer and which can give very accurate estimates of your circuit's performance. These simulation programs have very nice graphical user interfaces, so that it's easy to enter a circuit and to see plots of the circuit response. In addition, these programs are available free.

Circuit simulation history

Before we get into the details, it might be interesting to learn about how such programs developed. Not too long after computers became available, engineers developed simulation programs that would iteratively determine the response of electronic circuits. Our Excel program in Chapter 3, which calculated the RC circuit response, is actually an example of a crude iterative simulation program.

MSP430-based Robot Applications.
DOI: http://dx.doi.org/10.1016/B978-0-12-397012-1.00006-0

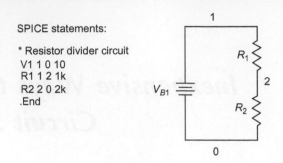

SPICE statements:

```
* Resistor divider circuit
V1 1 0 10
R1 1 2 1k
R2 2 0 2k
.End
```

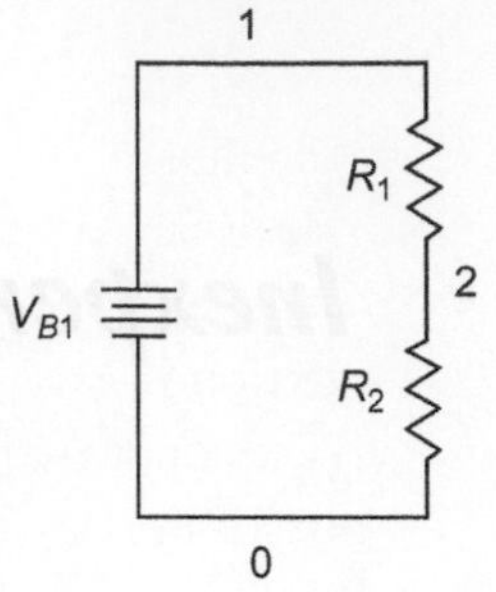

Figure 6.1
SPICE command list for the circuit at right.

The early simulation programs were mostly written in FORTRAN, which was the popular high-level programming language for scientific applications back then. One such program in particular, called SPICE (Simulation Program with Integrated Circuit Emphasis), written in the 1970s, became the de facto standard for circuit simulation.

Although it proved to be quite useful at the time, it seems clunky by today's standards. The simulator data entry is entirely text-based and, for all but the simplest circuits, the circuit designer has to make a hand drawing of the schematic and then number each node. The node numbering is arbitrary (except that the reference node has to be "0") but it's how the simulation program keeps track of the circuit. A SPICE command list for a simple resistor divider circuit is shown in Figure 6.1.

This simple example will return the voltages at nodes 1 and 2 with respect to node 0. SPICE is also capable of returning other useful information, such as how the circuit response changes with time or with frequency.

Modern circuit simulation programs

Virtually all circuit simulation programs available today have a graphical user interface (GUI). The user enters the circuit schematic using conventional symbols for the components (resistors, capacitors, etc.) and connects these together with lines that represent electrical connection. The result is a schematic readily readable to the user that also tells the simulation program how the circuit is connected.

There are a number of circuit simulation programs available for sale but, fortunately, there are also several good ones available free. One of these is TINA-TI, available from the Texas Instruments website. Another is LTSpice, available from the Linear Technology website. The examples that follow will use TINA-TI.

TINA-TI

Using the TINA simulator, we'll enter the circuit of Figure 3.13, the RC circuit (Figure 6.2). Note that, whereas the original RC circuit of Chapter 3 has a switch in series with a 5 V DC voltage, this circuit replaces that with a voltage step generator that goes from 0 V to 5 V when the simulation is run. The step function can be easily set up by double-clicking the VG1 icon. This opens the VG1 window and that then allows opening the Signal editor (the bottom box in Figure 6.3).

In the Voltage Generator box, the General Waveform function is chosen. This is a handy function for creating pulse trains in the simulations. As Figure 6.4 shows, the Signal Editor is set up so that the waveform has a 5 V, 50-microsecond high amplitude, a 0 V low amplitude, and a 500-microsecond pulse repetition period. The General Waveform parameter definition box also allows the user to enter rise and fall times. For the purpose of this example, those are made 1 nanosecond – that is, insignificant compared to the pulse width.

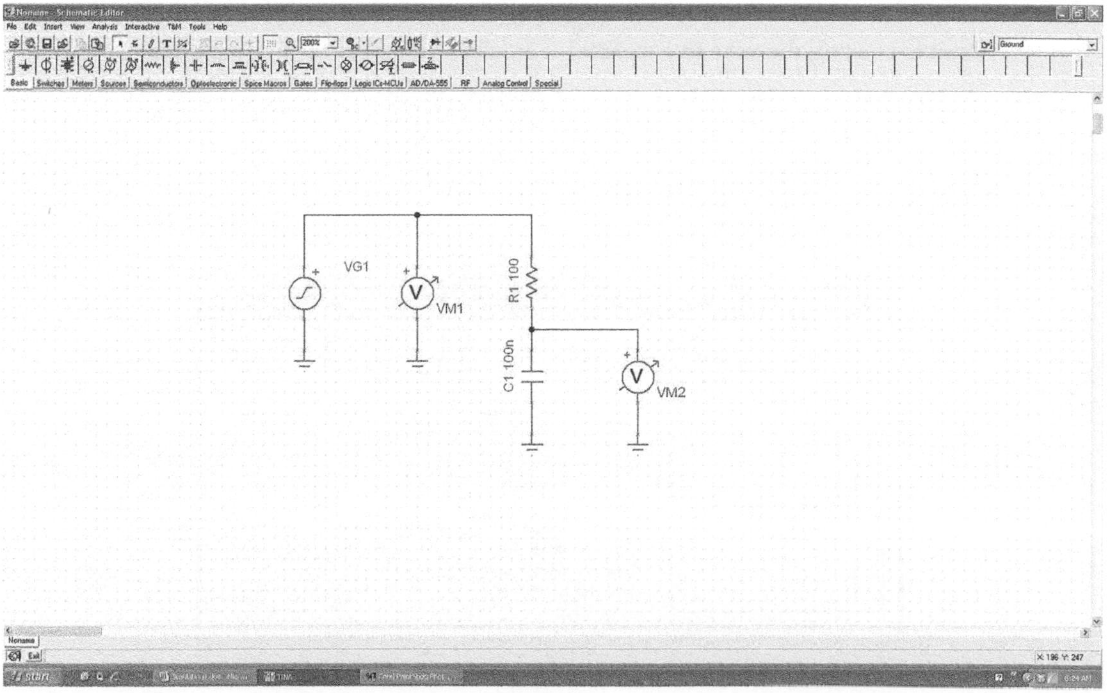

Figure 6.2
The RC circuit from Figure 3.13 entered into TINA simulation program.

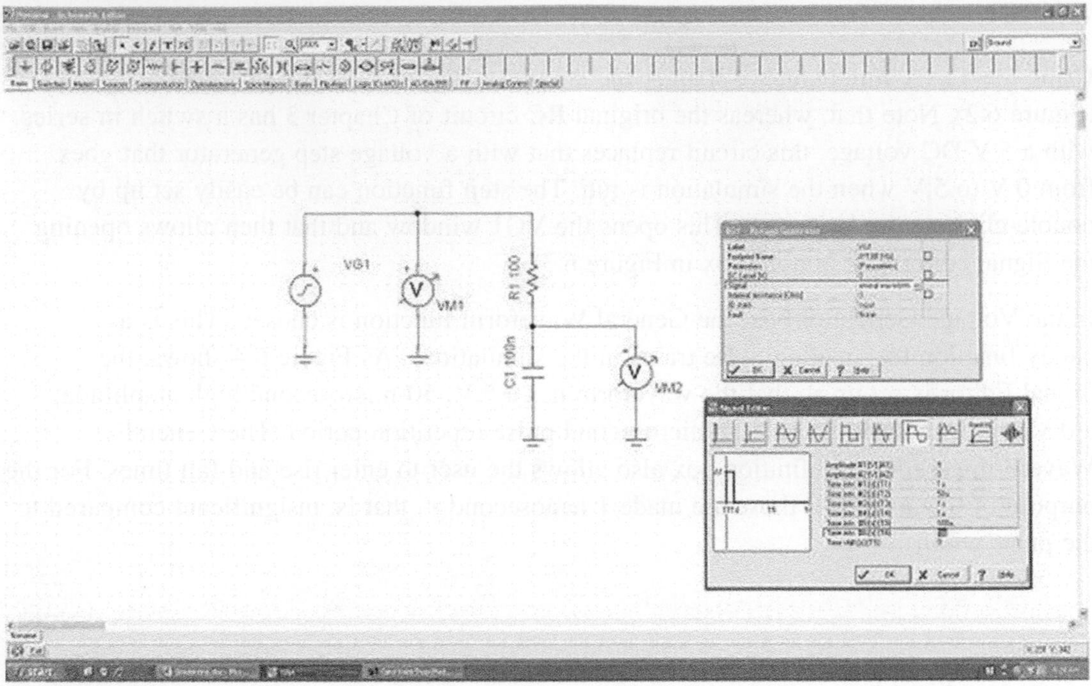

Figure 6.3
Double-clicking the voltage generator icon allows changes to the step voltage.

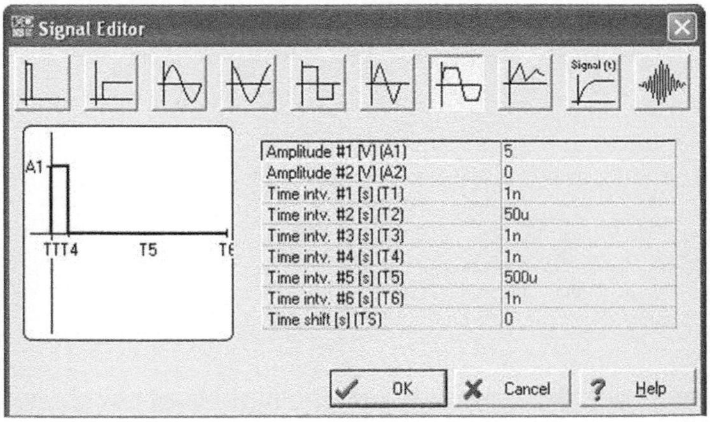

Figure 6.4
Signal Editor with parameters set for 10% duty cycle, 500 microsecond pulse train.

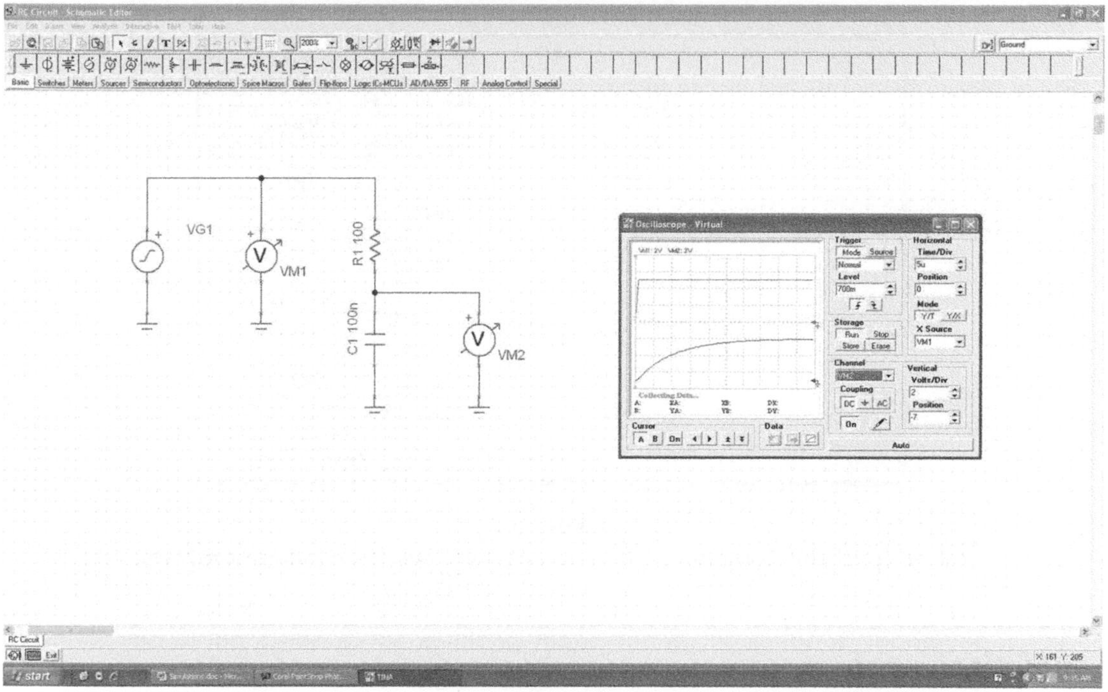

Figure 6.5
Selecting the Oscilloscope under the T&M tab allows the display of input and output.

Before setting up the display function, let's return to Figure 6.3 for a second. Note that there are two icons labeled VM1 and VM2. These are the Volt Meter icons. When we set up the virtual oscilloscope, they will serve as our means of identifying the parameter that we wish to observe.

In Figure 6.5, the Oscilloscope function is chosen under the T&M (Test and Measurement) tab at the top of the page. Here we want to trigger on the rising edge of the pulse created by VG1, so in the Trigger area of the Oscilloscope we choose Mode – Normal; Source – VM1; and Level – 700 mV (any level between 100 mV and 4.9 V would work). The Storage parameter is set to Run, the vertical sensitivity for the VM1 channel is set to 2 V/division, then the VM2 channel is chosen and set to 2 V/division sensitivity. Finally, the Horizontal Time/Div is set to 5 μ (that is, 5 μsec). To separate the two waveforms on the display, the Vertical Position for VM2 is set to −7.

Although that may seem like a lot of things to do to get the waveforms displayed, once you get used to it the process becomes quick and automatic. A simple circuit, like the one in this example, can be entered and displayed in minutes. A zoomed-in view of the oscilloscope is shown in Figure 6.6.

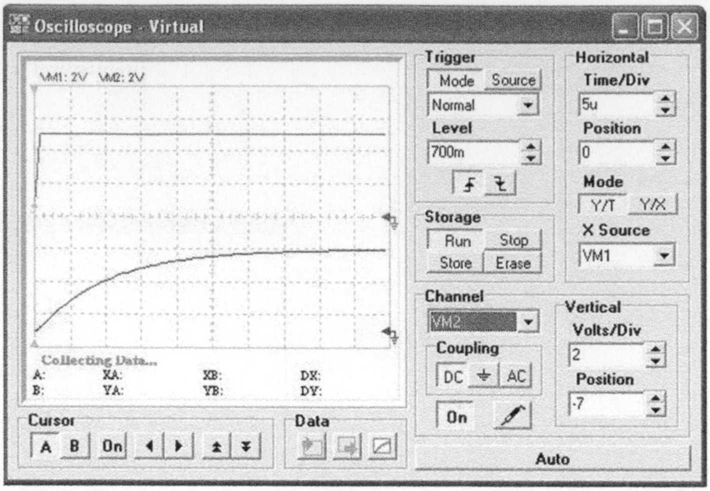

Figure 6.6
Close-up of the Oscilloscope function of the TINA-TI simulation program.

Simulations wrap-up

Circuit simulations are an easy, powerful way to "check your work". You can also play "what if" – trying different values of components to see what effect they have on circuit operation. As you do that, try to predict, based on your knowledge of circuits, what the circuit behavior will be.

We didn't talk about the full range of components available on simulators like TINA-TI, but they're fairly extensive. As you might expect, these free simulation programs have available integrated circuits that the company providing the simulator makes and not necessarily other manufacturers' parts. However, this is generally not a serious limitation if you're just starting out, since most manufacturers will have a wide selection of op amps and other basic building block parts.

CHAPTER 7

Computer Logic

Chapter Outline

The electronics revolution that has taken place over the last 50 years is the result of several factors coming together. Two of the most important of these are the ability to fabricate multiple transistors and other components on a single piece of silicon (the integrated circuit) and the ability to organize transistor circuits into digital devices.

The birth of integrated circuits

The invention of the integrated circuit (IC) is generally attributed to Jack Kilby, although Robert Noyce independently came up with the same idea shortly after Kilby. Jack Kilby was, in 1958, a young engineer at Texas Instruments.

It was already apparent to people in the industry at that time that, as electronic functions became increasingly sophisticated, a functional limit would be reached, due to the increased number of components. Electronics boxes would become not only a massive collection of

MSP430-based Robot Applications.
DOI: http://dx.doi.org/10.1016/B978-0-12-397012-1.00007-2

wires, but an expensive one as well, given the high cost of having skilled craftsmen solder each of those wires to their intended destinations. Printed circuit boards helped alleviate the problem somewhat, but nevertheless the size and cost of sophisticated electronics functions, such as the just-emerging computers, were huge as a result.

Kilby's invention allowed the transistors, resistors, and other components to be interconnected *within the integrated circuit.* As methods and equipment progressed, the number and speed of components integrated into a single device increased, to the point where today ICs with millions of transistors are quite common. The design of such large-scale ICs is not trivial but, once finished, they can be replicated at extremely low cost per IC. This was truly an essential step in the march to today's modern electronics revolution.

The advent of logic

The second field of development that brought about the modern electronics revolution is that of digital logic. The development of this field actually started long before modern electronic computers were invented.

Boole

In the mid-1800s an impoverished English shopkeeper's son by the name of George Boole, almost entirely self-taught in mathematics, developed the system that we discuss in this chapter, Boolean algebra. Using the constructs of this algebra, designers of complex modern computers are able to systematically describe their machines with mathematical equations that can then be manipulated so as to allow the reduction of the machine to its optimal collection of components.

Boole's algebra allowed logical propositions to be expressed as algebraic equations, but, beyond scholars of the day, it seemed to have little application. Some 80 years later it would prove to be extremely useful in designing digital circuits.

Shannon

Claude Shannon is best known for his contributions to the fields of digital communications and information theory. In fact, many consider him to be the father of these fields, having produced much of the theory that is used to this day to determine the limits of modern coding theory, encryption codes, and data compression. His groundbreaking work in communications, *A Mathematical Theory of Communications*, is available at a number of websites online.

But before Shannon had produced this work in communications, and while still a Master's student at MIT, he made a substantial contribution to modern electronics by realizing the potential of Boole's work in analyzing complex switching networks. His Master's thesis, *A Symbolic Analysis of Relay and Switching Circuits*, was written in 1937. In it, he described what is referred to today as a switching algebra, in which he demonstrated that the properties of bistable electrical switching circuits (relays back then but solid-state circuits today) can be represented by Boole's algebra.

Why is it called logic?

Back in Boole's day, the quest was for a means of applying mathematical rigor to logical propositions. Logic is a major branch of philosophy dealing with the truth of a particular statement. Complex statements, in which each of several propositions can be either true or false, must be evaluated to determine if the overall statement is true or not.

To illustrate, let's says that we're given a large collection of dogs and that our job is to separate the black terriers and white terriers from all the other dogs. In effect, we are testing the truth of the statement "This dog is a black or white terrier" for each dog considered. To visualize the problem, we could draw the *Venn diagram* of Figure 7.1. The dogs that we're interested in separating from the overall set of dogs are those represented by the darkest parts of the three circles that lie within the terrier circle. Note that some

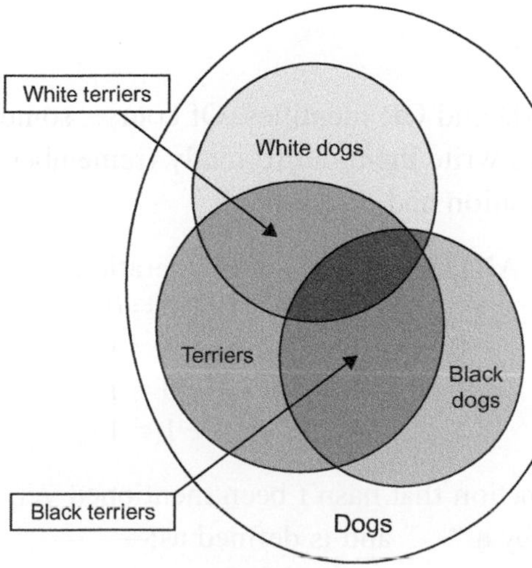

Figure 7.1
Venn diagram for the black and white terrier example.

terriers will be both black *and* white and that the darkest part of the diagram in the center represents the *intersection of those sets.*

Let's see if we can write our requirements statement as a mathematical equation. Let's set up a variable, w, which is 1 if the dog is white and 0 otherwise. Similarly, we'll set up a variable, b, for the black trait and a variable, t, for the terrier breed. Let's define some *operators*: " $\cdot$ " and " $+$ ". These aren't multiplication and addition, as you might have thought. Instead, let them stand for the words AND and OR, respectively. Now, we can describe a dog's inclusion in the designated group by the value of z, as in:

$$z = t \cdot (w + b) \tag{7.1}$$

We can think of this statement as saying "z is 1 if t is 1 AND if either w OR b is 1". Or we could say it in an equivalent, but more understandable, way by saying "The dog is included if it is a terrier AND if it is either white OR black".

If $z = 1$ the statement is true, that is, the dog is included, otherwise not. Let's try this out on some examples. A Jack Russell terrier is included in our group (that is, $z = 1$) because it is white (at least partly) and it is a terrier. A black Labrador retriever is not included ($z = 0$) because $t = 0$. An all-brown Norfolk terrier won't be included ($z = 0$) because, even though $t = 1$, both w and b are zero, making the quantity in parentheses in Eq. 7.1 equal to zero.

The variables just described, z, t, w, and b, are called *binary variables* because each takes on just one of two values, either 0 or 1. Equation 7.1 is a Boolean function of binary variables.

Some algebraic identities

First, let's write down the AND and OR identities. Of course, some of these were hinted at in the terrier example, but let's write them out formally (remember that " $\cdot$ " and " $+$ " refer to AND and OR, not multiplication and division):

AND Operation	OR Operation
$0 \cdot 0 = 0$	$0 + 0 = 0$
$0 \cdot 1 = 0$	$0 + 1 = 1$
$1 \cdot 0 = 0$	$1 + 0 = 1$
$1 \cdot 1 = 1$	$1 + 1 = 1$

Also, there's an important function that hasn't been mentioned yet, the NOT function. The NOT function is represented by a " $'$ " and is defined as:

$$0' = 1$$
$$1' = 0$$

Now, let's write some properties for a single binary variable in switching algebra:

$$x + x = x$$
$$x \cdot x = x$$
$$x + 1 = 1$$
$$x \cdot 0 = 0$$
$$x + 0 = x$$
$$x \cdot 1 = x$$
$$x + x' = 1$$
$$x \cdot x' = 0$$

The proof for the above eight properties is easy – simply substitute $x = 1$ or $x = 0$ into each statement and then use the AND, OR, and NOT identities. This is called *proof by perfect induction*, also sometimes referred to as *proof by exhaustion.*

Associativity, commutativity, and distributivity

Now let's look at some properties involving multiple binary variables. The first of these is associativity:

$$(x + y) + z = x + (y + z) \tag{7.2}$$

and

$$(x \cdot y) \cdot z = x \cdot (y \cdot z). \tag{7.3}$$

Again, you can prove these yourself using perfect induction – that is, just plugging in each combination of x, y, and z values (there are eight unique combinations to try for three variables).

Next is commutativity:

$$x + y = y + x \tag{7.4}$$

and

$$x \cdot y = y \cdot x. \tag{7.5}$$

Finally, distributivity is given by:

$$x \cdot (y + z) = x \cdot y + x \cdot z \tag{7.6}$$

and

$$x + y \cdot z = (x + y) \cdot (x + z). \tag{7.7}$$

The convention used for operation priority is the same as in ordinary algebra: $x + y \cdot z$ means $x + (y \cdot z)$ and not $(x + y) \cdot z$.

These properties of commutativity, associativity, and distributivity in Boolean algebra can, of course, be applied to functions of more than three variables.

Logic optimization

The power of switching algebra becomes especially apparent when the circuits become more complex. Using the properties already identified, as well as many other properties and methodologies developed over the years, very complicated logical expressions can be reduced to much simpler ones, resulting in huge reductions in the number of transistors that ultimately are needed to implement the function.

A very simple example of how powerful this reduction can be is the Absorption law:

$$x + x \cdot y = x. \tag{7.8}$$

Instead of needing both an AND and an OR function and needing both binary variables, x and y, it turns out that the expression, $x + x \cdot y$, can be reduced to just x!

Incidentally, you may have noticed that these properties seem to occur in pairs. That's not a coincidence. It's due to what is called the *Principle of Duality* (also referred to as De Morgan's theorem), which says that if, for a given property, we interchange the AND and OR functions and interchange the 0 and 1 constants, we arrive at the *dual* of the property. In the case of Eq. 7.8, that dual is:

$$x \cdot (x + y) = x. \tag{7.9}$$

Advanced topics in computer logic

Like the analog circuit theory covered earlier, the computer logic chapter of this book is limited to what is needed to get started in robotics. Specifically, there will be some simple logic circuits in the robotics design presented later and, most importantly, microcontrollers almost always include, in their instruction set, logical instructions that allow bits within the registers to be ANDed, ORed, or otherwise logically operated on. Understanding the basics of computer logic is therefore a must. However, this is where we must leave the theory for this book. The remainder of the chapter will look at circuit implementations of computer logic.

If you are interested in pursuing logic theory further, there are lots of books available. The ones that I've listed in the References section happen to be the ones that I own and am most familiar with. Many of them, like the references I list elsewhere in this book, are pretty old. However, if you go out to one of the book websites like Barnes and Noble, Amazon, etc., you'll find that there are lots of old, used books for sale, usually at very reasonable prices. Buying older textbooks is actually a great way to learn about things without spending a lot of money.

Electronic implementation of logic

Okay, so logical functions operate on 1s and 0s. But how do they do that? What do the circuits look like that make up the operations like AND, OR, and NOT? Figure 7.2 shows a circuit that performs a NAND function, which is an AND function followed by a NOT function:

$$\text{NAND}(x, y) = (x \cdot y)' \tag{7.10}$$

Let's analyze this circuit. Let's assume (we can verify this later) that current will flow through resistor R_1 under any circuit condition, either through one of the input diodes on the left or through the diode and transistor base-to-emitter on the right. When either x or y is at ground, then, if the current flows through that input diode, the point labeled "A" will be at 0.7 V (the forward drop of the diode). If point A is at 0.7 V, then the transistor is not on, since point A would have to be at around 1.4 V (the forward drop of the diode connected to the transistor base, plus the forward drop of the transistor's base-to-emitter voltage) for the transistor to be on. If the transistor is off then no current flows through R_2 and $z = 5$ V. This accounts for the first three table rows in Figure 7.2.

What about when both x and y are at 5 V? In that case, the current through R_1 flows through the right hand diode and the transistor's base-to-emitter. Point "A" will be at about 1.4 V, so the current through R_1 and therefore the current flowing into the transistor base will be around 3.6 V/4.7 kΩ ≈ 800 μA.

Now, is the transistor in its linear or in its saturated mode? Remember from Chapter 4 that in these situations we can just assume one or the other situation, then see if that actually makes sense. Let's assume the transistor is saturated. Then the output is at about 0.4 V, so the current through R_2, and therefore into the transistor's collector, is (5 V − 0.4 V)/1 kΩ = 4.6 mA.

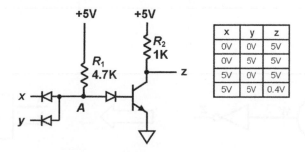

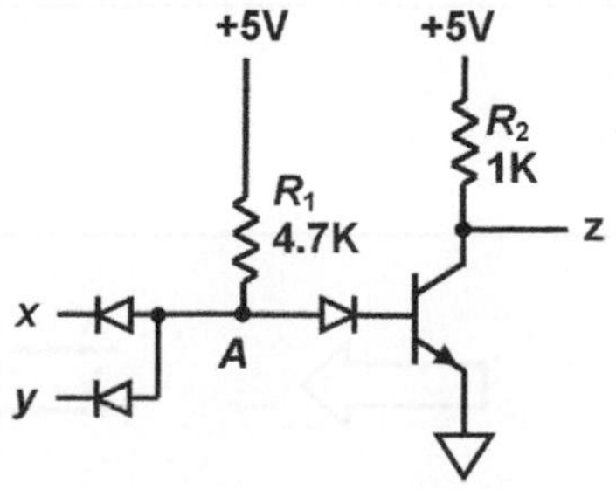

x	y	z
0V	0V	5V
0V	5V	5V
5V	0V	5V
5V	5V	0.4V

Figure 7.2
NAND circuit using diodes and transistors.

From Chapter 4, we know that, for the transistor to be saturated, we must have $I_C < \beta \cdot I_B$, where β is the transistor's current gain. Rewriting that inequality gives:

$$\beta > I_C/I_B = 4.6\ \text{mA}/800\ \mu\text{A} \approx 6. \tag{7.11}$$

That is, as long as the transistor has a β greater than 6, the transistor is saturated. Since virtually all transistors have a β greater than 6, it's safe to say that the transistor is saturated.

Now, let's take the table included in Figure 7.2 and assign logic values based on the criteria that any voltage greater than 4 V is a logic 1 and any voltage below 0.5 V is a logic 0. The table will then look like Table 7.1.

Such a table of input and output values for a logical function is called a truth table. Note that the Table 7.1 input values, *x* and *y*, in each row do indeed produce the given output value, *z*, when plugged into the NAND logic equation of Eq. 7.10.

The type of circuit implementation shown in Figure 7.2 is known as diode transistor logic, or DTL, and was a popular way to implement gates when logic integrated circuits were first introduced. It was later superseded by transistor-transistor logic (TTL), which has itself now been largely superseded by complementary MOSFET logic (CMOS). With advances in modern integrated circuits, the logic gate can often be thought of in terms of just its logical function, without much regard about how the circuit is actually implemented internal to the integrated circuit. These "gates" are therefore the lowest level of circuit operation that we need to concern ourselves with in designing the circuit.

For example, AND gate and NOT gate symbols are shown as in Figure 7.3. These two functions can be combined, of course, and the symbol for this is the NAND gate symbol. This is the symbol that represents the circuit of Figure 7.2 in a logic drawing.

Table 7.1: NAND function truth table for the Figure 7.2 example.

x	y	z
0	0	1
1	1	1
1	0	1
1	1	0

Figure 7.3
The AND gate and NOT gate can be symbolized as the NAND gate.

Similarly, we can create a *NOR gate* by logically cascading an OR gate and a NOT gate (incidentally, the NOT gate is usually referred to as an *inverter*). Why the discussion of NANDs and NORs, when the basic logic functions were introduced as AND, OR, and NOT? Two reasons account for this: one is that many circuit implementations (including the example in Figure 7.2) naturally implement NAND or NOR rather than the AND or OR. The second reason is the fact that any logic function can be implemented using only NAND gates or only NOR gates.

A logic example – the adder

Okay, let's build a logic circuit. We won't worry about what transistors or other physical components go into the makeup of the logic gate – that concern is conveniently handled for us by the manufacturer of the logic gate integrated circuits that we'll use. Instead, we'll just think of the signals in terms of their value, 1 (high voltage) or 0 (low voltage), and in terms of the logical functions operating on those signals.

One useful logic function we could build is an adder. This one will be the world's simplest adder, just one bit wide, but, hey, you've got to start somewhere. The adder will have two inputs – let's call them a and b – and two outputs – the sum, which we'll call s, and the carry, which we'll call c. Let's draw our function as just a box, like in Figure 7.4.

Since we're talking about the arithmetic addition of a and b, the sum, s, should be 0 if both a and b are 0 and it should be 1 if either a or b (but not both) are 1. In all three of those cases, there is no carry out of the adder, so $c = 0$. Now, what about when both a and b are 1? In that case, the arithmetic sum is 2, but our single-bit sum signal, s, can't represent a number that large. In that case, we have a carry out of the adder, so $c = 1$, and the sum bit will be 0.

Let's put the results of this discussion into tabular form (Table 7.2).

It's easy to show that the sum bit obeys the equation:

$$s = a' \cdot b + a \cdot b'. \tag{7.12}$$

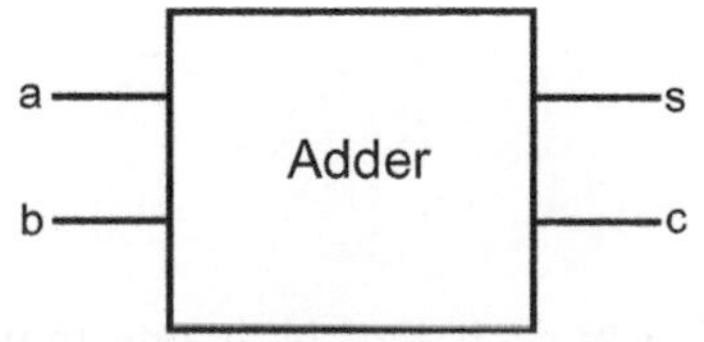

Figure 7.4
The Adder function symbol.

Table 7.2: Truth tables for the sum and carry outputs of the single-bit adder.

a	b	s
0	0	0
0	1	1
1	0	1
1	1	0

a	b	c
0	0	0
0	1	0
1	0	0
1	1	1

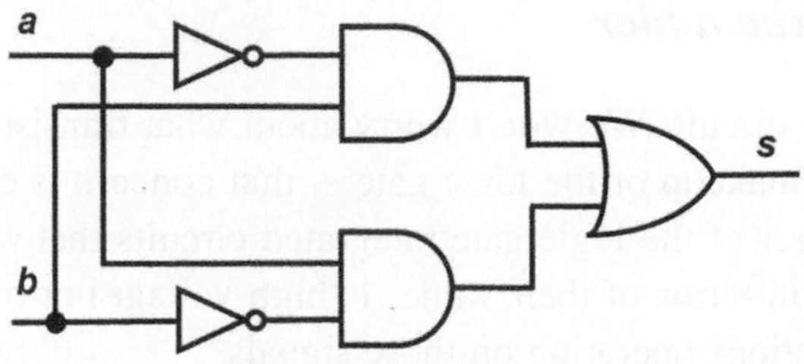

Figure 7.5
Logic drawing for the sum function of the adder example.

(Equation 7.12 is a logic equation, so " · " means AND and " + " means OR). Using the AND gate symbol and NOT gate symbol already introduced, as well as an OR gate symbol, the logic diagram for this function is shown in Figure 7.5.

Just to reinforce the statement made earlier, that any logic function can be built strictly from NAND gates, the sum function logic circuit is implemented entirely with NAND gates in Figure 7.6. Plug in values for *a* and *b* in the diagram to prove to yourself that the *s* values follow the truth table in Table 7.2.

Now let's go on to the logic implementation of the carry bit, *c*. Looking at the truth table in Table 7.2, it's clear that the carry bit is simply given by:

$$c = a \cdot b. \tag{7.13}$$

The carry bit logic circuit is therefore about as easy as it can get: just an AND gate (Figure 7.7).

The Exclusive-Or gate

The logical function of Eq. 7.12, $s = a' \cdot b + a \cdot b'$, is one that occurs so often in logic design that it is given its own gate symbol. Figure 7.8 shows the sum function when using an Exclusive-Or gate.

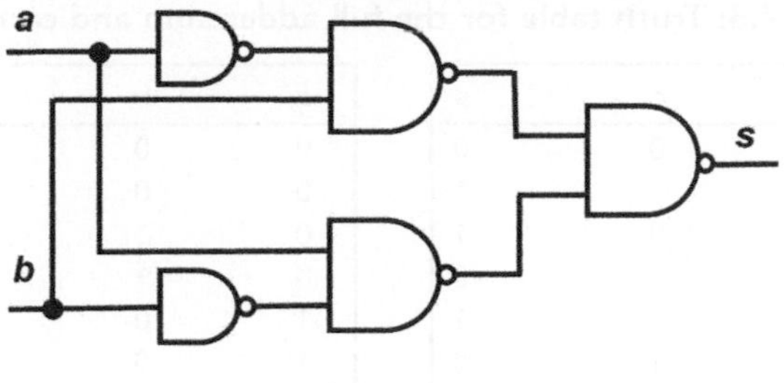

Figure 7.6
All-NAND version of the sum function.

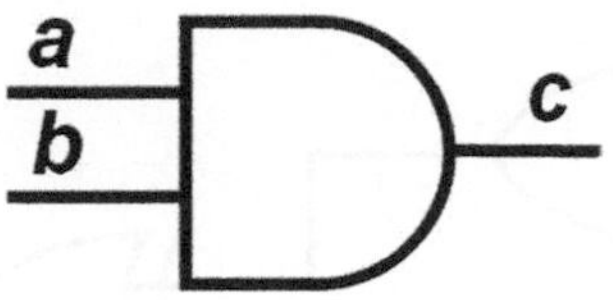

Figure 7.7
The carry function for the single-bit adder.

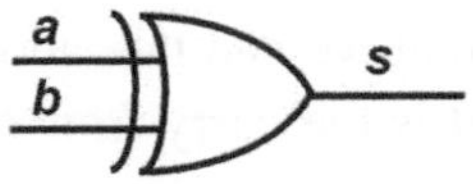

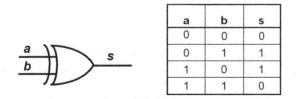

a	b	s
0	0	0
0	1	1
1	0	1
1	1	0

Figure 7.8
An Exclusive-Or implementation of the sum function.

An Exclusive-Or instruction is included in the MSP430 instruction set for performing this operation on bits within registers in the microcontroller.

The multiple-bit adder

The adder that we created in the previous section is known as a half adder. As you can see, it's not very useful, given that it can only add two numbers with a maximum value of 1. Let's at least try to extend the design to one that can handle two-bit numbers. To do that we'll need, for the next stage, what is called a full adder.

The full adder differs from the half adder in that it makes use of the carry bit from the previous stage. It therefore arithmetically adds the carry input to the arithmetic sum of *a* and *b*. The truth table for this function is shown in Table 7.3. Note that, because we now

Table 7.3: Truth table for the full adder sum and carry bits.

a_i	b_i	c_{i-1}	s
0	0	0	0
0	0	1	1
0	1	0	1
0	1	1	0
1	0	0	1
1	0	1	0
1	1	0	0
1	1	1	1

a_i	b_i	c_{i-1}	c_i
0	0	0	0
0	0	1	0
0	1	0	0
0	1	1	1
1	0	0	0
1	0	1	1
1	1	0	1
1	1	1	1

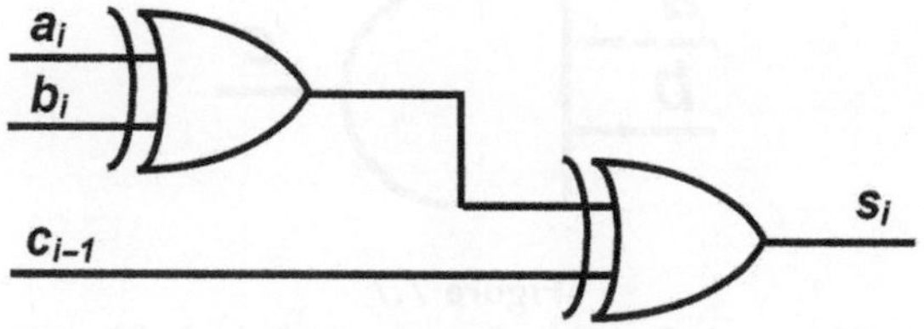

Figure 7.9
Sum logic implementation for the *i*-th bit.

have multiple bits, the subscript "i" is added, so that we can talk about the i-th bit. Note also that the carry input is labeled as c_{i-1}, since it is the carry from the previous bit.

Although the sum function may not appear to have an easy solution, we can use the fact that this function "adds the carry input to the arithmetic sum of a and b", as mentioned previously, and use this as a hint to implement this function as shown in Figure 7.9.

For the c_i output bit, the logic function can be determined just by observing that any time there are at least two bits that are 1, the output is 1. Therefore,

$$c_i = a \cdot b + a \cdot c_{i-1} + b \cdot c_{i-1}. \tag{7.14}$$

The logic diagram for the carry function is given in Figure 7.10.

We can combine the sum and carry functions of Figures 7.9 and 7.10 into a full adder logic function, as used for all of the parallel adder stages in Figure 7.11 except the least significant bit. The half adder function, which combines the sum and carry functions of Figures 7.7 and 7.8, is used for that adder stage.

Note that the n adder stages in Figure 7.11 are numbered 0 through $n-1$. This is a very common numbering scheme in computer applications and is something you'll have to get used to, since you're probably accustomed to starting the numbering with the numeral 1.

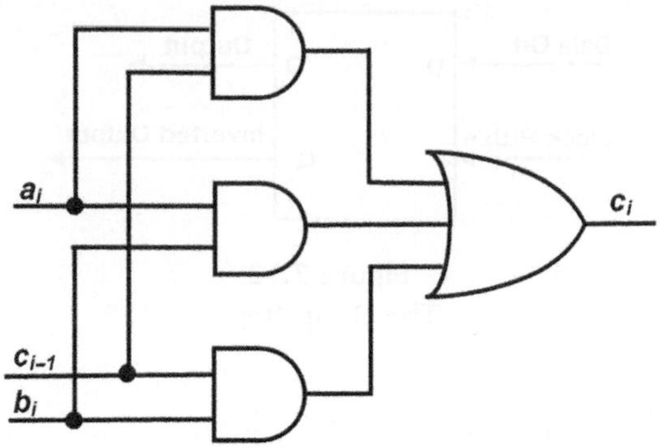

Figure 7.10
The carry function for the full adder.

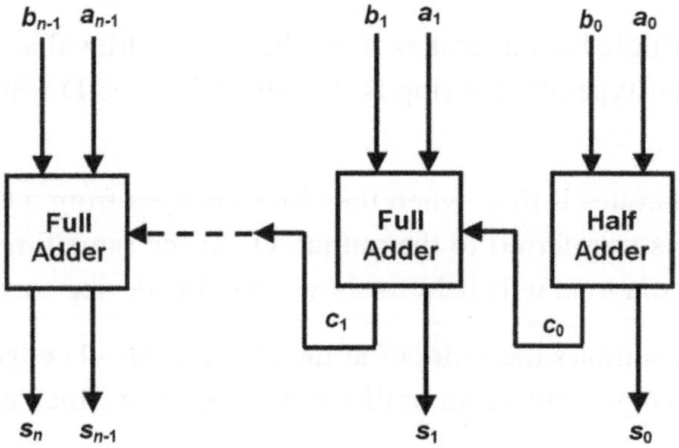

Figure 7.11
An *n*-stage parallel adder.

So this parallel adder adds two base-2 (binary) numbers, $a_{n-1}a_{n-2} \ldots a_1a_0$ and $b_{n-1}b_{n-2} \ldots b_1b_0$, and computes the sum of the two numbers, $s_ns_{n-1}s_{n-2} \ldots s_1s_0$.

Also note that this n-stage parallel adder produces $n + 1$ sum bits. The carry bit out of the most significant stage, the n-th stage, actually becomes the most significant sum bit, s_n. Why that's needed will become clearer in the next chapter, when we look at computer arithmetic.

Flip-flops and registers

One of the things that make digital circuits so powerful is the ability to remember. The most basic of all memory functions in a digital electronic circuit is the flip-flop.

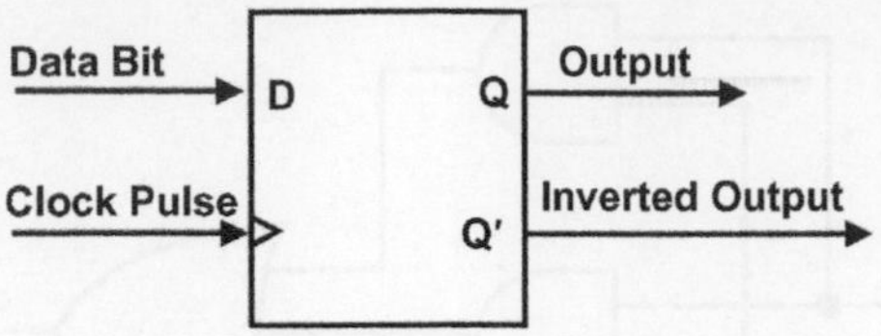

Figure 7.12
The D flip-flop.

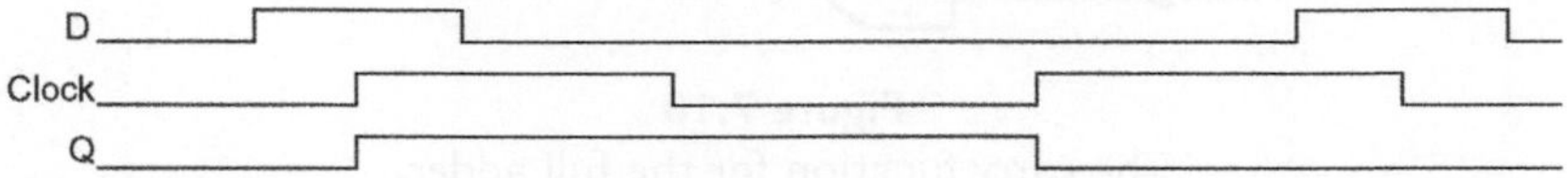

Figure 7.13
Timing diagram for the D flip-flop.

The flip-flop stores a single bit and makes it available for retrieval at some later time. One of the most popular types of flip-flop is the edge-triggered D flip-flop, depicted in Figure 7.12.

The way this device operates is that, when the clock changes from a 0 to 1, the value at the D input at that instant is transferred to the output, Q. At the same time, the inverse of Q is output at Q′. A timing diagram will help to show how this device works (Figure 7.13).

Note that the Q output samples the D input at the positive ($0 \rightarrow 1$) edge of the Clock input. The output does not change state again until the next positive Clock edge.

Registers

Just as half adder and full adder cells can be put together to form a parallel adder, flip-flops can be combined to form what is referred to as a register. Registers are quite common in microcontrollers, so we'll be talking about these more in subsequent chapters.

Logic chapter wrap-up

As with the last several chapters, this chapter is highly technical and may seem like quite a diversion if what you were ready for when you started this book was to build something. However, to do a good job of building the robot, you need the right tools. Some of those tools are the physical ones that you might think of first – pliers, meters, etc. But, for engineers, a lot of the tools that are needed are in their head – theory, troubleshooting techniques, circuit approaches used successfully in the past.

Successful design and troubleshooting is the result of using these more abstract tools to which you have access. Boole's story was included partly to illustrate this point. Boole had almost no formal training in mathematics beyond what his impoverished father could help him learn. He developed a brand-new field of mathematics, not by ignoring what others had done, but by expending considerable effort in first learning what they had already developed and building on that. Likewise, your success as an engineer, as a designer, or as a software developer relies on your ability to first learn what others have done, and then build on that.

Bibliography

[1] E.T. Bell, Men of Mathematics, Simon and Schuster, New York, 1937.
[2] T. Booth, Digital Networks and Computer Systems, John Wiley and Sons, 1971.
[3] Z. Kohavi, Switching and Finite Automata Theory, McGraw-Hill, 1970.
[4] F. Kuo (Ed.), Digital Electronics with Engineering Applications, Prentice Hall, 1970.
[5] M. Mano, Digital Logic and Computer Design, Prentice Hall, 1979.

CHAPTER 8

Computer Arithmetic

Chapter Outline

As we have seen from the logic chapter, early designers of computers quickly realized that performing logic operations and computations in a binary system, where each digit takes on only one of two values, 0 or 1, had real benefit. The mathematics is straightforward and well-developed and the precision required in the electronics is low, due to the fact that the electronics decisions require choosing between just two values.

Of course, humans generally perform their arithmetic in base-10. Compilers (used for high-level languages, such as C) and assemblers (for assembly language) usually allow the use of base-10 constants and make other accommodations to help the base-10-oriented

MSP430-based Robot Applications.
DOI: http://dx.doi.org/10.1016/B978-0-12-397012-1.00008-4

programmer. But microcontrollers not only add and subtract in base-2, they also manipulate single bits in many control applications. So base-2 arithmetic is a real skill that needs to be learned in order to make good use of microcontrollers.

Before diving into the base-2 numbering system, let's think about the base-10 numbering system (also referred to as the decimal system). Take, as an example, the number 192 decimal (you may also see this written as 192_{10}, although it's generally assumed that we mean decimal or base-10 when no base is written as a subscript). What we really mean by this string of digits is:

$$192 = 1 \times 10^2 + 9 \times 10^1 + 2 \times 10^0 = 100 + 90 + 2. \tag{8.1}$$

This might seem like a lot of work to express something that we take for granted in everyday decimal arithmetic. But thinking of the number in this way helps in understanding how base-2 (or any base) works.

Getting started with binary

Let's say that we are given the binary number 10100_2. The first thing you'll notice is that all digits are just 0 or 1. Why? Well, just as base-10 has no individual digits greater than or equal to 10, the base-2 numbering system has no digits greater than or equal to 2. So that means that the only digits used are 0 and 1. These *b*inary dig*its* are referred to as bits.

Before talking about how the binary system actually works, it's appropriate to discuss number organization inside a computer and some common terms. The word *byte* refers to a group of 8 bits. The term *word* refers to a group of bytes. For the MSP430 microcontroller, a word refers to 2 bytes, or 16 bits. There's nothing magical about groupings of 8 or 16 bits – it's just the standard that the computer industry has adopted. Indeed, early computers sometimes had 12 bits or other arrangements as their basic grouping.

In this chapter, the examples will all be for byte operations. This is simply for ease of understanding and ease of drawing. Keep in mind that in most computers, the operations will be on 16-bit and larger words.

Converting from binary to decimal

Now, let's get back to the example number, 10100_2. If this number is written as an 8-bit byte, we would include leading zeros and write the number as 00010100_2. Note that a byte can be used to represent integers between 0 and 255_{10}.

Bit Exponent (i)	7	6	5	4	3	2	1	0
Bit Weight (2^i)	128	64	32	16	8	4	2	1
Bit Value	0	0	0	1	0	1	0	0

Figure 8.1
Bit values of the binary number of Eq. 8.1 shown with their associated bit weights.

Now, let's write this binary number as powers of two, just as the decimal number, 192, was written as powers of ten:

$$00010100_2 = 0 \times 2^7 + 0 \times 2^6 + 0 \times 2^5 + 1 \times 2^4 + 0 \times 2^3 + 1 \times 2^2 + 0 \times 2^1 + 0 \times 2^0 = 16 + 4 = 20. \quad (8.2)$$

Figure 8.1 shows the individual bits under their associated power of 2. As you can see, conversion from binary to decimal is straightforward.

Converting from decimal to binary

What if we wish to convert from a decimal number to a binary number? The conversion process is a little more involved, but not bad. First, a verbal description is in order.

The process involves finding the largest power of two that is less than the decimal number being converted. That power of two is subtracted from the decimal number and the process is repeated until the difference between the collection of powers of two and the decimal number is zero. Writing the process out step-by-step gives:

1. Start with the largest power-of-two exponent (7 if the resulting binary number is to be a byte)
2. If the power-of-two number is less than the decimal number, mark a 1 in the position of the binary number corresponding to this power of two, otherwise go to step 5
3. Subtract the power of two from the original decimal number
4. If the difference is zero, the process is finished, else
5. Take the next largest power of two
6. Go back to step 2

Figure 8.2 is a flowchart of the process.

To demonstrate, let's look at the decimal number, 20. The first three powers-of-two, 128 (2^7), 64 (2^6), and 32 (2^5), are each greater than 20, so we mark zeros for each of these bits. The next power-of-two, 16 (2^4), when subtracted from the number, 20, produces a positive difference, 4, so the bit associated with that place is a 1. The next power-of-two, 8 (2^3), is greater than the difference, 4, so that bit is a 0. Finally, the next power-of-two, 4 (2^2), when subtracted from the difference, 4, produces a new difference that is zero, so the

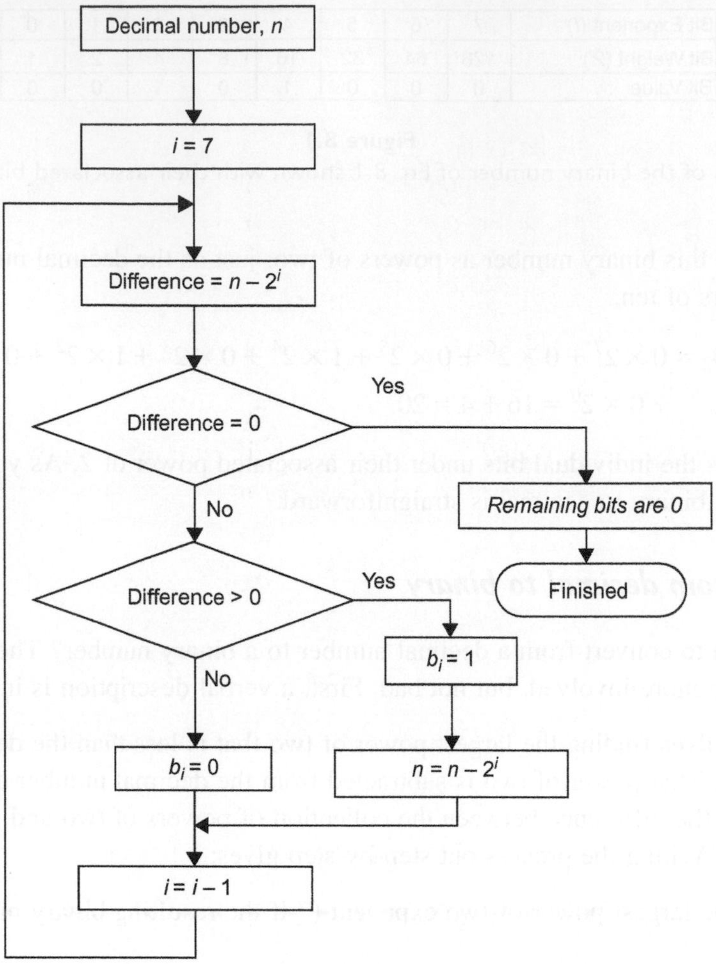

Figure 8.2
Flowchart of the decimal-to-binary (byte) conversion process.

process ends. The remaining two bits are zero. The binary equivalent to 20 is therefore exactly as shown in Figure 8.1.

Addition

We saw in the last chapter, on computer logic, what a parallel adder looks like. Each individual full adder cell of the complete parallel adder is pretty simple – there are three inputs, the carry-in, which we'll call c_{i-1} and the two data inputs, a_i and b_i. The two outputs are the carry-out, c_i, and the sum bit, s_i. If all three inputs to a cell are 0, both c_i and s_i are 0.

If one of the three inputs is 1 or if all three bits are 1, the sum bit, s_i, is 1. If two of the three bits are 1, s_i is 0. c_i is 1 if more than one input is a 1. Let's state that again in tabular form: see Table 8.1.

Let's try an example to help clarify how this works. Let's add the bytes 11010100_2 and 01011111_2. The first of these you should be able to show is 212_{10} and the second, 95_{10}. Let's write these two numbers as the a and b input operands:

$$a_7a_6a_5a_4a_3a_2a_1a_0 = 11010100$$

$$b_7b_6b_5b_4b_3b_2b_1b_0 = 01011111.$$

When we add these two numbers, we do so exactly as we would when adding two decimal numbers together – that is, we start with the least significant digits and work our way to the left. Figure 8.3 shows the process. As you follow through the process, use Table 8.1 if you aren't sure about the sum and carry output values.

Note that whereas the two addends can both be contained in 8-bit bytes, the sum is a 9-bit number. This should not be a surprise. Remember that an 8-bit binary number can represent decimal numbers in the range 0 to 255, so it is possible for a sum to be as great as 510.

Table 8.1: Truth table for the full adder cells of a parallel adder.

Inputs			Outputs	
c_{i-1}	a_i	b_i	s_i	c_i
0	0	0	0	0
0	0	1	1	0
0	1	0	1	0
0	1	1	0	1
1	0	0	1	0
1	0	1	0	1
1	1	0	0	1
1	1	1	1	1

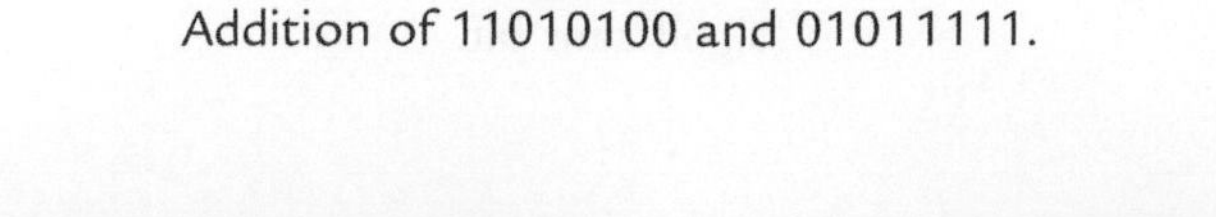

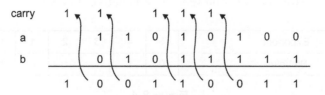

Figure 8.3
Addition of 11010100 and 01011111.

If you remember back to the parallel adder example at the end of the computer logic chapter, the most significant full adder stage had two sum bits that it produced – one from the normal sum output and the other that would normally be the carry output. That was the reason for that unusual last stage.

Now, let's check to see if the addition went as expected. The sum is the 9-bit number 100110011_2, which we can show to be $256 + 32 + 16 + 2 + 1 = 307$. Given that we previously computed the decimal equivalents for the two addends to be 212 and 95, we can see that the result is correct.

Fractional numbers

So far we've dealt strictly with integers. But there's no reason that we can't have bits that represent fractions. For example, let's say that, instead of wanting to represent numbers in the range from 0 to 255 with a single byte, we decide to represent numbers in the range 0 to 63.75, where the two least significant bits represent increments of 0.25. Our powers of two could range from 5 down to -2, rather than from 7 down to 0. A number like 27.25 would be represented by the bit sequence 01110101_2 as shown in Figure 8.4.

When the binary number includes powers of two that are negative like this, the number is said to have a *binary point* to the right of the power-of-0 bit (in the example, this is also to the left of the second least significant bit). Another term for fractional numbers like this is *fixed-point numbers*.

If this idea of representing fractional numbers seems a little confusing, don't worry – we're not going to be using fractional numbers in the robot project. Its inclusion in this chapter is just to introduce you to a topic that you will likely encounter in the future.

Negative numbers

So far we've talked only about non-negative numbers. For example, we've talked about a byte representing numbers in the range 0 to 255. But the real world contains lots of data with negative values. How do we represent these negative numbers?

Exponent	5	4	3	2	1	0	–1	–2
Bit Value	0	1	1	1	0	1	0	1

Figure 8.4
The number 27.25 represented by a byte with binary point to the left of the second least significant bit.

Sign-magnitude representation

One way this could be done is to limit the number of bits used for the magnitude to seven bits (assuming byte-sized operations) and use the most significant bit as a sign bit. This would allow the representation of numbers in the range −127 to +127. $0111\ 1111_2$ would be our representation for +127 (the most significant bit = 0 meaning positive sign) and $1111\ 1111_2$ would represent −127. One little quirk in this system is that the decimal number 0 can be represented by either 0000 0000 or 1000 0000.

The system just described is called sign-magnitude. Its biggest advantage is that it feels very natural for us to use such a system. Why is that? Well, it turns out that this is exactly how we deal with numbers in our everyday decimal calculations. The number −25, for example, is made up of two parts: a two-valued sign symbol (which can take on the value of either " + " or " − ") and a positive magnitude; in this example, it is "25". We humans have learned to make good use of this sign-magnitude system.

The disadvantage of this system in computers can be seen when we consider what is required to add two numbers in such a system. If the two numbers have the same sign, then we just add the two magnitudes and assign the result the same sign as the two addends have. However, if the sign bits of the two addends are different, we subtract the smaller from the larger number, then assign the sign bit according to the sign of the larger addend.

So, in order to perform the simple operation of addition, the computer, if using sign-magnitude, must:

- compare the addend sign bits
- based on that comparison, perform either an addition or subtraction
- after performing the computation, decide the sign bit of the result.

Although this may not seem like a big deal, such extra steps can significantly increase computing time.

Complements

We saw in the logic chapter that a bit can be changed to its opposite value and the function name for this is NOT. It is also called inversion or complementation. When this operation is performed on an entire byte or word, it is called a *one's complement.* The one's complement for the number 00010100_2 is 11101011_2. An important complementation process for computer arithmetic is the *two's complement.* This is just the one's complement with 1 added to the number. So the two's complement of our example number is:

$$\text{Two's complement of } 00010100_2 = 11101011_2 + 1 = 11101100_2 \qquad (8.2)$$

Two's complement arithmetic

A better numbering system than sign-magnitude and one that is almost universally used in computers' internal processing is the 2's complement system. Here a positive decimal number from 0 to 127 is represented with a 0 in the most significant bit and the seven least significant bits containing the magnitude, just as was done with the sign-magnitude numbering system. However, negative numbers are represented by the 2's complement of the magnitude. So to create the binary representation for -107, we first find the binary representation for 107 (that is, $0110\ 1011_2$), then find the 2's complement of this number:

$$\text{Binary representation of } (-107) = \text{2's complement of } (0110\ 1011_2)$$
$$= 1001\ 0100_2 + 1 = 1001\ 0101_2.$$

Okay, so what's so great about that? Well, it turns out that adding two numbers in this system becomes simple, regardless of whether the addends are positive or negative. Let's say we want to add the binary equivalent of 95 to -107. The addition, shown in Figure 8.5, is performed just like the addition that was performed in the first addition example of the chapter.

What is this number in decimal? The sign bit, being a 1, tips us off that this is a negative number. To determine the magnitude of the number, just take the 2's complement, which is $00001100_2 = 12_{10}$. Thus, this is the 2's complement version of -12_{10}, which is the correct answer.

Note that, while we had to do a little arithmetic gymnastics to come up with the decimal answer, that was only necessary to make the answer understandable to humans. The computer is perfectly happy with the result of the Figure 8.5 example just as it is, and no further processing is required to use the result in subsequent operations.

One additional note about 2's complement – since this system has only one binary representation for the decimal number 0, then, for a single byte, that leaves 255 other binary numbers that can be represented. If we decide to have 127 positive numbers (1 through 127),

carry			1	1	1	1	1	
a (−107)	1	0	0	1	0	1	0	1
b (+95)	0	1	0	1	1	1	1	1
	1	1	1	1	0	1	0	0

Figure 8.5
Two's complement addition of -107_{10} and 95_{10}.

then that means we have *128* negative numbers that can be represented (−1 through −128). It might sound odd that the computer can represent a slightly larger range of negative numbers than positive numbers, but that's how all computers employing the 2's complement system work.

Overflow

One issue that the programmer needs to be aware of is overflow. Overflow is simply the outcome of having only a finite set of numbers that can be represented. Say, for example, that we have a 2's complement system and that we are representing each number with a single byte, so that the numbers range from -128_{10} to $+127_{10}$. If we attempt to add +68 ($0100\ 0100_2$) and +82 ($0101\ 0010_2$), the result is as shown in Figure 8.6.

The fact that the most significant bit is 1 tells us immediately that the number is negative, an impossibility since both addends are positive. What has happened is that two numbers were added that produced a result (+ 150) that is beyond the +127 maximum that the single-byte numbering system can handle.

Fortunately, almost all computers include internal hardware that monitors these operations and can alert the program that such an error has occurred.

Subtraction

As you've probably guessed, subtraction is a very straightforward operation in a computer employing 2's complement arithmetic. The computer simply forms the 2's complement of the subtrahend (the operand being subtracted) and adds that to the minuend (the operand from which subtraction is taking place). This entire operation is taken care of internal to most computers. The programmer simply uses the computer's Subtract command (we'll talk about that in subsequent chapters) and its internal hardware forms the 2's complement automatically for the subtrahend.

carry	1							
a (+68)	0	1	0	0	0	1	0	0
b (+82)	0	1	0	1	0	0	1	0
	1	0	0	1	0	1	1	0

Figure 8.6
Overflow from addition.

Multiplication

That takes care of addition and subtraction, so what about multiplication and division? These operations are more difficult to perform. Some microcontrollers, including some MSP430 microcontrollers, have an internal multiply function built in. Such an internal multiply function is referred to as a hardware multiply. In such a case, the programmer simply includes the multiply instruction in the program in the same manner as an addition or subtraction instruction would be used. The instruction typically executes in just a few microcontroller clock cycles.

Note that, with any type of multiply, the number of bits required for the result is twice the number of bits used to represent the two inputs.

Software multiply

Most inexpensive microcontrollers, including the ones that we will use in the robot designs in this book, do not include a hardware multiply. In that case, the programmer has to use some other method of multiplying. The algorithms that can be used to implement a multiply when a hardware multiply is not available are called software multiply algorithms.

Multiplication by iterative addition

One method that can be used to implement a software multiply is the one that your teacher probably taught you when you first learned what multiplication was – you can simply add one of the numbers to itself over and over, as specified by the second number. For example, 23×3 can be computed by adding 23 to itself 3 times: $23 \times 3 = 23 + 23 + 23 = 69$.

This method actually works well and is a relatively fast way of computing the product, as long as one of the numbers is small. The method is slow when the two numbers are both large. Nevertheless, if compute time is not a big issue, this is a very straightforward way to compute the product.

Multiplication by shifting

A much more common method of performing a software multiply is to use an "add-and-shift" algorithm. This is also a method that you were taught in grade school. Think about how you multiply two decimal numbers on paper. Let's say that we need to know 347×192. The way you compute this is illustrated in Figure 8.7.

The top number is multiplied by a single digit of the bottom number, starting with the least significant digit of the bottom number. As you proceed to each successive digit in the bottom number, you write the interim product one digit to the left. So the second interim product, 2987, is shifted left one digit from the previous interim product, 686.

```
        3 4 3
      × 1 9 2
    ---------
        6 8 6
      2 9 8 7
      3 4 3
    ---------
    6 5 8 5 6
```

Figure 8.7
Decimal multiplication example.

```
              1 0 0 0 1 0
        ×       1 1 0 1 1
            -------------
              1 0 0 0 1 0
            1 0 0 0 1 0
        1 0 0 0 1 0
      1 0 0 0 1 0
      -------------------
      1 1 1 0 0 1 0 1 1 0
```

Figure 8.8
Binary multiplication of 34_{10} and 27_{10}.

Why the shifting? Well, remember how, at the beginning of the chapter, we broke 192 down into its actual powers-of-ten representation. So, when you multiply 343 by 9 in this example, you're really multiplying 343 by 90, right? The second product, 2987, is really 29870. Your grade school teacher left the trailing 0 off to avoid confusing the children. Likewise, the third interim product is really 34300.

In exactly the same way, we can multiply two binary numbers, using add-and-shift. Let's say that we wish to multiply 27_{10} by 34_{10}. What we have in binary is shown in Figure 8.8.

Let's look at how this works. We start with the least significant bit of the lower number (11011). Since it's 1, the upper number (100010) is written down. We shift left one bit for the next interim product, then check the next least significant bit of the lower number. Since this is also a 1, we write the upper number down again. Continuing this process, we find upon examination of the next bit of the lower number that it is 0, so no interim product is written down (the interim product is 0). We do, however, shift left one bit for the next interim product.

Note the simplicity of an algorithm based on binary add-and-shift. The interim products in this process consist of the upper number multiplied by either 1 or 0. So the whole process is just a series of adds and shifts. In a later chapter, a binary multiplication program is included to demonstrate how this is done on the MSP430 microcontroller.

Multiplication by powers of 2

In some instances, it is only necessary to scale a signal within a microcontroller by a power of 2. When this happens, a simple way to perform the multiplication is to shift left for each power of two. For example, to multiply a variable by 8, you can just shift left 3 times ($2^3 = 8$).

Division

Division can be performed in a manner similar to multiplication. However, we are able to avoid the need for division in the robot projects in this book so, to limit the scope of the book to just the tools that we need, division methods, other than powers-of-two scaling, aren't covered in this book. If you should decide to pursue division algorithms on your own, the Sifferlen and Vartanian text is a good source. In addition, there are a number of division algorithms that can be found by searching the Internet.

When a division algorithm is implemented, the programmer must decide what to do with any remainder that occurs. For example, 5/4 produces a result of 1 with remainder 1 or the result can be expressed as 1.25 (01.01 in binary).

Division by powers of 2

When we just want to divide by a power of 2, a shift right works in a manner similar to the multiply by powers of 2. To divide by 8, for example, a number can be shifted to the right three bits ($2^{-3} = 1/8$).

One thing to be aware of with division by powers of 2 through the use of right shifts is that potentially useful information is being shifted out. This is equivalent to the problem that you have dividing integers in base-10. For example, when you divide 7 by 4 in base-10, the quotient is 1 but there is a remainder of 3. Similarly, if we are dividing 0111_2 by shifting two bits to the right (thereby dividing by 4), the result is 01_2 but we have shifted the lowest-order two bits (11_2) out of the register being shifted. In both the decimal and the binary cases (as well as any base system we might use) it can increase accuracy to round to the nearest integer (in this example, making the result 2_{10}) rather than truncating as was done.

Hexadecimal and octal

Computer compilers and assemblers, the programs that turn your microcontroller program into the bits that actually tell the microcontroller what to do, generally allow numbers to be expressed in your microcontroller program as either base-2 or base-10 (binary or decimal). The decimal representation is provided to make things easier for the human: in this case, you.

Once you've gotten familiar with microcontrollers, you may find that another base is useful, that being base-16, usually referred to as hexadecimal. The nice thing about hexadecimal is that a single hexadecimal digit (programmers often refer to this as a "hex" digit) represents exactly four bits. So, unlike conversion between binary and decimal, conversion between binary and hexadecimal is simple.

The first 10 digits of the hexadecimal numbering system are exactly the same as in decimal, 0 through 9. The symbols for the six largest hexadecimal digits, representing 10_{10} through 15_{10}, are A through F, respectively. A 16-digit binary number like $0100\ 1001\ 1111\ 1110_2$ is written in hexadecimal as 49FE. As you can see, it's much easier to remember 49FE than it is to remember $0100\ 1001\ 1111\ 1110_2$.

Octal numbers are base-8 representations, so valid digits are 0 through 7. Octal has the advantage of representing long strings of binary numbers in a more human-friendly way and, like hexadecimal, conversion between it and binary is very straightforward (three bits of binary to a single octal digit). However, octal is rarely used these days.

Floating-point arithmetic

If you've done any C or C++ (or any other high-level) programming on a "regular" computer, by which I mean a PC or Mac or something like that, then you've probably encountered floating-point arithmetic. In floating-point arithmetic, each number is represented by a sign, a mantissa, and an exponent – basically what we refer to as scientific notation.

While floating-point arithmetic is very useful and simplifies the job of the programmer considerably, it is hardware- and/or time-intensive, and therefore is generally not used with inexpensive microcontrollers. The applications that will be presented in the remainder of this book achieve their results without the use of floating-point arithmetic.

Bibliography

[1] T. Booth, Digital Networks and Computer Systems, John Wiley and Sons, 1971.
[2] M. Mano, Digital Logic and Computer Design, Prentice Hall, 1979.
[3] T. Sifferlen, V. Vartanian, Digital Electronics with Engineering Applications, Prentice Hall, 1970.

CHAPTER 9

Introducing the MSP430 Microcontroller

Chapter Outline

What is a central processing unit and how does it work?

The central processing unit (CPU) is the real brain of any computer. If you have ever looked inside a PC or Mac, the CPU is the big square integrated circuit on the motherboard or one of the other circuit cards. It almost always has a heatsink mounted to it with lots of fins to help dissipate the considerable heat that it generates. Many CPU heatsinks also include a small fan with this assembly to further cool things down.

The MSP430 microcontrollers also have a CPU, although it is, as you probably guessed, less sophisticated than the CPU in your home computer. So just what is this CPU? Figure 9.1 is a block diagram of the TI MSP430 CPU.

At first this might look like a complicated structure. Keep in mind that most of the figure just depicts the 16 registers that the MSP430 can use for its arithmetic and logic operations. The real guts of the processor is the “ALU”, the arithmetic logic unit. This part of the processor performs the arithmetic (add, subtract) and logic (AND, OR, Exclusive Or, etc.) operations that we talked about in the last couple of chapters. It also sets certain

MSP430-based Robot Applications.
DOI: http://dx.doi.org/10.1016/B978-0-12-397012-1.00009-6

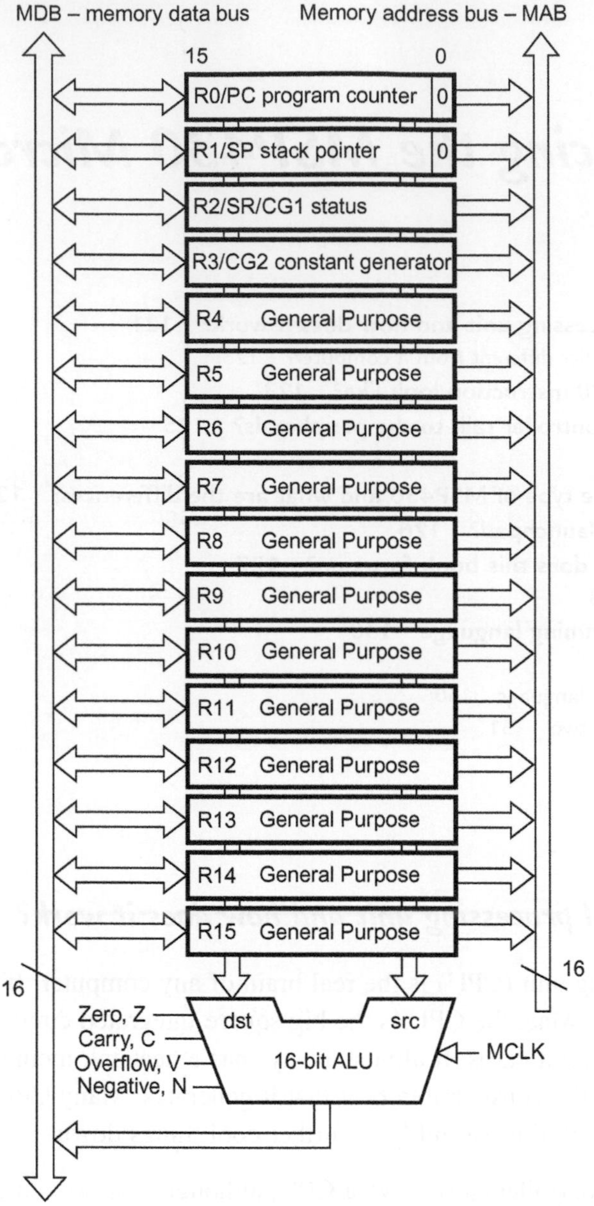

Figure 9.1
MSP430 CPU *(figure courtesy of Texas Instruments).*

"condition bits" – Zero (Z), Carry (C), Overflow (V), and Negative (N). In addition, it determines where to go for the next instruction (called the "instruction fetch") and where to find operands other than registers ("data fetch"). These fetches are made possible by the address and data busses on the right-hand and left-hand sides of the figure, respectively.

You may have noticed that Figure 9.1 doesn't show any memory. Technically, the memory is not part of the CPU. In a home computer, the memory is located in a separate set of integrated circuits from the CPU as well.

How is a microcontroller different from a computer?

Both the microcontroller and the computer contain a CPU for processing instructions. The microcontroller *is* a computer, just not as sophisticated as the one in what we normally refer to as a computer. However, what really sets the microcontroller apart from conventional computers is the fact that the microcontroller contains much more than just a CPU.

As already mentioned, your home computer has separate integrated circuits for the memory, but the microcontroller contains the memory on the same integrated circuit as the CPU. Of course, this means that the microcontroller will have to get by with considerably less memory than your home computer. Whereas home computers typically have in excess of a gigabyte of random access memory (RAM), the microcontroller typically has just a few kilobytes of memory. In fact, as "minimalist" as that might sound, it's actually worse. Much of the microcontroller's memory is typically "flash" memory, which isn't intended to change during execution of a program (or at least not change often). This flash memory is similar to the memory that you have inside a memory stick. The actual RAM that is available for frequent change is typically less than a kilobyte in these microcontrollers.

In addition to the on-board memory, the microcontroller integrated circuit has lots of built-in peripherals. Most MSP430 microcontrollers have multiple analog-to-digital (A/D) converters built in. We haven't talked about A/D converters yet but, as you've probably guessed, these are the devices that convert analog voltages from sensors and other circuits into digital numbers for the processor to use.

Another peripheral included in most MSP430 microcontrollers is one or more timers. These are counters that keep track of the time so that the user can precisely set delays, determine when an event occurred, etc. Like the A/D converter, the timer will be covered a little later in this book.

Digital input and output registers are included for controlling and sensing logic values. Some of this digital input and output (I/O) is in the form of parallel registers. Serial digital I/O is also included, in the form of the serial peripheral interface (SPI) and the inter-integrated circuits (I^2C) interface.

What does an MSP430 instruction look like?

Let's say we wanted to add the contents of two registers, R7 and R8. What would the instruction look like that would tell the MSP430 CPU to perform this operation?

15	14	13	12	11	10	19	8	7	6	5	4	3	2	1	0
Op-code				S-Reg				Ad	B/W	As		D-Reg			

Figure 9.2

An MSP430 double-operand instruction format *(courtesy of Texas Instruments).*

This section of the chapter is not intended to teach MSP430 assembly language programming. That will happen in the next chapter. What we want to do here is just to look at a simple instruction, see what it looks like in the CPU's native language (called machine language), and see how it looks in assembly language.

The first word of an MSP430 *double-operand* instruction has the format shown in Figure 9.2. (There are also *single-operand* instructions, such as rotating the bits of a single register, and those have a different format.)

Sometimes there will be one or two words that follow this instruction, depending on the addressing mode used. The example instruction was deliberately chosen to have the simplest addressing mode, called Direct Mode, and this will require only a single word for the instruction.

Note that there is something called an op-code in the first four bits of the instruction. This designates the general type of instruction to be executed – in this case, an Add. If we go to the "CPU" chapter of the *MSP430x2xx Family User's Guide* (URL is given in the References section) we find that the op-code for the Add instruction is 0101_2. The *source* register (called S-Reg in Figure 9.2) is the first of the two registers, in this case R7. So the next four bits, bits 8 through 11, are 0111_2. The destination register is R8, so bits 0 through 3 are 1000_2. The bit called *Ad* in Figure 9.2 designates the destination addressing mode and the two bits called *As* designate the source addressing mode. For the Direct addressing mode, these bits are all 0. Finally, the B/W stands for Byte/Word. The MSP430 has the flexibility of operating on either a single byte of data or on a 16-bit word of data. Because we are operating on a 16-bit word, this bit will be 0.

Okay, so to add the contents of register R7 to the contents of register R8, then store the sum in register R8, we would use an instruction: 0101 0111 0000 1000_2. That string of bits is the *machine code* and it tells the computer (the MSP430) exactly what to do. The MSP430 immediately gets R7, adds its contents to the contents of R8, then stores the result in R8. Unfortunately, that string of bits is quite difficult for the programmer to remember. Its hexadecimal equivalent, 5708_{16}, is pretty hard to work with as well.

To get around this, the assembly language equivalent instruction, *Add R7,R8*, is used. This has the advantage that it is easy for the human programmer to interpret. And it can be easily converted to the machine language equivalent using an Assembler, which is a program written for the purpose of translating assembly language into the machine language code that the MSP430 needs.

How does the microcontroller talk to the peripherals?

One of the nice things about any microcontroller is that the peripherals – the A/D converters, the timers, serial and parallel I/O, etc. – already have an interface designed in for you. In most cases, these peripherals can be operated in many different modes. For example, the MSP430 timer can be operated to count up to its maximum count (65535), or count up to some preset number, or alternate between counting up and counting down. The timer can be set up to "capture" a certain event, by which is meant that it will store its counter contents at the time of the event. And the timer can use a number of different "clocks" as its source of counting.

The way all of this is set up in the MSP430 is through a group of microcontroller registers dedicated to the purpose of tailoring the timer to your specific needs. These registers are typically set up at the beginning of the program but can be modified throughout execution of the program as needed. Each of the various peripherals in the MSP430 has its own dedicated set of registers for such setup, which we will talk about in detail in the subsequent chapters covering those parts of the MSP430.

Interrupts

One of the very powerful features of most computers, including the MSP430, is the use of interrupts. Interrupts allow your program to perform "event-driven" tasks. What does that mean? Well, let's say that your program has a main task that it performs, maybe some kind of number-crunching routine that it executes. But every so often, there is some other task that needs to be attended to – for example, maybe some other device has just sent, via the serial interface on the microcontroller, some new information.

There are two ways you could handle this. One is to write the program so that it stops the main task every so often and checks the serial interface to see if any new information has been received. This sort of checking, without knowing beforehand if anything needs to be done, is called "polling". Its biggest disadvantage is that the program will have to be written so that it periodically stops what it's doing and checks the status of the peripheral – often without anything being there.

The other way this can be handled is to make the interface task "event-driven". In that case, you would set up the peripheral, in this example the serial interface, to generate an interrupt upon being filled with new information needing to be read. Your program would continuously perform the main number-crunching task until it received an interrupt.

Upon receiving the interrupt, it would store the location in the number-crunching part of the program that it last executed (so that it knows where to return) and would then jump to some prearranged interrupt service routine (ISR). In this example, the ISR might retrieve the information received through the serial interface and store it at some memory location for later use. The ISR would then return to the program location in the main number-crunching task and continue with that task until another such interrupt occurred.

Notice how much more efficient this is? The program never has to do periodic checks of the serial interface for possible transmissions. It knows that only when an interrupt occurs is there information to be retrieved from the serial interface.

Now, after having talked up the advantages of interrupts, I've got some bad news for you. Interrupts can be a little tricky to use. Although they are very powerful and are something that I incorporate in many of my programs, I consider them to be an advanced feature, not something for novices to attempt. As a result, they are considered outside the scope of this book. All the sample programs in this book have been deliberately written to avoid interrupts. I mention them here because they are definitely something you will want to learn to use as you develop your microcontroller skills. But for now, we're going to say goodbye to interrupts.

Is there more than one type of MSP430 and what are the differences?

There are literally hundreds of different MSP430 configurations. Each comes with different combinations of peripherals and different amounts of Flash memory and data memory. Some have hardware multiply capability, some have high-resolution A/D converters, etc. The focus in this book will be the simplest of the MSP430 microcontrollers which are compatible with the MSP430 LaunchPad product.

What is the MSP430 launchpad?

The LaunchPad MSP-EXP430G2 is a very powerful little TI board that you can plug your MSP430 microcontroller IC into (Figure 9.3). You can use it just to program and debug your microcontroller program, or you can design it into your robot as we will do in this book. Figure 9.4 shows an example of a kit that I've developed for extending infrared remote controls to other rooms (not part of the robot projects in this book). The development was made much easier by the fact that I only had to design the extra analog circuits on the board that is connected to the LaunchPad – the digital hardware was already designed, debugged, and ready to go, in the form of the LaunchPad board.

The LaunchPad connects through a USB interface to a PC and includes all the development software needed to turn the PC/LaunchPad into a complete development system. One of the

Figure 9.3
The LaunchPad MSP-EXP430G2 board.

really nice things about this development system is the price – for $4.30 you get the development board, the development software, two MSP430 microcontroller ICs plus connectors.

Which MSP430 types does this book focus on?

With so many different types of MSP430 microcontrollers out there, the task of picking the right one for your application might seem daunting. However, there are several things that will help narrow the choices. First of all, you will want to choose an MSP430 that is in a Dual-In-Line (DIP) package. There are two reasons for this. First, the LaunchPad board is designed for DIP versions of the MSP430. The LaunchPad board and its associated software development system are such powerful and useful tools, especially for beginning MSP430 projects, that we should do what we can to use these tools.

Second, packages much smaller than a DIP are hard to work with. An example is shown in Figure 9.5. It is the MSP430F6736. It has lots of great features – a 32 × 32 multiplier,

Figure 9.4
LaunchPad board with additional board connected.

three 24-bit A/D converters, relatively large amounts of program and data memory, lots of peripherals, etc. But those pins on the package are spaced only about 1/40 inch apart! That's great if you're trying to cram this microcontroller into some really tiny product, which is exactly what designers do with this part. But for a beginner, successfully soldering the part to a circuit board is close to impossible. These parts are usually installed by a professional technician or, more commonly, by a machine capable of such precision.

MSP430 pinout

Fortunately, TI has the same pinout for all 14-pin DIP MSP430 devices and for all 20-pin DIP MSP430 devices. Not only that, but the pins on the two sizes of DIPs are set up so that

Figure 9.5
A 100-pin MSP430F6736 package.

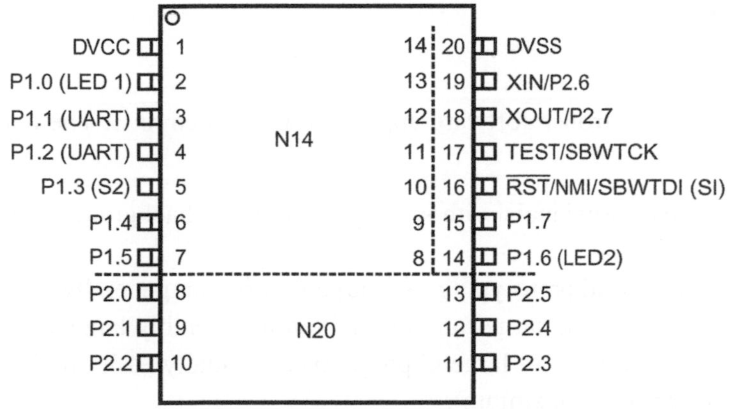

Figure 9.6
Pinout for 14- and 20-pin MSP430 DIP packages *(courtesy of Texas Instruments).*

either of them can be used in the same 20-pin socket of the LaunchPad. To see how this works, take a look at Figure 9.6. The 20-pin device has the same functions on pins 1 through 7 as the 14-pin device. And the 20-pin device has the same functions on the top 7 pins of the right-hand side as the 14-pin device. The extra six pins on the 20-pin device are just six pins of the digital "Port 2" register, P2.0 through P2.6.

As a result, you can plug either a 14-pin or a 20-pin version of the MSP430 into the 20-pin socket of the LaunchPad – just remember that, for a 14-pin version, the device is plugged in using the upper part of the socket.

Choosing the programming language

At this point, you're ready to program the MSP430, but before doing that, you'll have to decide what programming language to choose. You have two choices – the language C or assembly language. C is a "high-level language". What does that mean? Well, unlike assembly language, where each instruction corresponds to exactly one machine-language instruction and where we think in terms of registers and memory addresses, in a high-level language the correspondence between the high-level language instruction and its machine-level instructions is no longer one-to-one. In most cases, a C programming instruction will correspond to several machine-level instructions. When we get to the chapter on C programming, some examples of C programs, along with their disassembled machine language equivalents, will help show this difference.

Deciding between assembly language and C is actually a controversial question, something like the Mac vs. PC question or digital audio vs. vinyl LPs question. You'll need to decide for yourself where to come down on this issue. Some background on the choices might help in making that decision.

The case for C

The majority of microcontroller software appears to be written in C. The reasons are probably:

1. Familiarity – most individuals with some programming background have been exposed to C.
2. High productivity, rapid prototyping – a large C program generally consists of many fewer instructions than the same function written in assembly language.
3. Maintainability, debugging – large C programs are usually easier to fix (debug) or to change at a later date (maintain).
4. Portability – programs written in C can be used on multiple platforms: for example, a program written for the PC in C can be "ported" to the Mac by simply recompiling the program using a Mac C compiler.

The case for assembly language

Those individuals who program microcontrollers in the microcontroller's assembly language can claim these advantages:

1. Faster execution
2. Smaller program size
3. More visibility into the way the microcontroller actually works
4. Less abstraction

Deciding between the two

It might appear that C wins hands down. Certainly, if you were writing a brand-new word-processing program for a PC or Mac and had only these two choices, you would pick C. After all: (1) the effort to write a word-processor program would be huge and any productivity improvement a big help; (2) likewise, anything that would help make the debugging easier would be important; and (3) portability between different types of computers would be essential, since you would want to sell your word processor to as many different customers as possible.

On the other hand, the speed at which this hypothetical word processor executes wouldn't be terribly important, since most modern PCs and Macs have extremely high CPU clock speed. Program size for the word processor also wouldn't be terribly important, given the huge capacity of RAM and hard drive space on modern PCs and Macs. So there's no question, a word-processor program (or any other *huge* application written for a PC or Mac) would be written in C, given just C and assembly language as the two programming choices.

But, of course, that's not what you'll be doing with the microcontroller. Let's think about the important issues associated with the microcontroller in the control types of applications for which we'll be using it:

1. The available program memory is tiny – flash memory for the MSP430 DIP versions ranges in size from just 512 bytes for the MSP430G2001 up to 16 kB (16384 bytes) for the MSP430G2553.
2. Execution speed is often very important – the robot must react quickly to real-world changes.
3. A large part of any microcontroller program will be setting up the peripherals correctly and the choice of language has no bearing on that problem.

Given all of the evidence, it seems that assembly language is indeed a good choice for many applications. Even if you plan to write your programs in C, it is helpful to understand how the lower-level assembly language works. In subsequent chapters, we'll present program examples for both languages and you can then decide for yourself.

Clocks

A CPU's control of all the activity within it is dependent on a "clock". This is a square wave occurring at some constant frequency. The edges of this clock trigger the registers within the CPU to change state. In the MSP430, the master clock (MCLK) is used to clock the CPU. MCLK can be driven by several different square-wave generators: the digitally controlled oscillator (DCO), a crystal-controlled oscillator (LFXT1CLK, which has extremely accurate timing), or a very low-frequency oscillator (VLOCLK).

There is also a sub-main clock (SMCLK), which can be used to drive peripherals. The square-wave generators that can source its clock signal are the same as for the MCLK: the DCO, the LFXT1CLK, and the VLOCLK. This allows the programmer to run one or more of the peripherals at a clock speed different from the CPU.

There is also an auxiliary clock (ACLK), which can be used as a peripheral clock and which can be sourced by LFXT1CLK or VLOCLK. Having this third source of clocking for peripherals gives the programmer further flexibility in choosing peripheral clock speeds.

Finally, some of the peripherals have available to them their very own source of clock. For example, the ADC10 A/D converter has, in addition to the MCLK, SMCLK, and the ACLK choices, a fourth choice, the ADC10OSC, which runs at about 5 MHz, the maximum frequency at which the ADC10 can reliably change. This can be useful if, for example, the other clocks are set up for relatively slow speeds (to save power) but you need to convert analog voltages fast, or when you are running the other clocks at very high frequency, for example, 16 MHz, which is too high for reliable ADC10 operation.

Does all of this clocking stuff seem confusing? It confused me when I was first learning how to use the MSP430 microcontrollers. The entire MSP430 clocking system is extremely powerful, allowing the programmer lots of choices, but with that flexibility comes complexity. We will simplify things in this book by observing a few constraints on clock choice:

- the MCLK will always be sourced from the DCOCLK
- the SMCLK will always be sourced from the DCOCLK (so MCLK and SMCLK frequencies should be identical)
- we won't use the ACLK
- we won't use any dedicated peripheral clocks except for the ADC10OSC clock.

Bibliography

[1] J. Davies, MSP430 Microcontroller Basics, Elsevier-Newnes, 2008.
[2] MSP430x2xx Family User's Guide. Literature Number: SLAU144H.
[3] MSP430 IAR Assembler Reference Guide, second ed., January 2003.
[4] MSP430G2452 Datasheet. Literature Number: SLAS722B, December 2010; revised March 2011.
[5] C. Nagy, Embedded Systems Design Using the TI MSP430 Series, Elsevier-Newnes, 2003.
[6] Available from: <www.ti.com>.

CHAPTER 10

Getting Started with MSP430 Assembler

Chapter Outline

Now that we've got some idea how the MSP430 works, it's time to look at how assembly language programming works for this microcontroller. This chapter introduces the way in which registers and memory are addressed and introduces the MSP430 instruction set.

MSP430-based Robot Applications.
DOI: http://dx.doi.org/10.1016/B978-0-12-397012-1.00010-2

The TI MSP430x2xx family user's guide

One document that you will absolutely need is the TI *MSP430x2xx Family User's Guide.* This is available as a free download from the TI website. It is document slau144h.pdf (http://www.ti.com/lit/ug/slau144i/slau144i.pdf). The *User's Guide* goes into great detail about the instructions, how the peripherals work, and how the microcontroller is organized. The document is several hundred pages long, so you probably won't want to make a hard copy of the entire document, but an electronic copy of the guide kept handy will be an invaluable resource.

The MSP430 datasheets

Each of the MSP430 versions has a data sheet available from Texas Instruments as a download. For example, the MSP430G2452 is described in the TI document SLAS722E, *Mixed Signal Controller - MSP430G2x52/MSP430G2x12* (http://www.ti.com/lit/ds/symlink/msp430g2452.pdf).

Registers and memory

Take another look at the first figure back in Chapter 9, showing the MSP430 CPU. As you can see, there are 16 registers within the CPU. There is also what is called an address bus (referred to in the figure as MAB, for memory address bus) and a data bus (MDB, for memory data bus). Every instruction that you execute involves either the 16 internal registers, or an internal memory location, or both (some instructions will involve peripheral control registers, but these are actually memory locations).

Special registers

Although there are 16 registers in the MSP430 microcontroller, not all are available for general use. The first four have special uses.

Program counter

Referring again to the CPU diagram in Figure 9.1, the first register, *R0*, is the *Program Counter* (often abbreviated PC). What's that? It's a register that exists in every CPU. It's the way that the CPU keeps track of where it is in the program (referred to as *program control*). When the program starts, the PC is loaded with the program's starting memory location. As each new instruction is executed, the PC is incremented by the number of bytes needed to take the CPU to the next instruction in program memory. If the program encounters a "jump" instruction, the CPU figures out how to modify the PC so that program control proceeds to the new instruction memory location.

Stack pointer

A very powerful feature of most CPUs is the subroutine, which we'll discuss later in this chapter. A subroutine is a set of instructions that can be "called" to perform some task, with execution then returning to the program that called the subroutine. The *stack pointer* (register *R1*) is the register that allows the CPU to remember where it was in the calling program so that, after performing the subroutine task, it can return to that point in the calling program.

Status register

In addition to modifying the bits of the register or memory location explicitly referred to in each instruction, the CPU sets or resets some "status" bits when it executes instructions. For example, executing an instruction like ADD R6,R7 adds together the contents of registers *R6* and *R7* and replaces the contents of *R7* with this sum. We can write this symbolically as:

$$R6 + R7 \rightarrow R7.$$

However, in addition to modifying the contents of R7 (the explicit result), the CPU changes the value of four "status" bits upon execution of this instruction. These are the overflow bit (V), the negative bit (N), the zero bit (Z), and the carry bit (C). The *MSP430x2xx User's Guide* explains how each instruction affects these four bits. Note that not every instruction affects all four bits and a few instructions affect none of the four bits. As we go through the instruction set, you'll see why these status bits are useful.

Before leaving the topic of the Status Register, let's take a look at the other bits in the register (Figure 10.1). SCG1, SCG0, OSC OFF, and CPU OFF provide ways to turn parts or all of the microcontroller off. Yikes! Why would you want to do that? Well, for some applications, there are periods where the microcontroller isn't doing anything and turning off the microcontroller's clocks provides a way to significantly reduce its power during those intervals. But how does it turn itself back on? The answer to this is almost always some type of interrupt that causes the microcontroller to "snap out of it". Now, since we've sworn off interrupts for the projects in this book, we can ignore these bits. As with interrupts in general, however, this is something that you'll want to consider as you get more comfortable with the MSP430 and its more advanced features.

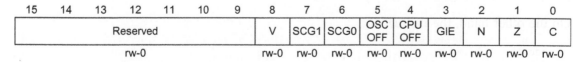

15–9	8	7	6	5	4	3	2	1	0
Reserved	V	SCG1	SCG0	OSC OFF	CPU OFF	GIE	N	Z	C
rw-0	rw-0	rw-0	rw-0	rw-0	rw-0	rw-0	rw-0	rw-0	rw-0

Figure 10.1
The MSP430 Status Register *(courtesy of Texas Instruments).*

There is one more bit in the mix that we haven't talked about, the GIE bit. This is the General Interrupt Enable bit. It needs to be set any time we want an interrupt to occur from any of the microcontroller peripherals. Again, since we are doing without interrupts this go-round, we can simply ignore this bit (it is reset to 0 when the microcontroller first powers up). So the only bits that we will be interested in within the Status Register are the V, N, Z, and C bits.

Constant generator register

R3 is a Constant Generator register. This provides a way to use commonly occurring constants, such as 0, 1, 2, etc., with slightly less program memory or slightly less execution time. It's a nice feature, but it's not something that you'll need to be concerned with using at this point.

General-purpose registers

The remaining registers, R4 through R15, are general-purpose registers, which will be extremely useful in implementing our programs.

Addressing memory and registers in assembly language

Byte vs. word instructions

Most of the instructions can operate on either a single byte or on a two-byte word. For example, adding the 16-bit contents of R6 to the 16-bit contents of R7, and replacing the contents of R7 with the 16-bit sum of these two numbers, can be accomplished with:

```
add      R6,R7
```

or

```
add.w      R6,R7.
```

The extra suffix, ".w", simply tells the Assembler that this instruction is to operate on word-long data. The default version of the instruction is the one that operates on word-long data, so the two statements are equivalent.

On the other hand, the instruction:

```
add.b      R6,R7
```

just adds the low-order byte of R6 to the low-order byte of R7. The result (the sum of these two bytes) is stored in the low-order byte of R7. A 0 is stored in the high-order byte of R7.

These byte-data instructions can be useful in some instances but they can also be a little confusing. As such, we will avoid them in this book. Like interrupts, they are a tool for the advanced user but one that can be ignored for now. The one exception to this will be with certain peripheral control bits, which, in some instances, are contained in single-byte memory locations and *must* be addressed with the ".b" version of an instruction.

Addressing words in memory

Remember that 16-bit words are two bytes. And memory location addresses refer to bytes, not words. So when we store a 16-bit number at some memory location, for example, 0200_{16}, that number takes up both locations 0200_{16} and 0201_{16} (the low-order byte goes in 0200_{16} and the high-order byte in 0201_{16}). So if we store a number at location 0200_{16} and we want to store another number at the next available memory location, we need to remember that this next available memory location is 0202_{16}, not 0201_{16}.

Hexadecimal memory address conventions

By now, you're probably realizing that a lot of constants will be expressed in hexadecimal. Up to this point, we've used the subscript "16" to indicate base-16. However, for programming purposes the convention is to use other ways of notating hexadecimal. For MSP430, a hexadecimal number can be indicated by adding an "h" at the end of the number, for example, 0200h, or by adding a "0x" at the beginning of the number, for example, 0x200.

Also, note that a hexadecimal number beginning with a hexadecimal digit A through F should be preceded by a 0. In general, there is no restriction on having any number of leading zeros in a constant.

Memory addressing modes

Register mode

The instruction operands can be addressed in several ways. The instruction examples given so far used the *Register Mode* – just plain, old registers R0 through R15. Of course, lots of the time we'll want to address memory locations (including the peripheral control words) and for this we'll need some other addressing modes.

Indexed mode

The *Indexed Mode* allows memory locations to be addressed using the registers as pointers. The Indexed Mode uses the contents of the specified register, along with an offset, to point to the memory location. This can best be illustrated with an example.

Let's say that we need to add the two numbers at memory locations 0200h and 0202h. Assume that R8 already contains the value 0200h. We could then perform the instruction:

```
add     0(R8),2(R8).
```

The microcontroller's CPU takes the contents of R8 (0200h). It adds the offset, but since this is 0, the final address is just 0200h. The contents of 0200h (and 0201h) are now the *source operand*. The CPU then takes the contents of R8 (again) and adds the destination offset of 2 to it to get 0202h, so the contents of 0202h (and 0203h, since this is a word-long-data instruction) are the *destination operand*. The contents at these two word locations are then added and the sum is stored at 0202h (and 0203h), replacing the previous content, which was the original destination operand.

Special versions of the Indexed Mode – the Symbolic Mode

What if the register used for Indexed Mode is the Program Counter (PC)? Then the offset determines where, relative to the current instruction, the memory location is. This is called the Symbolic Mode. Fortunately the offset is computed by the Assembler, so this instruction is much easier to use than it sounds.

The Symbolic Mode is often used for jump instructions. Let's say that our program gets to the bottom of a loop and we want it to jump back to the top (labeled Top in this example). The assembly language program would look like this:

```
Top

    instruction #1
    instruction #2
    instruction #3
    jmp             Top
```

The Assembler figures out the actual number of bytes that need to be subtracted from the current contents of the PC to get program control back to the program location labeled Top.

Special versions of the Indexed Mode – the Absolute Mode

The Absolute Mode is a form of the Indexed Mode in which the register contents are forced to zero – that is, only the offset is used to form the address so that the offset becomes the actual address. This address mode is used mainly to address peripheral memory locations. In the assembly language program you precede the memory label with an "&" symbol. For example, to read the 8 bits of the Port 1 digital I/O and store it in memory location 0200h, we could write:

```
mov.b     &P1IN,&0200h.
```

The indirect and the indirect autoincrement addressing modes

For the source operand, there are additional addressing modes. The more common of these is the Indirect Addressing Mode, an example of which is:

```
mov     @R6,R8
```

which moves the contents of the memory location pointed to by R6 into R8. But doesn't the Indexed Mode do the same thing when used with a zero offset? Well, yes. In fact, the ADD example given a little earlier could be rewritten as:

```
add     @R8,2(R8).
```

Not only can you use this last statement in place of the earlier ADD syntax, but the Assembler will actually substitute this Indirect Addressing Mode for the earlier Indexed Addressing Mode, should you write your program using the Indexed Addressing Mode for the source operand. The reason for this is that, although they produce equivalent results, the Indirect Addressing Mode is more compact, requiring just one word of instruction, instead of the three words that the Indexed Addressing Mode requires.

The Indirect Autoincrement Addressing Mode is exactly like the Indirect Addressing Mode, except that it increments the pointing register by 1 or 2 (depending on whether the instruction is for byte-long data or word-long data). This can be useful when loading data from a table (that is, a list of bytes or words in memory). The program can execute a MOV instruction that loads the byte or word from memory and, without using an extra instruction to increment the table pointer, be ready to load the next memory location.

Immediate addressing mode

The Immediate Addressing Mode is a form of the Indirect Autoincrement Addressing Mode that uses register R0 (the Program Counter, PC). It is used to insert constants into the program memory. For example, let's say that you want to store the number 3 in register R7. To do this, an instruction like:

```
mov     #3, R7
```

could be used. The actual machine-language instruction would consist of the first word, that informs the CPU that this is a MOV, that the source operand is the PC, that the addressing mode is Indirect Autoincrement, and that the destination operand is register R7. The second word would contain the number 3. The PC would automatically increment the PC so that it is pointing to the next instruction after the execution is completed.

Simplifying addressing

Does all of this addressing stuff seem confusing? It certainly did to me when I was first learning about it. The examples in this book will follow some guidelines that will make addressing a little simpler:

- No explicit use of the Indirect Addressing Mode
- No explicit use of the Indirect Autoincrement Addressing Mode
- Minimal use of the Indexed Addressing Mode
- Jump instructions will always use the Symbolic Addressing Mode. You don't really even have to remember that – just remember that if you want to jump to a location, you simply use the label associated with that location, with no preceding symbol before the location name.
- Call instructions will always use the Immediate Addressing Mode. However, you don't need to remember that either. Just remember that anytime you wish to use a Call instruction, use the syntax: CALL #*label*, where *label* is the starting location of the subroutine that you are calling.
- Memory locations corresponding to peripheral control registers will always by addressed using the Absolute Addressing Mode. But you don't have to remember that. Just remember that anytime you address a memory location associated with a peripheral control register, precede the register name with a "&". For example, to move the number FF_{16} into the Port 1 I/O output, you would use the instruction: MOV #0FFh, &P1OUT.

Once you get the hang of how addressing works, you can always graduate to the more advanced features that we're avoiding here.

Instruction set

The *CPU* chapter of the *MSP430x2xx User's Guide* contains a good detailed description of each assembly language instruction. The purpose of this section is simply to give a short introduction to these instructions.

Let's first list all of the instructions available for use with the MSP430. Table 10.1 gives that list.

Instructions not used in this book

The first thing to notice is that some of the instructions in Table 10.1 are grayed out. These are instructions that won't be used in this book. In some instances, these are instructions related to interrupts, which we are deliberately avoiding in this book. So that means we don't need to concern ourselves with DINT, EINT, RETI, and a bunch of others.

Table 10.1: MSP430 instruction set.

Instruction	What does it do?	Symbolic explanation
ADC(.B) dst	Add C to destination	dst + C→dst
ADD(.B) src,dst	Add source to destination	src + dst→dst
ADDC(.B) src,dst	Add source and C to destination	src + dst + C→dst
AND(.B) src,dst	AND source and destination	src .and. dst→dst 0
BIC(.B) src,dst	Clear bits in destination	not.src .and. dst→dst
BIS(.B) src,dst	Set bits in destination	src .or. dst→dst
BIT(.B) src,dst	Test bits in destination	src .and. dst 0
BR dst	Branch to destination	dst→PC
CALL dst	Call destination	PC + 2→stack, dst→PC
CLR(.B) dst	Clear destination	0→dst
CLRC	Clear C	0→C
CLRN	Clear N	0→N − 0
CLRZ	Clear Z	0→Z
CMP(.B) src,dst	Compare source and destination	dst − src
DADC(.B) dst	Add C decimally to destination	dst + C→dst (decimally)
DADD(.B) src,dst	Add source and C decimally to dst	src + dst + C→dst (decimally)
DEC(.B) dst	Decrement destination	dst - 1→dst
DECD(.B) dst	Double-decrement destination	dst - 2→dst
DINT	Disable interrupts	0→GIE
EINT	Enable interrupts	1→GIE
INC(.B) dst	Increment destination	dst +1→dst
INCD(.B) dst	Double-increment destination	dst + 2→dst
INV(.B) dst	Invert destination	.not.dst→dst
JC/JHS label	Jump if C set/Jump if higher or same	
JEQ/JZ label	Jump if equal/Jump if Z set	
JGE label	Jump if greater or equal	
JL label	Jump if less	
JMP label	Jump PC + 2 × offset→PC	
JN label	Jump if N set	
JNC/JLO label	Jump if C not set/Jump if lower	
JNE/JNZ label	Jump if not equal/Jump if Z not set	
MOV(.B) src,dst	Move source to destination	src→dst
NOP	No operation	
POP(.B) dst	Pop item from stack to destination	@SP→dst, SP + 2→SP
PUSH(.B) src	Push source onto stack	SP - 2→SP, src→@SP
RET	Return from subroutine	@SP →PC, SP + 2→SP
RETI	Return from interrupt	
RLA(.B) dst	Rotate left arithmetically	
RLC(.B) dst	Rotate left through C	
RRA(.B) dst	Rotate right arithmetically 0	
RRC(.B) dst	Rotate right through C	
SBC(.B) dst	Subtract not(C) from destination	dst + 0FFFFh + C→dst
SETC	Set C	1→C
SETN	Set N	1→N
SETZ	Set Z	1→C

Continued

Table 10.1: MSP430 instruction set.—cont'd

Instruction	What does it do?	Symbolic explanation
SUB(.B) src,dst	Subtract source from destination	dst + .not.src + 1 → dst
SUBC(.B) src,dst	Subtract source and not(C) from dst	dst + .not.src + C → dst
SWPB dst	Swap bytes	
SXT dst	Extend sign 0	
TST(.B) dst	Test destination	dst + 0FFFFh + 1
XOR(.B) src,dst	Exclusive OR source and destination	src .xor. dst → dst

Other grayed-out instructions are simply instructions that aren't needed in these projects. For example, there are a number of instructions to set or reset condition bits and those won't be used. The POP and PUSH instructions allow a programmer to load and store information on the CPU's stack, a sometimes-useful but advanced feature. So POP and PUSH are on our do-not-use list.

As with everything else that's done in this book in the interest of simplification, it's not that these instructions aren't useful or that you won't be using them in the future. It's just that this whole project will be difficult enough and a little simplification is a good thing at this point.

Instructions by category

Most of the instructions can be categorized as arithmetic, logic, rotation, or program control. Rather than go through each of the instructions, we'll go through these categories, since many of the instructions within a category behave similarly. To find a detailed description for a particular instruction, consult the *MSP430x2xx Family User's Guide.*

The move instruction

The MOV instruction doesn't really fit with any other category and it's used so often that it deserves special mention. As you've probably guessed from earlier examples in the chapter, this instruction is used to move data from one register or location to another register or location. The source operand with this instruction (that is, the first of the two operands and the one from which data is moving) can be a constant. For example:

```
MOV     #7,R9
```

moves the constant 7 into register R9.

Note that the CLR instruction is identical to moving a constant of 0 into a register or location. We can write this equivalency as:

```
CLR     R7     ⇔     MOV     #0,R7
```

Arithmetic instructions

The arithmetic instructions from Table 10.1 are as follows:

1. The addition instructions:

```
ADC(.B) dst
ADD(.B) src,dst
ADDC(.B) src,dst
INC(.B) dst
INCD(.B) dst
```

2. The subtraction instructions:

```
SBC(.B) dst
SUB(.B) src,dst
SUBC(.B) src,dst
DEC(.B) dst
DECD(.B) dst
CMP(.B) src,dst
TST(.B) dst
```

Note that, in addition to the three subtract and two decrement instructions, there are the compare (CMP) and test (TST) instructions. The CMP instruction, like the SUB instruction, subtracts the source operand from the destination operand. But unlike the SUB instruction, which takes this result and stores it in the destination location or register, the CMP does not change the contents of either the source or destination. So what's the point of such an instruction? Well, it turns out that CMP does set the condition bits, the same as the subtract instructions, and this can be very useful in deciding whether one operand is greater than, equal to, or less than a second operand.

The TST instruction is exactly like the CMP instruction except that the comparison is always between the constant 0 and an operand. We can write this equivalency as:

```
TST    R7    ⇔    CMP    #0,R7
```

Where are the multiplication and division instructions?

The answer is: there aren't any, at least not on the inexpensive versions of the MSP430 microcontrollers that we're using in these projects. The more sophisticated versions of the MSP430 have a hardware multiply (meaning that the multiply is built into the microcontroller CPU and executes as a single instruction) but, remember, we're avoiding those more sophisticated versions because:

- they're more expensive
- they come in tiny packages that most beginners will find impossible to work with
- they can't be used with the TI LaunchPad experimental board.

So we make a trade-off. We use inexpensive microcontroller versions that cost a buck or two and come in nice DIP packages. And, in return, we have to figure out ways to do without hardware multiplication and division (actually, hardware division is not available even in the more sophisticated versions of MSP430).

One alternative to a hardware multiply instruction is to develop a multiplication subroutine. We'll talk about subroutines a little later in this chapter, but basically it's just a chunk of code that any part of your program can access to do special functions like a multiply. The multiply subroutine will be given as an example in the next chapter.

Division can similarly be achieved by a subroutine written for that purpose.

Logic instructions

The logic instructions include the following:

```
AND(.B) src,dst
BIC(.B) src,dst
BIS(.B) src,dst
BIT(.B) src,dst
INV(.B) dst
XOR(.B) src,dst
```

Note that the BIS (Set Bits) instruction is really just a logical "OR" instruction. Its most common use is as a means of setting desired bits in the destination operand. For example:

```
BIS     #3,R8
```

sets the two least significant bits of R8 to a 1, while leaving the remaining bits of R8 unchanged.

In a similar fashion, the BIC (Clear Bits) instruction *resets* bits in the destination when the corresponding bits in the source are set to 1. For example:

```
BIC     #3,R8
```

resets the two least significant bits of R8 to a 0, while leaving the remaining bits of R8 unchanged.

BIT is equivalent to AND except that, whereas AND takes the bit-by-bit AND between bits in the source and destination operands and *replaces* the destination operand's contents with this result, the BIT instruction just takes the AND of the two operands and does not store the result anywhere. Its sole purpose is to set condition bits based on the outcome of the operation. This is similar to the relationship of CMP and SUB.

Rotation instructions

The rotation instructions include the following:

```
RLA(.B) dst
RLC(.B) dst
RRA(.B) dst
RRC(.B) dst
SWPB dst
```

The SWPB instructions swap the two bytes of the designated word and can be thought of as rotating left (or right) 8 bit positions. RLA and RLC are the rotate-left instructions. Likewise, RRA and RRC are the rotate-right instructions.

Program Control instructions

Program Control instructions include:

```
JC/JHS label
JEQ/JZ label
JGE label
JL label
JMP label
JN label
JNC/JLO label
JNE/JNZ label
CALL dst
RET
```

Program Control instructions cause the CPU's usual sequential execution of instructions to be changed so that the Program Counter points the CPU to some other location, rather than the next sequential instruction. For example, in the program snippet

```
            CMP     R7,R8
            JZ          ZeroResult
            ADD     R9,R10
                        .
                        .
                        .
  ZeroResult
            SUB     R11,R12
```

the program executes the ADD instruction as the next instruction after the JZ instruction if R7 and R8 are not equal, but executes the SUB instruction as the next instruction if R7 and R8 are equal. Each of the jump instructions (those beginning with the letter "J") depends on the condition of the status bits as to whether the jump is taken or not. The only exception to this is the JMP instruction, which is an *unconditional* jump.

Implementing subroutines with the CALL and RET instructions

The remaining two instructions in the program control group are the CALL and the RET. These are the instructions used to implement *subroutines*. Get used to using these – they can be very useful. The CALL instruction is like a JMP instruction, in that program control is transferred to the location identified by the label in the instruction. But unlike the JMP instruction, program control eventually returns to the instruction following the CALL instruction. A very simple example, in which the contents of registers *R9* and *R10* are set equal to zero in a subroutine, is this:

```
         .
         .
         .
  call     #ZeroR9R10
  next instruction
         .
         .
         .
ZeroR9R10
  clr      R9
  clr      R10
  ret
```

When the CPU executes the CALL statement, it changes the contents of the Program Counter so that program control is transferred to the location corresponding to the label *ZeroR9R10*. The program executes the instructions that it encounters there until it executes a RET (return) instruction. When it executes the RET instruction, it transfers program control back to the instruction immediately following the CALL instruction.

But how does it remember where the CALL instruction was? This is where the *stack* comes into play. So, just what is a stack?

How the stack memory enables subroutine calls

When the CALL instruction executes, it does more than just change the contents of the Program Counter to the location associated with the label. Before it changes the Program Counter contents, it first writes the existing Program Counter contents into a section of data memory called the stack. The particular memory location that the PC contents are written to is determined by the Stack Pointer (referred to as SP or alternatively as register R1).

When the RET instruction is executed at the end of the subroutine, the CPU knows that it must then go to the location that the Stack Pointer is pointing to and replace the PC contents with this value, which then causes the execution at the instruction labeled *next instruction* to occur in the above example.

If all we ever did was to make these single calls to a subroutine followed by a return, then we wouldn't need a *stack* of memory locations – we could just get by with one dedicated memory location that would hold the return address. But a called subroutine can call another subroutine, which can call a third subroutine, and so on. This is called nesting and a simple example is shown in Figure 10.2. Here the first subroutine initializes R6 to $0F_{16}$ and then immediately calls a second subroutine, which calls a third subroutine.

In this case we will need three memory locations, each one holding the return address to get back to the location from which it was called (keep in mind that a memory location referred to here is actually a word location – that is, two bytes).

To take care of this situation, the stack exists. When the first subroutine is called, the contents of the Program Counter are "pushed" onto the stack (which just means that the PC data is stored at the location pointed to by the Stack Pointer). The Stack Pointer is decremented by two (remember, there are two bytes to an address) just prior to this push. When the second subroutine call occurs, the Stack Pointer is again decremented by two and the PC is again pushed onto the stack. When the third subroutine call occurs – same thing – the SP is decremented by two and the PC is again pushed onto the stack.

```
                    .
                    .
                    .
                call    #FirstLevel
                next instruction
                    .
                    .
                    .
FirstLevel
                mov     #0Fh,R6
                call    #SecondLevel
                ret
                    .
                    .
                    .
        SecondLevel
                        call    #ThirdLevel
                        add     R6,R7
                        ret
                    .
                    .
                    .
                ThirdLevel
                                clr     R7
                                ret
```

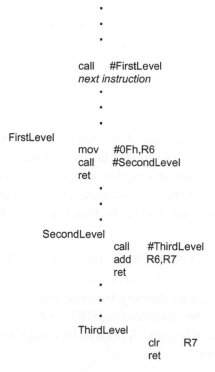

Figure 10.2
Nested subroutine example.

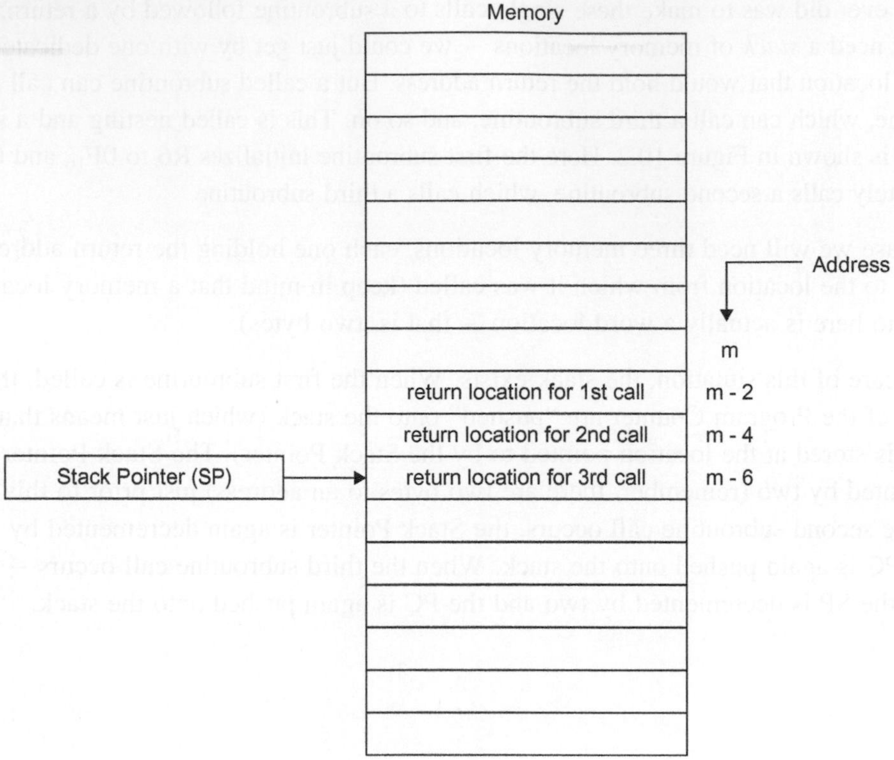

Figure 10.3
Stack memory operations.

As the RET instructions are each executed, the return addresses stored on the stack are "popped" off the stack in last in first out (LIFO) basis. That is, the first RET instruction causes the most recently stored address on the stack to be loaded into the PC. The process is just the reverse of the stack pushes – with each RET executed, the PC address is loaded, followed by incrementing the Stack Pointer by two. Figure 10.3 is a diagram of the stack in memory for this example.

As a matter of convention, the MSP430 Stack Pointer is initialized to be just above the data memory, so that the stack occupies the top of data memory. For example, an MSP430 with 128 bytes of data memory has data memory between addresses 0200_{16} and $027\,F_{16}$ and the Stack Pointer would therefore be initialized to 0280_{16}. Of course, if you use parts of data memory to store program variables, you'll need to be careful not to use locations which might become part of the stack.

Like some of the other operations in this chapter, the subroutine calls and returns probably seem more complicated than they actually are. The fact is that the stack operation is handled for you automatically by the CPU and most of the time you won't even need to think about that.

Why use subroutines?

Why even have subroutines? There are at least two reasons. One is that there are often tasks common to two or more sections of your program. A common subroutine can be called by each of these program sections, resulting in reduced code. A second and possibly less obvious reason to have subroutines is that it helps develop more structured, simpler-to-understand-and-maintain programs. A main program, with each major task as a subroutine call, is much easier to follow than a long "spaghetti-code" program.

The lowly NOP instruction

Well, that's all the instructions in Table 10.1 that needed to be mentioned except for one – the oddest instruction of the set – the NOP. NOP stands for No Operation, and that's what it does – nothing, or almost nothing. What it does do is to waste one clock cycle. This can actually be a useful thing if you need to get the timing of some event just right. Later, when we get the ultrasonic transmitter going, we'll make use of this instruction.

Bibliography

[1] J. Davies, MSP430 Microcontroller Basics, Elsevier-Newnes, 2008.
[2] MSP430x2xx Family User's Guide. Literature Number: SLAU144H.
[3] MSP430 IAR Assembler Reference Guide, second ed. January 2003.
[4] MSP430G2452 Datasheet. Literature Number: SLAS722B, December 2010; revised March 2011.
[5] C. Nagy, Embedded Systems Design using the TI MSP430 Series, Elsevier-Newnes, 2003.
[6] Available from: <www.ti.com>.

CHAPTER 11

Running Assembly Language Programs

Chapter Outline

The last chapter was about how assembly language works – about registers and addressing memory, and about the MSP430 instruction set. To actually run the assembly language program (or a C program for that matter) we need to learn to use the development system provided by Texas Instruments, and that's what this chapter is about. TI offers two development systems, the Code Composer Studio and the IAR Embedded Workbench. We use the IAR system in this book.

As the focus example for this chapter, we'll develop a software multiply program, which you may find useful in its own right, but which will also provide us with plenty of opportunities to see how the development system works and to further understand the MSP430 assembly language.

MSP430-based Robot Applications.

DOI: http://dx.doi.org/10.1016/B978-0-12-397012-1.00011-4

There will be further examples, in subsequent chapters, to look at what assembly language instructions are needed to make the various MSP430 peripherals work, but for now we'll just focus on the strictly software task of creating a multiply function.

Getting started

The first thing to do, of course, is to get the LaunchPad going. As soon as you plug in the LaunchPad board to the USB cable and connect it to your PC, it will start its existing program, which is probably a toggle between the red and green LED. Verify that this works, then download the IAR Embedded Workbench software from the TI website, per the instructions that come with the LaunchPad.

Now, for the projects in this book, the preferred MSP430 version is the MSP430G2452. That device is usually included with the LaunchPad. If not, order the part from a distributor like DigiKey or request a sample from TI (be sure to order the *DIP* version of the part). Okay, so we'll assume you now have the MSP430G2452. Assuming your LaunchPad came with some version of the MSP430 other than the MSP430G2452 installed in the on-board socket, this is a good time to practice changing DIPs on the LaunchPad board. Carefully pry the originally installed MSP430 DIP out of the socket, using a small screwdriver or some other small tool. Easy does it on this – don't try to pry the IC out all at once or you'll end up with bent pins on the IC. Gently pry up, first one end, then the other, until the IC is finally ready to "let go".

Install the MSP430G2452. Make sure you have the IC oriented correctly – the IC end with the indentation should be pointing toward the USB-socket-end of the LaunchPad board (see Figure 9.3). The IC initially will probably seem too wide to fit into the socket. If that's the case, very gently bend the leads towards one another. Then insert one side of the DIP into the socket and then gently push the other side's leads into the sockets. When everything seems aligned, push the IC down. Push gently and make sure that no pins are being bent as the IC is pushed into the socket.

Downloading the code examples

Look on the TI LaunchPad website for "MSP430 Code Examples" and download the code examples. There are different sets of code examples for different MSP430 versions. Download the examples for MSP430G2xx2. Open the folder titled "Assembly (IAR)". Make a new folder on your computer under "My Documents". Call the folder "IAR Programs" or something like that. Then transfer the programs from the "Assembly (IAR)" folder to your "IAR Programs" folder.

Among the Code Example programs is one called "msp430g2xx2_1.s43". This is a very simple program that blinks just the red LED. We'll use this program shortly.

Creating the project

Okay, let's create the project. For our purposes in this book, the project is just your program, plus some additional hardware definition for the IDE (the Integrated Development Environment, a fairly common acronym that you'll see from time to time). To create the project, open the IAR Embedded Workbench IDE application that you have downloaded, go to the tab labeled "Project" at the top of the page, click on that tab to pull down the menu choices, then click on "Create New Project . . .". This will bring up a small window that has "Empty Project" highlighted. Click OK. At this point a window pops up that is looking for the name of the project. Let's call this project "First", so just type that in for the File Name.

You should now see, in the left-hand side of the IAR Embedded Workbench IDE window, the project name in boldface letters (Figure 11.1).

Pull down the *Project* menu again and select *Options*. A window like the one in Figure 11.2 will appear.

In this window, click on the button under *Device*. Select *MSP430Gxxx Family*, then MSP430G2452.

Next, click on the Library Configuration tab at the top of this window, and in the dropdown menu under *Library*, select *None* (Figure 11.3).

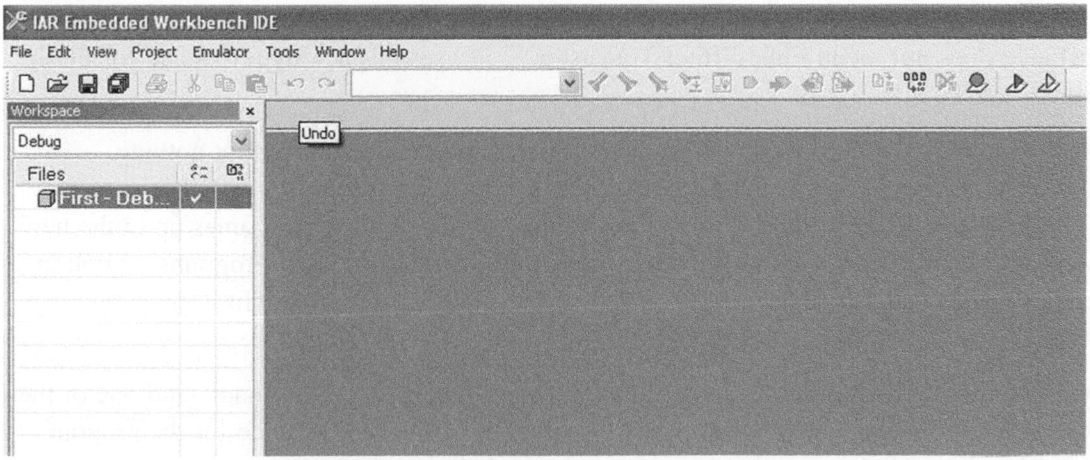

Figure 11.1
The IDE window.

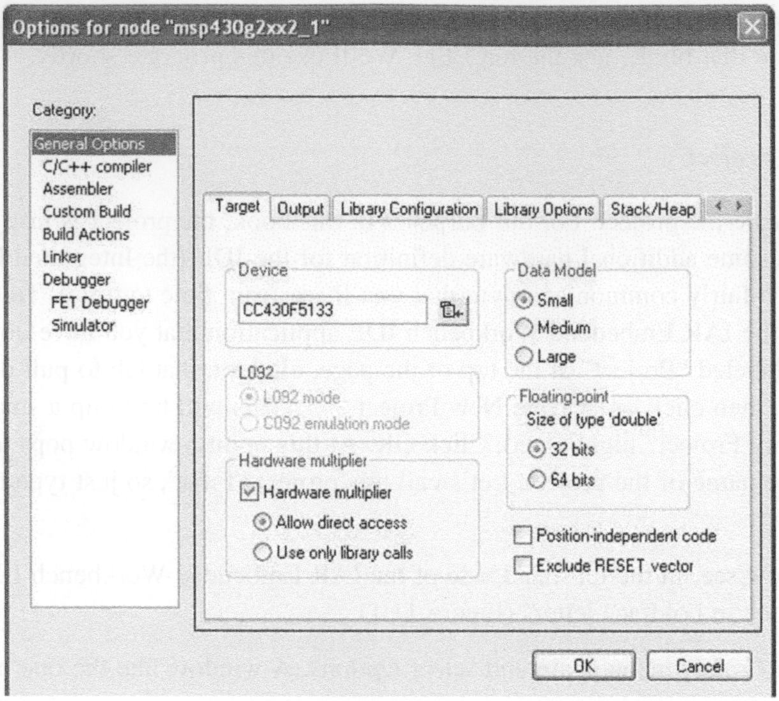

Figure 11.2
The Project Options window.

Finally, select Debugger under "Category" on the left-hand side. Under "Driver", select "FET Debugger (Figure 11.4). Now click "OK" at the bottom of the window. (FET in this instance does *not* mean Field-Effect Transistor – TI uses this acronym to stand for Flash Emulation Tool.)

At this point, you should still see the project name in boldface letters, but nothing underneath the project name. Left-click once on the project name to highlight the name. Then go to the Project tab and select "Add Files...". The folder that comes up in the new window that opened should be the one that you just created – "IAR Programs". Look in there for the "msp430g2xx2_1.s43" program, left-click once to highlight it, then click "Open" at the bottom of the window.

At this point, you should see some additional items under the project name and one of these will be your program. Double-left-click the program name and the code for the program should pop up. Notice that the program is nice and short – just what we need to get started.

Let's figure out what this program does before trying to use it.

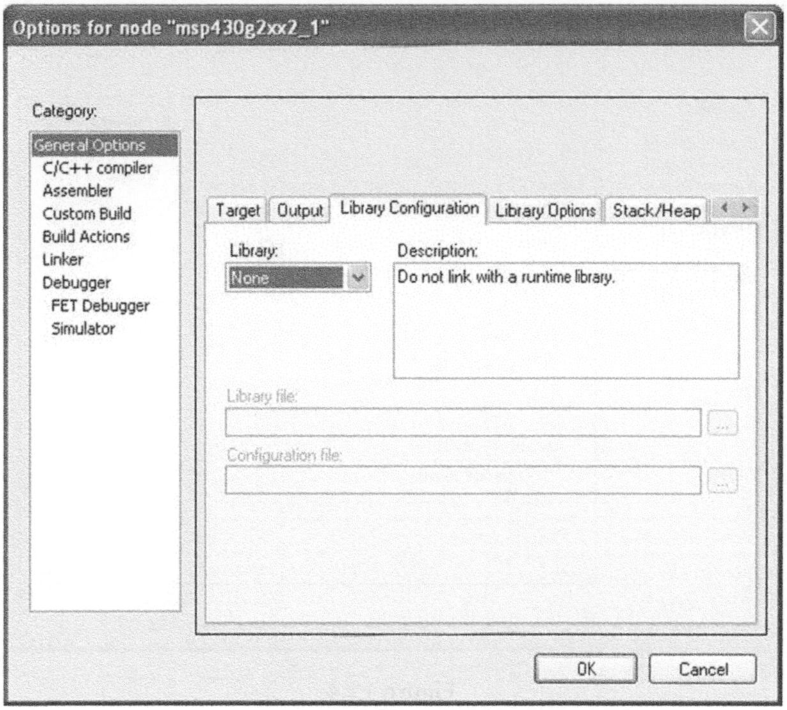

Figure 11.3
Configuring the Assembler library.

What does the Include statement do?

Notice the *Include* statement in the program. It should read

```
#include      "msp430g2452.h".
```

What is the ".h" file that this refers to? This is a text file, called a header file, which contains the information that the Assembler needs to figure out what peripheral names correspond to what memory locations. For example, the instruction:

```
xor.b     #001h,&P1OUT     ; Toggle P1.0
```

refers to the Port 1 digital I/O output register, P1OUT. According to the MSP430G2452 datasheet, P1OUT is memory location 021h. So the Assembler needs to know that when it sees P1OUT, the machine code should refer to memory location 021h. The Assembler knows this association because of the information in the ".h" file.

The #include statement as well as the ORG statement that we'll discuss next are examples of what are called Assembler directives. These are instructions to the Assembler. They are not program instructions. They are included to help the Assembler figure out how to

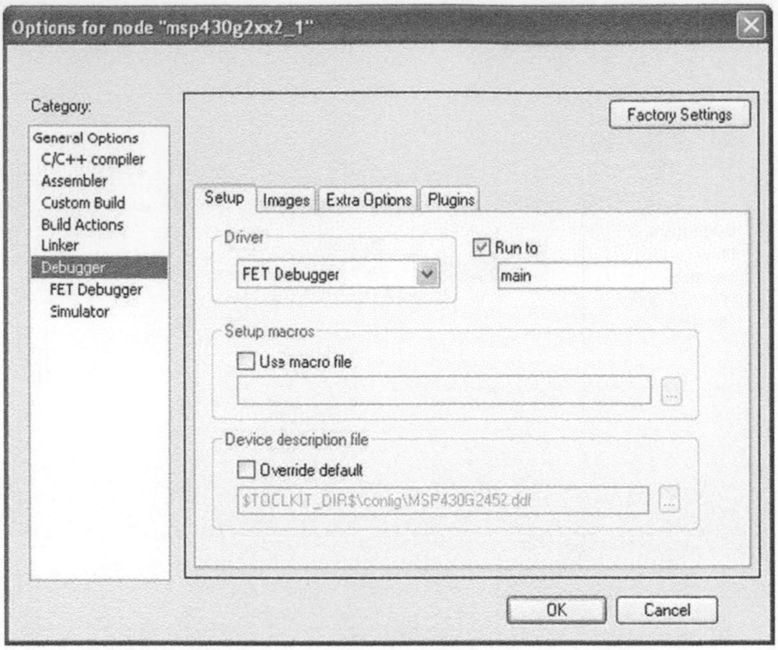

Figure 11.4
Selecting "FET Debugger".

assemble the program instructions. You can see all the directives available for the Assembler by going to the Help tab on IAR Embedded Workbench and opening the "Assembler Reference Guide". However, the directives needed for the example programs in this book are all covered in the book.

What's an ORG statement?

The next statement is

```
ORG     0F800h
```

This is another Assembler directive. It tells the Assembler that the program instructions which follow are to be placed in the microcontroller's memory starting at address $F800_{16}$.

The blinking LED program

The program "msp430g2xx2_1.s43" is shown in Figure 11.5.

The first program instruction is one that will be in every program we write in this book. It does a couple of things. First it establishes where program control is to be transferred in the event of a microcontroller reset event. Second, it initializes the Stack Pointer.

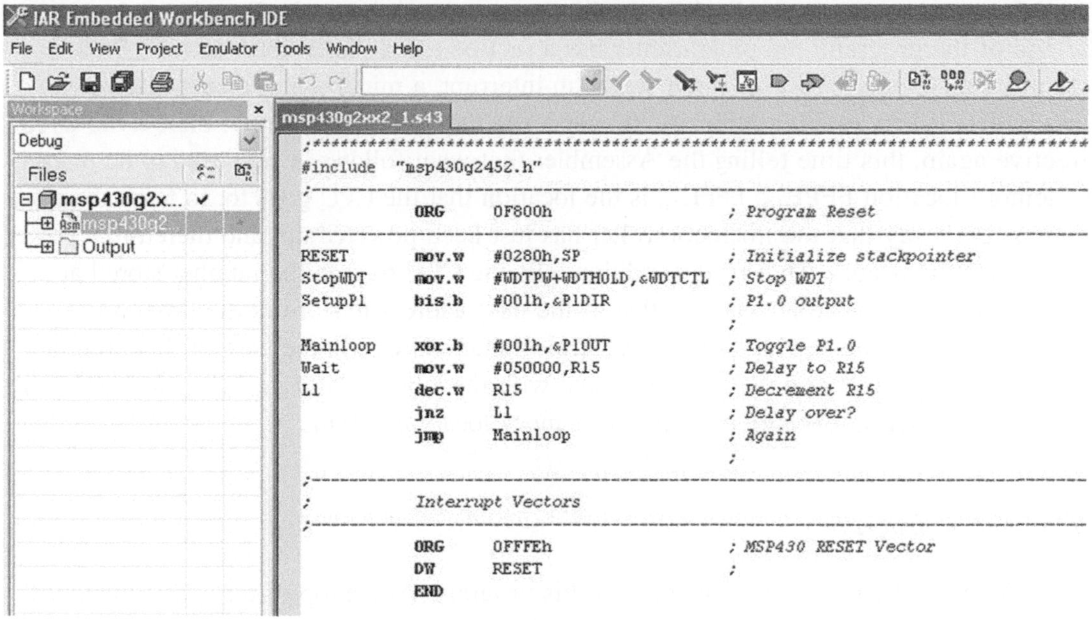

Figure 11.5
Program msp430g2xx2_1.s43.

The second instruction deals with an MSP430 peripheral that we haven't talked about – the Watchdog timer. Watchdog timers can be a useful safeguard to keep programs from going into unexpected parts of memory, but they are an extra complication that we choose to do without for now. The programmer who created "msp430g2xx2_1.s43" wisely turns this peripheral off with this instruction.

The next instruction sets up the least significant I/O bit for output. We'll have more to say about such instructions in a later chapter on the subject of digital I/O ports.

The "XOR" instruction is the one that causes the least significant bit of the Port 1 I/O register to change state each time this instruction is executed. So if the bit is 1 (which illuminates the LED) this instruction will make the bit 0 after execution, and vice versa.

The next three program instructions create a delay. R15 is loaded with the number 50000_{10} (there is no "h" after the number or "0x" preceding the number so this is a decimal, not a hexadecimal, constant). The "DEC" and "JNZ" instructions then create a loop. Each time the "DEC" instruction is executed it decrements R15 by 1. The "JNZ" (Jump if Not Zero) instruction keeps returning program control to location "L1" until the R15 contents are finally 0, at which time program control drops through to the "JMP" instruction that returns program control to "Mainloop". This delay is what causes the LED to stay in each state for a relatively long time.

There is one more thing to talk about in this program, and that is the Reset Vector. Note at the end of the program, the programmer has a section titled "Interrupt Vectors". This is actually a misnomer. A Reset is not really an interrupt, a minor point except that we've already decided to avoid interrupts with respect to the robot. At any rate, there's the ORG directive again, this time telling the Assembler that what follows is supposed to be placed at memory location $FFFE_{16}$. $FFFE_{16}$ is the location that the CPU goes to, to fetch the reset address. Let's say that the microcontroller has just been powered up, and therefore a power-on reset (POR) occurs. That reset condition tells the CPU to grab the number stored at $FFFE_{16}$ (the reset "vector") and use that as the next address to store in the Program Counter, thereby transferring program control to that address. In the case of this program, the number happens to be the one associated with the label RESET, which is at $F800_{16}$. So at reset, the program execution begins at memory location $F800_{16}$.

Running the program

Let's complete the preparation of running this program. At the top of the IAR Embedded Workbench IDE page, left-click on the "Tools" menu and select "Options..." (Figure 11.6). On the left-hand side of the "IDE Options" window that has popped up, select "Stack". Unselect any checked boxes and then click "OK". This just gets rid of some annoying, and not helpful, warnings that would otherwise occur when we try to run the program.

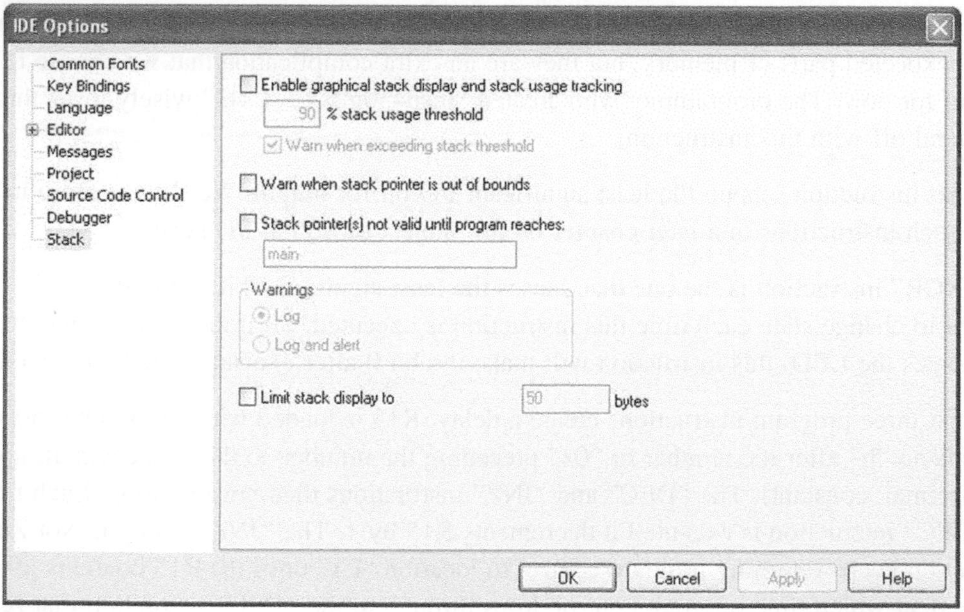

Figure 11.6
Clearing the boxes for the stack.

At the top of IAR Embedded Workbench IDE, now left-click "Project" and in the pull-down menu select "Rebuild All". A window will pop up that says "Save Workspace As". The workspace is an abstract construct that holds your project plus, maybe, some other projects. In all the examples that we do in this book, there will only be one program per project and one project per workspace. Nevertheless, we have to have a workspace and we have to give this workspace a name. Let's call the workspace "First" as well. This performs the assembly of your program. Now left-click on "Project" again and select "Download and Debug". After a few seconds of download, you should see a display that looks like Figure 11.7, with the first instruction highlighted in green.

At this point your program is ready to run. Pull down the "Debug" menu at the top of the screen and select "Go". You should see the red LED on the LaunchPad board blink on and off.

What do I do if this doesn't work?

What do you do if you get stuck getting this example to work? That's actually the reason that this first program is a short one written by engineers at Texas Instruments. TI has a technical support group that helps in getting you going in the event that you get stuck. Just go to their site at www.ti.com and follow the links to their technical support group.

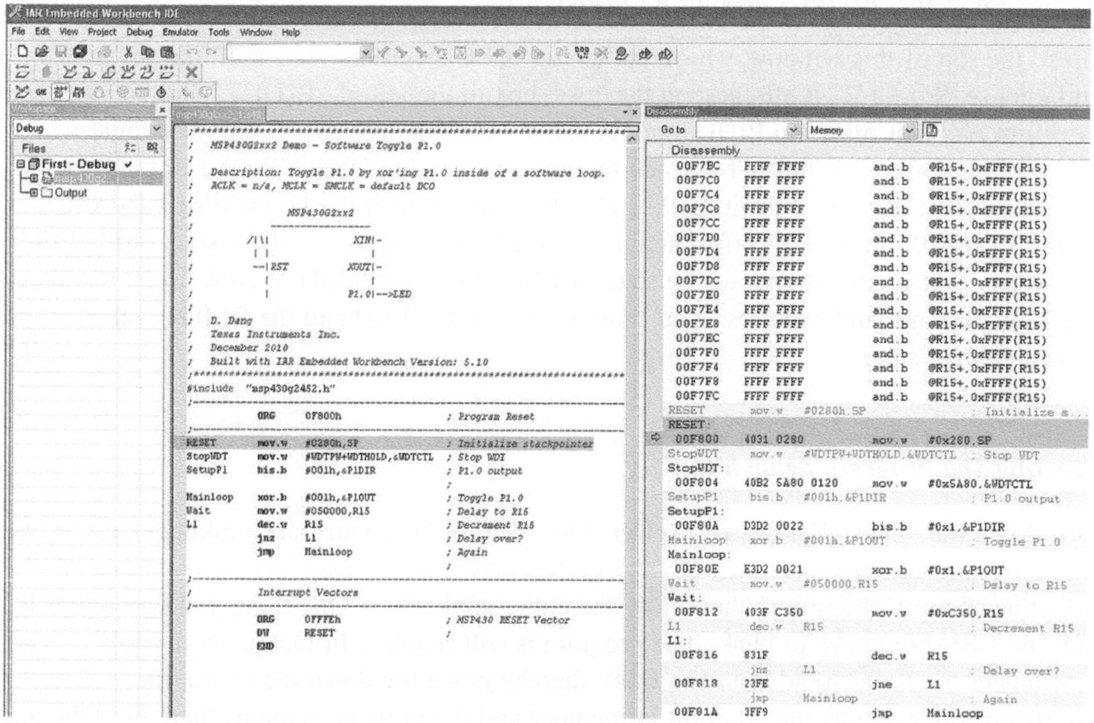

Figure 11.7
Screen after "download and debug" – program is ready to run.

Getting familiar with IAR Embedded Workbench IDE

Okay, we're not done with this first example yet. While we've got it running, let's get familiar with some of the features of IAR Embedded Workbench IDE. Assuming you selected "Go" in the pulldown menu under "Debug", now select "Break" in that menu. You'll see a green-highlighted line, which is the instruction that will be executed next. Press F11 and you will see that that next instruction is executed. Stepping through instructions one-at-a-time like this is called single-stepping.

Now let's let the development system step through several instructions and arrive at some desired location – in this case, just before executing the XOR.B instruction. To do this, mark the instruction by left-clicking on the instruction line somewhere before the "x" in "xor". You should see the cursor blinking on that line. Now go to the "Debug" menu and choose "Run to Cursor". The green highlight should now be highlighting that instruction, indicating that it is next to be executed. "Run to cursor" causes the microcontroller to execute all the instructions between the last break and the instruction that you marked with the cursor.

Now let's take a look at some of the internal microcontroller values. Pull down the "View" menu and choose "Registers". At the right-hand side of the screen will be a window showing registers R0 (PC) through R15 (Figure 11.8).

Go to the top of this window, where the words "CPU Registers" are, pull this menu down and choose "Port 1/2". Left-click on the " + " box to the left of "P1OUT" and you will see the eight bits that make up P1OUT individually listed. P1 through P7 are meaningless for this program, since we haven't defined them as outputs, but P0 (we'll refer to this as P1.0, to indicate that it is the 0-th bit of Port 1) defines the state of the LaunchPad's red LED. When P1.0 is 0, the LED is off; when it is 1, the LED is on. Look at the state of the LED and verify that it corresponds to the state of P1.0 (P0 on the display). Now hit F11 so that the XOR.B command executes. You should see both the LED and the P1.0 bit value on the screen change state.

Starting the program again later

Pull down the "Debug" menu and select "Go". The LED should start blinking again. Now exit IAR Embedded Workbench IDE by pulling down the "Debug" menu again and selecting "Stop Debugging", then close the entire application. One thing you will notice is that the LED continues to blink. The program is still running. In fact, even if you disconnect the LaunchPad from the USB, thereby powering down the board, the microcontroller retains the program in memory and the program remains there until the next time you perform a "Download and Debug" within the IDE (the program memory is a form of nonvolatile memory called flash memory).

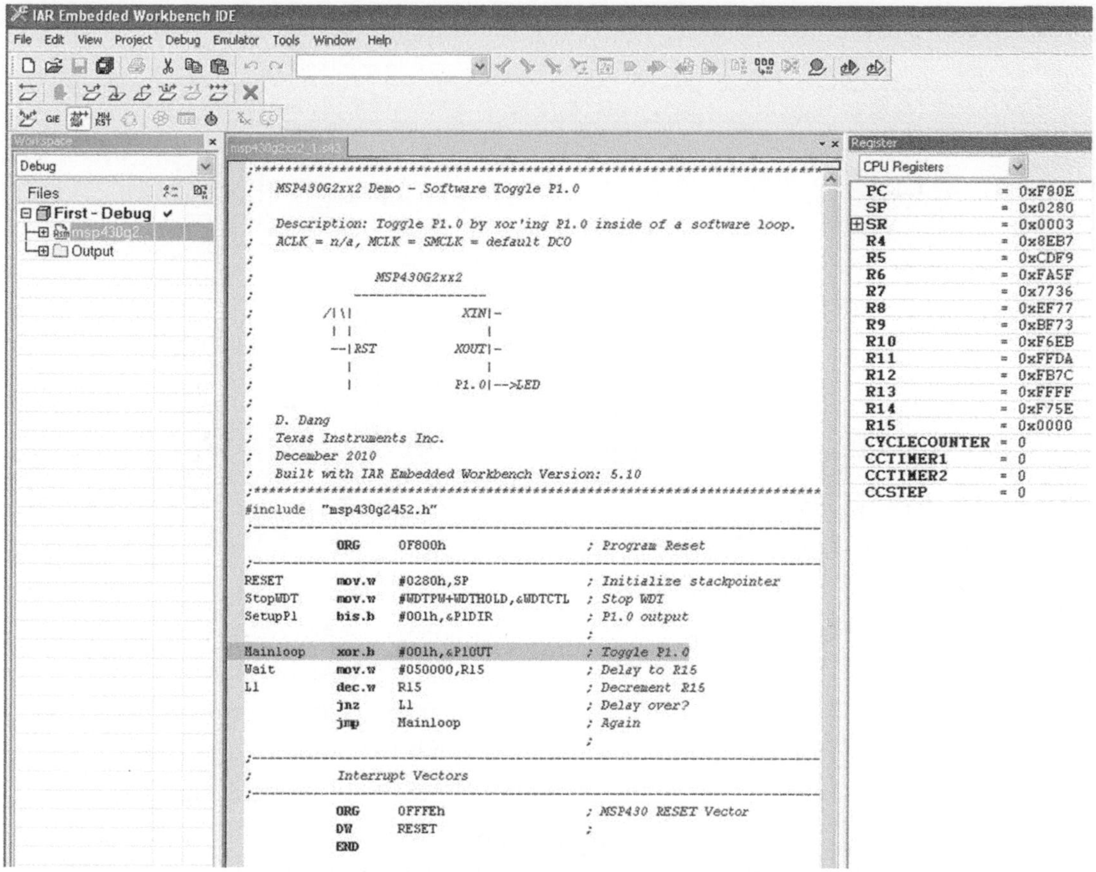

Figure 11.8
Register window.

Now, let's say that you want to bring the program up with the IAR Embedded Workbench IDE at some point in the future, perhaps to make modifications to it. How do you do that? Remember that you named the Workspace "First". So, after opening the IDE, you go to the "File" menu, select "Open", then "Workspace". When the "Open Workspace" window pops up, type "First" for the File Name, then click the "Open" button. This brings up the Project, which was also called "First", which includes the program called "msp430g2xx2_1.s43".

Modifying the LaunchPad board

If you've followed this first programming example to this point, you have verified that your LaunchPad board is working. Now it's time to make a modification to that board. It turns out that Texas Instruments included something on the board that we don't need and that will interfere with the robot projects that will be developed.

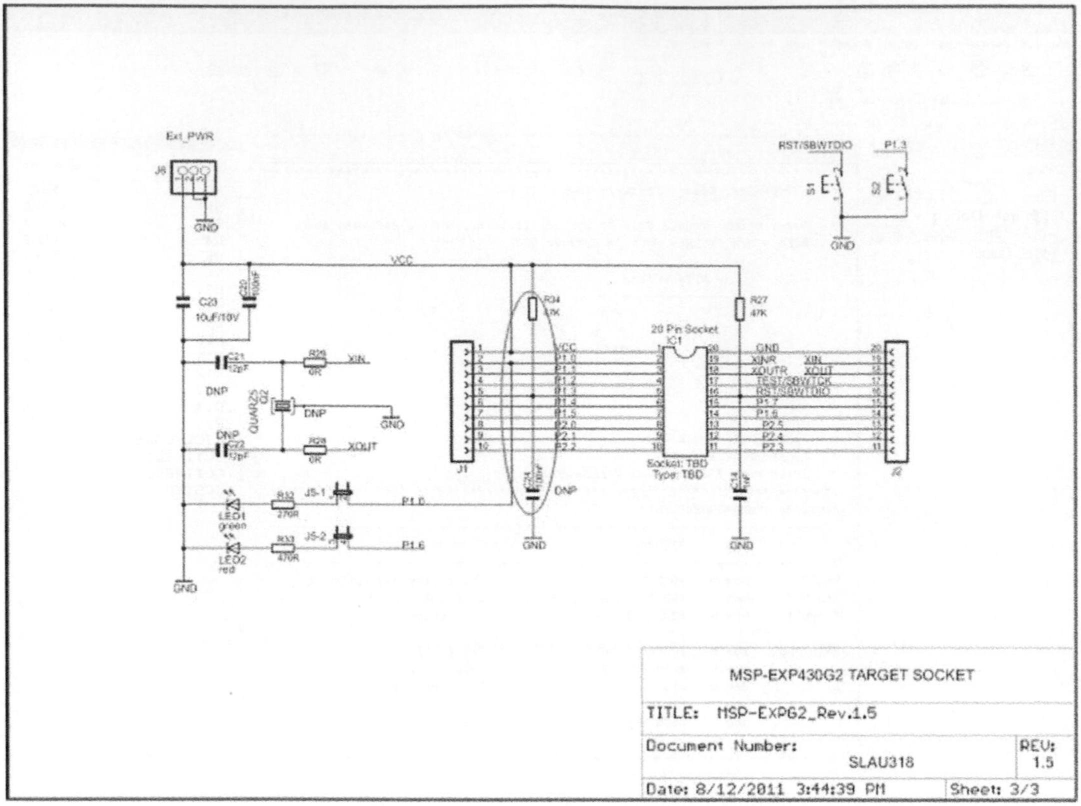

Figure 11.9
Partial LaunchPad schematic, showing resistor/capacitor connected to P1.3 *(courtesy of Texas Instruments).*

If you look closely at the schematic sheet copied from the LaunchPad User's Guide (Figure 11.9), you'll see two components connected to node "P1.3". One of these is a 47 kΩ resistor, R34, and the other is a 0.1 μF capacitor, C24. These are included on the board because they help to "debounce" switch S2. Debouncing is sometimes needed because a switch like S2 is a mechanical device that flexes as it is pressed, possibly causing it to "bounce" as it makes contact and thereby repeatedly make and break contact. That can confuse the microcontroller program (depending on how the program is written) and an easy fix is to add a resistor/capacitor, like R34/C24.

However, we're not going to be using switch S2 and the capacitor, C24, will potentially alter signals coming into P1.3. Some versions of the LaunchPad board that I've used leave R34 and C24 off (note that the schematic shows "DNP" for C24, which is a note to the board builder, meaning "Do Not Place").

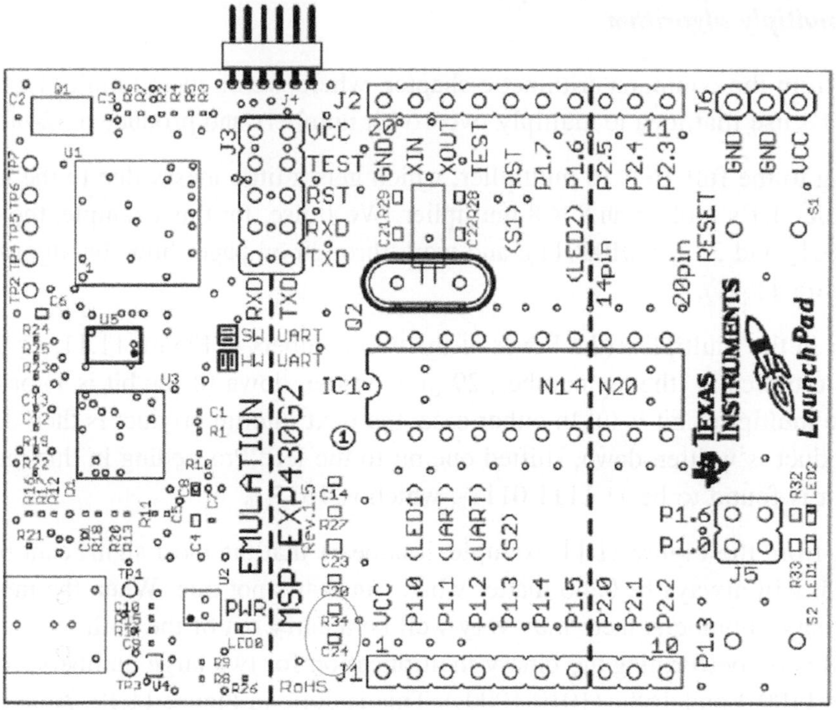

Figure 11.10
Board components for the LaunchPad showing C24 and R34 *(courtesy of Texas Instruments)*.

If the components are missing on your LaunchPad board, then no further action is required. To see if they're on the board, compare your board to the board drawing in Figure 11.10. If the components are on the board then you'll have to remove them. This isn't too hard because the components are very small surface-mounted parts. Any broad-tipped soldering iron should be able to heat both sides of the component simultaneously, and you can just push the component off the board with a small screwdriver or other small tool.

A second programming example – the software multiply

At this point, let's develop a program that's longer than the first and not part of the TI examples group. This will be a software multiply program. It will be set up as a subroutine, so we'll get some practice at how to use them in assembly language. Although this algorithm won't be used in the robot project in this book, it serves as a good way to demonstrate the development of a software program. Later, in the sections on peripherals, we will develop the assembly language programs for use with the robot.

The basic multiply algorithm

Remember from the computer arithmetic chapter, where we discussed multiplying by shifting? Let's use that idea to multiply two 16-bit numbers and produce a 32-bit result.

Before going to the full 16×16 multiplier, which gets a little messy due to the large number of bits, let's look at an 8×8 multiplier. We'll use, for this example, the numbers $29_{10} = 011101_2$ and $35_{10} = 0100011_2$, and work through on paper how the algorithm should behave (Figure 11.11).

For each bit in the multiplier (the lower of the two numbers in Figure 11.11, equivalent to 35_{10}), the multiplicand (the top number, 29_{10}) is written down (if the bit is 1) or 0 is written down (if the multiplier bit is 0). In either case, the next interim product is then determined, and that product is written down, shifted one bit to the left. Proceeding in this manner, the final product is found to be 11 1111 0111_2, which is 1015_{10}.

Notice that, from the Figure 11.11 example, it appears that the most significant bit of the result, bit 15, will always be 0, no matter what. But that's not true. When the multiplier operates on large numbers, there may very well be a carry out of the 14th bit, into the 15th bit. Verify this by performing the binary multiplication for two large numbers; for example, 252_{10} (1111 1100_2) and 168_{10} (1010 1000_2). This is done in Figure 11.12. Note that, rather than try to add together all the numbers at once, in this example the first two non-zero numbers are added and that interim sum is then added to the final non-zero number.

As expected, the product of these two numbers, 1010 0101 0110 0000_2, is the decimal number 42336_{10}.

The unsigned 16×16 multiply subroutine

Scaling this algorithm to a 16×16 multiplication with 32-bit result isn't too difficult. We will just need to do 32-bit operations for the adds and for the rotated multiplicand.

```
                0 0 0 1 1 1 0 1
              × 0 0 1 0 0 0 1 1
                ---------------
                0 0 0 1 1 1 0 1
              0 0 0 1 1 1 0 1
            0 0 0 0 0 0 0 0
          0 0 0 0 0 0 0 0
        0 0 0 0 0 0 0 0
      0 0 0 1 1 1 0 1
    0 0 0 0 0 0 0 0
  0 0 0 0 0 0 0 0
-------------------------------
0 0 0 0 0 0 1 1 1 1 1 1 0 1 1 1
```

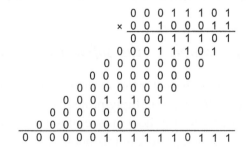

Figure 11.11
Multiply and shift for the binary equivalents of 29_{10} and 35_{10}.

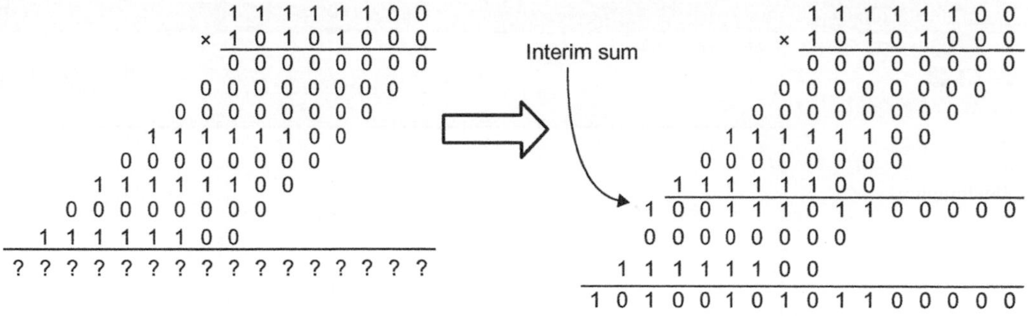

Figure 11.12
Multiply and shift for the numbers 252_{10} and 168_{10}.

Writing the program

Repeat the process of creating a project. We'll call both the Workspace and the Project *Multiplying in Assembler*. After saving the project to your IAR Programs folder or whatever you've named that folder, set the Options for the project:

- The MSP430G2452 is selected
- "None" is selected for the library
- FET Debugger is selected for Debugger

Also, under Tools→Options, make sure the boxes are unselected for the Stack. Now you're ready to start writing the program.

The program file

To start creating your program file, select from the File tab at the top of the IDE page, New→File. The right side of the IDE screen should be a big white area and the title at top left should say "Untitled1" or something like that. Let's give this program a name. From the File tab, select Save As. The Save As window comes up. Let's call the program "Multiplying in Assembler", as well. Don't hit save yet. One slight annoyance of the IAR Embedded Workbench IDE is that it assumes that you're saving a C program, so it wants to give the program a ".c" extension. You have to manually go to "Save as type" at the bottom of the Save As page and select "Asm Files (*.s;*.msa;*asm)". Then go to your File name and add the extension ".s43". At this point, the file name should read "Multiplying in Assembler.s43".

Notice that the title of the program in the IAR Embedded Workbench IDE window is "Multiplying in Assembler.s43" but that at the far left of the IDE page, where the project name is displayed, there is nothing underneath the project name (see Figure 11.13). This is because the file has not yet been *added* to the Project. So let's do that.

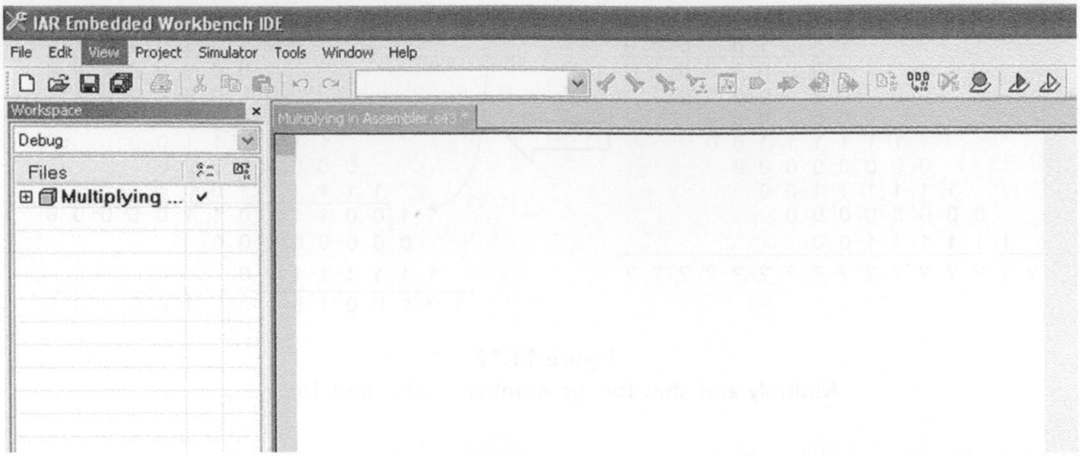

Figure 11.13
IAR Embedded Workbench IDE with program titled "SoftwareMultiply".

To add the file, go to the Project tab and select "Add Files…". The displayed Add Files window should show the contents of your IAR Programs folder. Select "Multiplying in Assembler.s43". This program will now appear underneath the "Multiplying in Assembler" project name in the IDE window. Now it's time to start the programming – we'll start generating the program within the "Multiplying in Assembler.s43" window.

Figure 11.14 shows the start of the program. This is just the start of things – a kind of template, but it's the kind of program shell that you'll use for almost any program you write, so let's analyze it.

Notice that the program starts out with some comments describing the program. It's a good idea to say a little about what the program is intended to do, etc. Following this is the "include" statement that alerts the Assembler to the fact that an MSP430G2452 microcontroller is the version being used. The ORG statement tells the Assembler that the program will start at location $E000_{16}$ (this is the start of program memory in the MSP430G2452). This is followed by the Stack Pointer initialization, which is the first program instruction and is assigned the RESET label, which will be stored at the Reset Vector. Also, the watchdog timer is turned off.

Let's skip over the "call Mult" instruction, the "jmp" instruction and the "Mult" subroutine for the moment and go down to the ORG Assembler directive at the bottom. As in the first example program in this chapter, this one takes the address of the first program instruction (the stack pointer initialization instruction) and stores that address at location $FFFE_{16}$. The Define Word (DW) directive tells the Assembler that the address associated with the RESET label is what should be stored at $FFFE_{16}$. If we had chosen to label the first

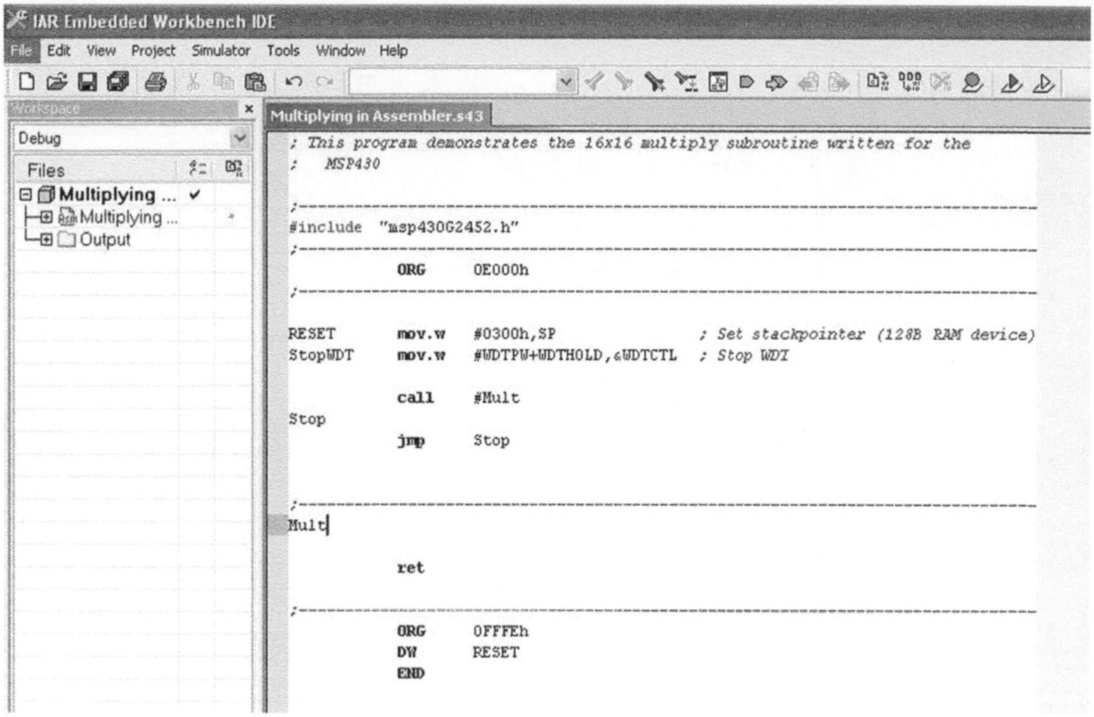

Figure 11.14
The "Multiplying in Assembler" program "shell".

program instruction "Tangiers", then the DW directive would have the label "Tangiers" in it – but "RESET" is much more descriptive.

Now, let's return to the "call Mult" instruction. This instruction, as we've seen, transfers program control to the instruction labeled "Mult". Note that in this particular program, the label "Mult" actually is on the line above the first instruction of the subroutine. This is perfectly legitimate, as far as the Assembler is concerned, and it may help in the readability of your program.

At this point, the subroutine does absolutely nothing. It just returns program control to the instruction following the "call" instruction. Of course, eventually the complete multiply subroutine will be located within this subroutine area.

Note the "jmp" instruction that follows the "call" instruction. What is its purpose? Well, in this particular program, we just want the product of two 16-bit numbers, after which we want the program to halt (that is, do nothing). The "jmp" instruction serves this purpose by jumping back to itself continuously.

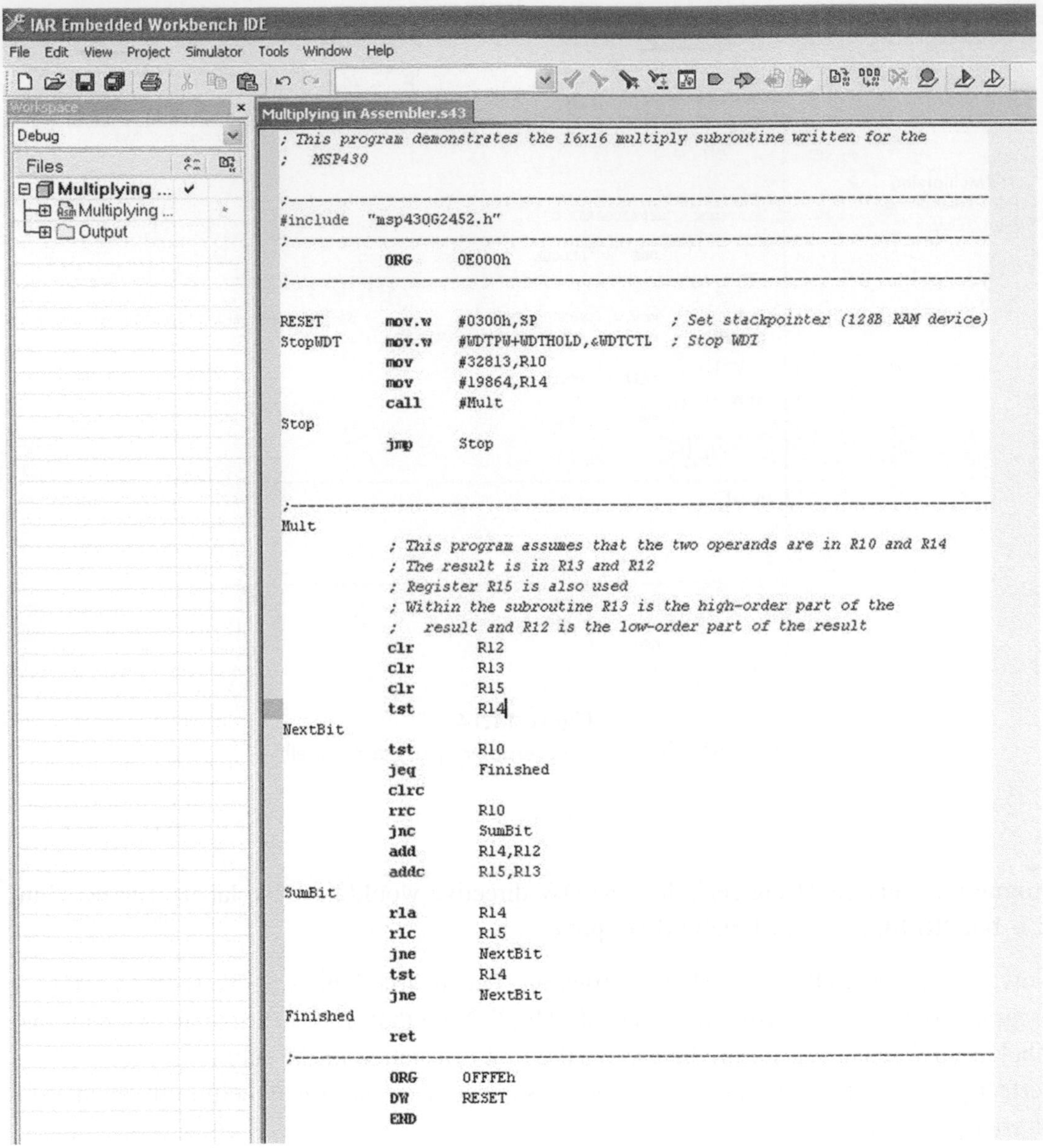

```
; This program demonstrates the 16x16 multiply subroutine written for the
;   MSP430

;------------------------------------------------------------------------------
#include "msp430G2452.h"
;------------------------------------------------------------------------------
            ORG     0E000h
;------------------------------------------------------------------------------

RESET       mov.w   #0300h,SP                ; Set stackpointer (128B RAM device)
StopWDT     mov.w   #WDTPW+WDTHOLD,&WDTCTL   ; Stop WDT
            mov     #32813,R10
            mov     #19864,R14
            call    #Mult
Stop
            jmp     Stop


;------------------------------------------------------------------------------
Mult
            ; This program assumes that the two operands are in R10 and R14
            ; The result is in R13 and R12
            ; Register R15 is also used
            ; Within the subroutine R13 is the high-order part of the
            ;   result and R12 is the low-order part of the result
            clr     R12
            clr     R13
            clr     R15
            tst     R14
NextBit
            tst     R10
            jeq     Finished
            clrc
            rrc     R10
            jnc     SumBit
            add     R14,R12
            addc    R15,R13
SumBit
            rla     R14
            rlc     R15
            jne     NextBit
            tst     R14
            jne     NextBit
Finished
            ret
;------------------------------------------------------------------------------
            ORG     0FFFEh
            DW      RESET
            END
```

Figure 11.15
16 × 16 software multiply subroutine.

Now, let's fill in that big blank of a subroutine. Let's start with some comments. Comments are an excellent way to not only document what the subroutine does, but to formulate your thoughts and strategy. Okay, let's use registers (rather than memory locations) for all the variables. Let's put the multiplier in R10 and the multiplicand in R14. R13 and R12 are the registers for the result (R13 is the high-order word).

The resulting subroutine is shown in Figure 11.15. Note that this is the algorithm that Figure 11.10 depicts. Note that the multiplicand in R14 is shifted left one bit each time through the loop and that a second register (R15) is then needed to receive the bits being shifted out to the left of R14.

Note also, in Figure 11.15, that the two operand registers, R10 and R14, are initialized with the desired input numbers prior to the subroutine call.

Getting more familiar with IAR Embedded Workbench IDE

Once you've got this program entered, assembled, and running, get familiar with the troubleshooting aspects of the IDE. For example, pull down the "View" tab and select registers. You should then see, on the right-hand side of the screen, the values of the 16 MSP430 registers. Now single-step through the program (using the F11 key) and observe how the registers R7 and R8 change with each successive iteration through the loop.

Bibliography

[1] J. Davies, *MSP430 Microcontroller Basics*, Elsevier-Newnes, 2008.
[2] MSP-EXP430G2 LaunchPad Experimenter BoardUser's Guide. Literature Number: SLAU318B, July 2010; revised March 2012.
[3] MSP430x2xx Family User's Guide. Literature Number: SLAU144H.
[4] MSP430 IAR Assembler Reference Guide, second ed. January 2003.
[5] MSP430G2452 Datasheet. Literature Number: SLAS722B, December 2010; revised March 2011.
[6] C. Nagy, *Embedded Systems Design using the TI MSP430 Series*, Elsevier-Newnes, 2003.
[7] Available from: <www.hti.com>.

CHAPTER 12

Programming the MSP430 in C

Chapter Outline

This chapter assumes the reader already has some familiarity with the C programming language. If that's not the case, there are lots of books available to teach the language. One very good guide to C programming is the IAR reference guide, which you can obtain by simply pulling down the Help menu within the IAR Embedded Workbench IDE. You may also want to defer your MSP430 C programming for some later time, since, for several reasons including microcontroller timing, the final MSP430 robot program in this book will be written in Assembler. Just keep in mind that, should you skip this chapter now, chances are you'll need C or some other high-level language in the future.

Let's start off with the blinking LED program. As we did in the last chapter, we'll target the MSP430G2452 as our MSP430 version-of-choice. Go out to the LaunchPad website (http://e2e.ti.com/group/msp430launchpad/w/contents/1130.aspx), find the Code Examples link, left-click on the "MSP430G2xx2" link and open. Select the "C" link and copy "msp430g2xx2_1.c" to the "IAR Programs" folder that you created in the last chapter.

Now create a New Workspace and then a New Project, as was done in the last chapter. This time, however, the project's Options will be a little different. We'll still choose MSP430G2452 under Device, but for Library Configuration, choose "Normal DLIB". In the Debugger window, under "Driver", choose "FET Debugger". Let's call the project "Blinking LED in C". Select Add File from the "Project" tab and, in the window that opens, type the program name. The program name, given to it by the TI programmer, is "msp430g2xx2_1.c" (Figure 12.1).

Choose "Rebuild All", which compiles the program, and then select "Download and Debug", which sends the machine code from your personal computer to the MSP430.

MSP430-based Robot Applications.
DOI: http://dx.doi.org/10.1016/B978-0-12-397012-1.00012-6

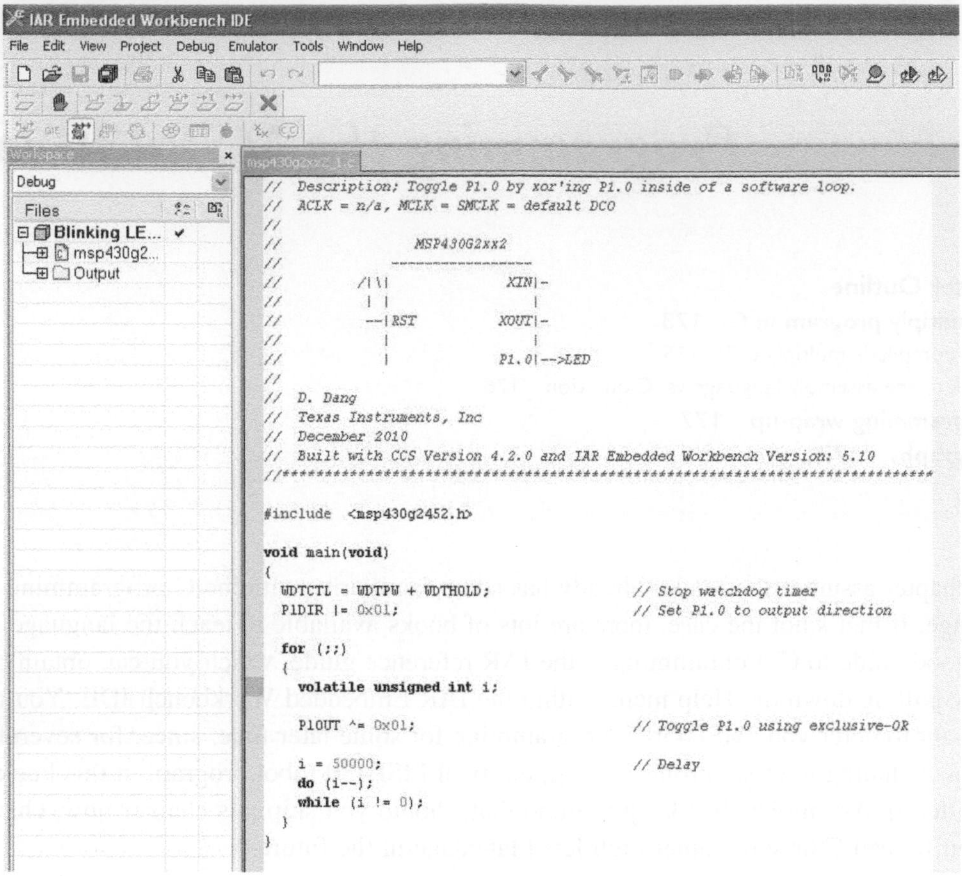

Figure 12.1
The blinking LED program written in C.

Note that, even though the program is written in a high-level language, it still must be converted to machine code for the microcontroller to understand it.

Actually, this is a good opportunity to see one of the features of the IAR Embedded Workbench IDE application called Disassembly. To do this, pull down the "View" menu at the top of IAR Embedded Workbench IDE and select "Disassembly" (assuming that the Disassembly window isn't already in view). Figure 12.2 shows what this display should look like.

Note several things about the Assembler equivalent set of instructions that the C compiler creates:

- The stack pointer is initialized without the C program explicitly invoking that action.
- The initialization instructions (for the watchdog timer, etc.) actually have a one-to-one correspondence between C and the equivalent Assembler.

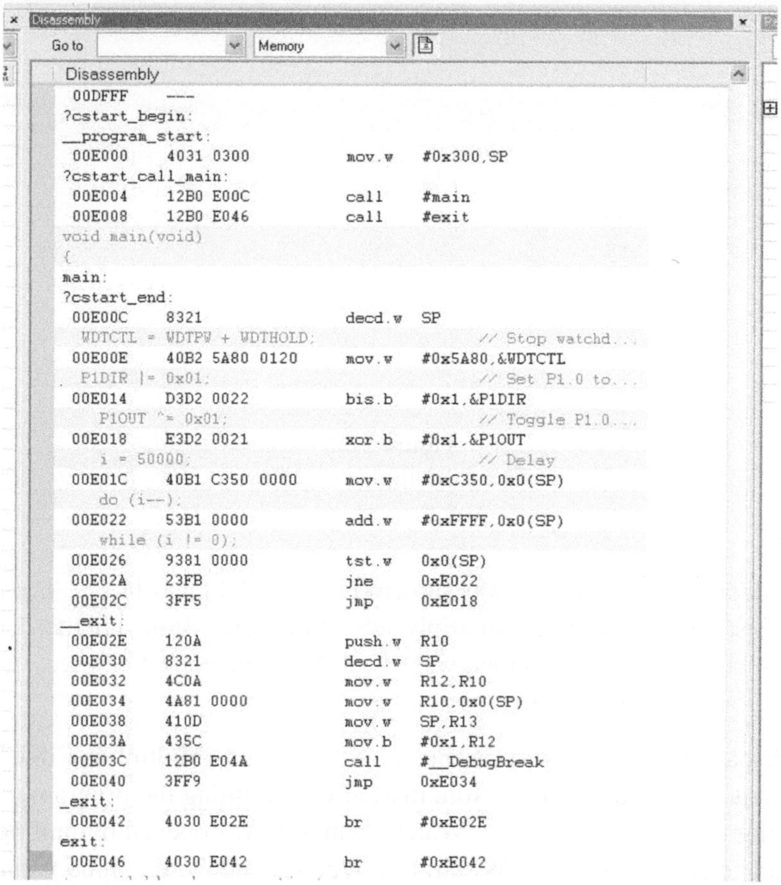

Figure 12.2
The Disassembly window for the blinking LED C program.

- The absence of parameters in the "for" loop make this a "forever" loop.
- The set of brackets – the left one after the "for" statement and the matching right one after the "while" statement – mean that the program body corresponding to the "for" loop is the sequence of instructions enclosed within these parentheses.
- The stack is actually used as the storage location for the variable "*i*".

Now pull down the "Debug" menu and select "Go" and verify that the red LED on the LaunchPad board is blinking.

The multiply program in C

Let's continue the parallel track of writing programs in C that perform the same functions as we earlier produced in assembly language. In the last chapter, a 16 × 16 multiply

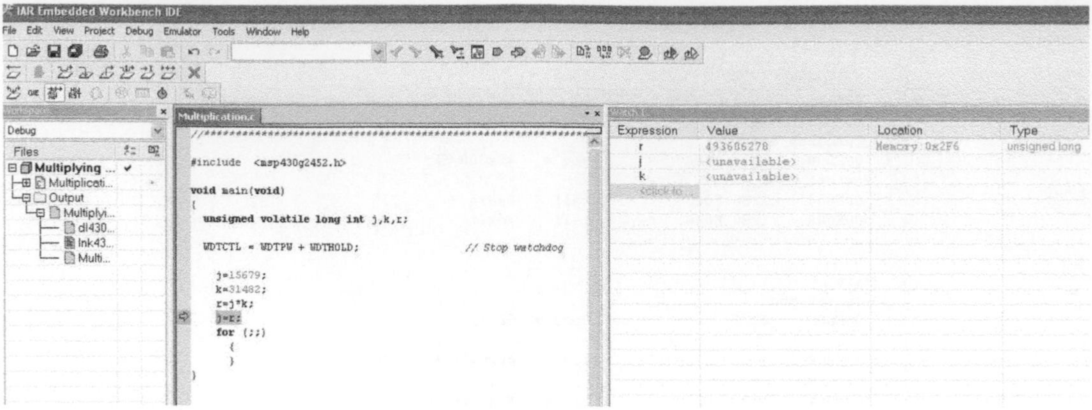

Figure 12.3
The C multiply program.

subroutine, producing a 32-bit result, was shown (Figures 11.14 and 11.15). The full subroutine, consisting of an unsigned multiply subroutine called from a higher-level subroutine that took care of signed input and output, was about 30 instructions. Let's see what is required in C. Figure 12.3 shows the program.

Both the Workspace and the Project, in this case, are called "Multiplying in C" and the program is called "Multiplication.c". Note that, after compiling the program ("Rebuild All") and downloading to the LaunchPad, a "Watch" window was opened by pulling down the View menu and selecting "Watch". Within this Watch window, the three variables, "*r*", "*j*", and "*k*" are entered into the "Expressions" column. Then the program is single-stepped (using the F11 key), until the dummy "*j* = *r*" command is executed. As each of the statements above this instruction is executed, you should see, in the "Value" column of the "Watch" window, the decimal value of the variable.

The three variables – the multiplier, multiplicand, and product – are all declared to be 32-bit integers, even though we only require 16-bit multiplier and multiplicand. This is because the C standards that the IAR Embedded Workbench IDE compiler is based on specify that the multiplication instructions be set up this way. If you try to declare "*j*" and "*k*" as 16-bit integers (that is, "unsigned volatile int" instead of "unsigned volatile *long* int"), leaving "*r*" as a 32-bit integer, the program returns just the low-order 16 bits of the result.

Okay, so what's the lesson with this program? Well, the C version does the entire multiply function by using just a single C instruction – pretty impressive. To get more insight into what the compiler actually does to make this happen, close the "Watch" window and open View→Disassembly (Figure 12.4).

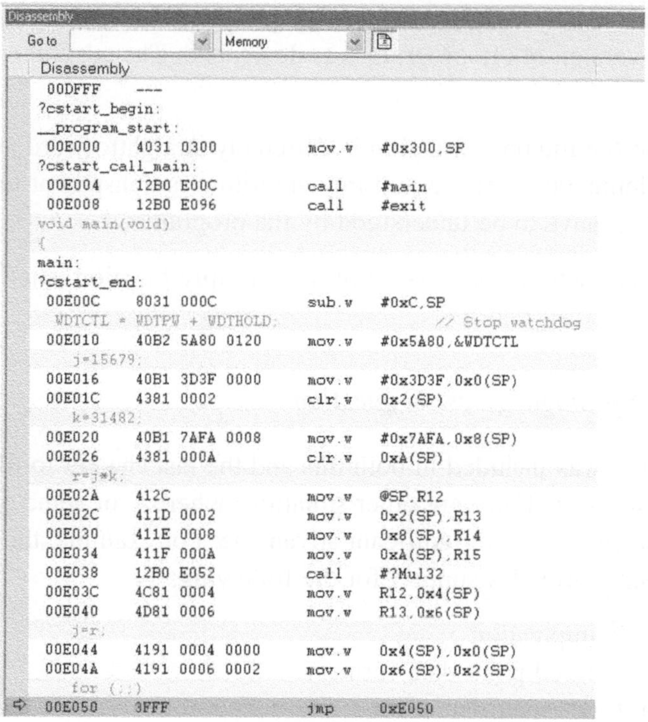

```
Disassembly
Go to                    Memory
Disassembly
 00DFFF     ---
?cstart_begin:
__program_start:
 00E000     4031 0300           mov.w    #0x300,SP
?cstart_call_main:
 00E004     12B0 E00C           call     #main
 00E008     12B0 E096           call     #exit
void main(void)
{
main:
?cstart_end:
 00E00C     8031 000C           sub.w    #0xC,SP
  WDTCTL = WDTPW + WDTHOLD;              // Stop watchdog
 00E010     40B2 5A80 0120      mov.w    #0x5A80,&WDTCTL
    j=15679;
 00E016     40B1 3D3F 0000      mov.w    #0x3D3F,0x0(SP)
 00E01C     4381 0002           clr.w    0x2(SP)
    k=31482;
 00E020     40B1 7AFA 0008      mov.w    #0x7AFA,0x8(SP)
 00E026     4381 000A           clr.w    0xA(SP)
    r=j*k;
 00E02A     412C                mov.w    @SP,R12
 00E02C     411D 0002           mov.w    0x2(SP),R13
 00E030     411E 0008           mov.w    0x8(SP),R14
 00E034     411F 000A           mov.w    0xA(SP),R15
 00E038     12B0 E052           call     #?Mul32
 00E03C     4C81 0004           mov.w    R12,0x4(SP)
 00E040     4D81 0006           mov.w    R13,0x6(SP)
    j=r;
 00E044     4191 0004 0000      mov.w    0x4(SP),0x0(SP)
 00E04A     4191 0006 0002      mov.w    0x6(SP),0x2(SP)
    for (;;)
 00E050     3FFF                jmp      0xE050
```

Figure 12.4
Disassembly of the C multiply program.

Note that, in the Disassembly, the C instructions are shown in the lightly grayed-out lines, with the compiler-generated assembly language commands shown below that. Let's examine the multiplication instruction; that is, "*r* = *j***k*". We see that the Assembler instructions amount to retrieving the two numbers that had previously been pushed onto the stack and then calling the routine, "?Mul32". ?Mul32 is the unsigned multiply subroutine.

The compiler's multiply call

When we observe the disassembly of the C program, we get some insight into how the compiler works. The compiler, after parsing the instruction "*r* = *j***k*", recognizes that the "*" means that a multiply is being requested. Based on the definition of *j*, *k*, and *r* as unsigned, long variables, it further knows that the version of the multiply subroutine that handles 32-bit, unsigned numbers is the one that must be used. Thus, the C compiler is able to interpret this instruction as a "call" to a specific type of multiply subroutine. The compiler "pastes" the multiplier subroutine code into the assembly language code that it generates.

Now contrast this software multiply with the one presented in the last chapter, in which an assembly language version of the program was developed. The advantages of the C version are obvious:

- The inclusion of the multiply function is extremely straightforward and readable.
- The actual implementation of the multiply algorithm is transparent to the programmer and does not even have to be understood by the programmer.

The main disadvantage with the C version of the multiply function is its increased execution time.

Back to the assembly language vs. C question

The multiply function was included in both this and the last chapter to illustrate the advantage that C *can* have. There are other situations where C or some other high-level programming language can offer significant advantage. For example, the C/C++ programming language provides support for the following:

- Character string manipulation
- Floating-point and fixed-point data formats
- Iteration control (do..while, while, for)
- Input/output functions
- Standard mathematical functions (in addition to multiplication)
- Arrays
- Classes

That last bullet, "Classes", is a special type of C variable, something like an array, but in which the elements may have different types.

While these functions are of great use in many applications, their potential utility diminishes considerably when the processor becomes a simple microcontroller for our planned robot. For example, there are no character strings within the robot applications that we'll be discussing. Floating-point representation (that is, scientific notation, with sign, magnitude, and exponent) is an extremely powerful way of calculating and is probably what you use most on your personal computer or your calculator, but floating-point is just too time-consuming for our little microcontroller applications. The input/output functions can be of considerable use in sending and receiving information between computer and display, hard drive, or printer – but our microcontroller will communicate with none of these.

Even the multiply function used as an example in these chapters will not be used for the robotics applications in this book. Instead, the microcontroller programs that we will be developing will change bits within registers, count down intervals, control A/D converters, etc. – simple tasks that require approximately the same amount of programming effort whether the program is written in C or Assembler. Indeed, the first example in this chapter and its Assembler equivalent example in the last chapter illustrate this point.

C programming wrap-up

We have touched very lightly on the subject of programming the MSP430 in C. This light touch occurs for two reasons: (1) programming of the actual robot will be done in Assembler in this book, due mainly to the need for some tight timing control that requires the use of assembly language; and (2) to really cover C and C++, either as a general programming topic or as it relates to the MSP430 microcontroller, is a topic that can easily take an entire book by itself. Nevertheless, any reader who plans to work with the MSP430 is wise to consider C as a potential programming language in the future.

Bibliography

[1] HowStuffWorks. The Basics of C Programming. <http://www.howstuffworks.com/c.htm>.

[2] J. Hubbard, *Schaum's Outline of Theory and Problems of Programming with C++*, McGraw-Hill, 2000.

[3] IAR C/C++ Compiler Reference Guide for Texas Instruments' MSP430 Microcontroller Family.

CHAPTER 13

System Clocking for the MSP430

Chapter Outline

Back to hardware!

The last several chapters gave the basics of MSP430 software and, previous to that, we talked about electronics at a general level. Now it's time to look at specific aspects of the MSP430. In the "Introducing the MSP430 Microcontroller" chapter, we discussed briefly the clocking of the MSP430. In this chapter, we'll discuss clocking in more detail.

System clocking in a computer

Virtually all computers have system clocks. This is usually a square-wave signal that is distributed throughout the computer. You've no doubt heard of personal computers having clock speeds of a certain number of GHz. Such computers therefore have a clock cycle of less than 1 nanosecond – an amazing speed, considering that the earliest computers had clock speeds of only around 100 kHz.

One thing to note – on most computers, an instruction can require more than one cycle to execute, so even though a clock might have a period of 1 nanosecond, the instruction might take several nanoseconds to complete. Nevertheless, this is extremely fast and accounts for why many complex operations performed by a personal computer appear to occur almost instantly.

MSP430-based Robot Applications.
DOI: http://dx.doi.org/10.1016/B978-0-12-397012-1.00013-8

The MSP430, like most microcontrollers, has a more modest clock speed. Top speed for the versions that we will use is 16 MHz and then only if the microcontroller's supply voltage is at its maximum.

But why does a computer need a clock at all? What is the purpose of this clock? The answer to this question is that the computer logic inside a computer's CPU does not change its state instantaneously. Each transistor of each gate of the memory, registers, and the CPU's arithmetic/logic unit (ALU) has its own "transition time" required to change state. These transition times are the result of parasitic capacitances and inductances in the transistors and other parts of the circuit. Even with ideal (zero) inductances and capacitances, the different signals in the microcontroller must travel at finite speed (the speed of light) over different path lengths.

So, let's say, for example, that our microcontroller needs to perform a compare (CMP) between a word in its data memory and a word in a register. Remember that such an instruction sets or resets the microcontroller's Zero (Z) condition bit (among other condition bits) based on whether the two operands are equal to one another. The path from the bits in data memory may arrive earlier or later than the bits in the register, and during this "transition time" the input to the Z bit flip-flop may bounce around between 1 and 0. All values for Z are suspect during this interval. Only after some period of time are we sure that both operands have reached their final values. The microcontroller must therefore have a clock that keeps the Z flip-flop output bit from changing until all signals have settled. The clock period must be long enough to guarantee that the time from when the CMP instruction is started (on one clock edge) to the time the Z bit is used (and assumed stable) in the next instruction is less than the clock period.

The MSP430 system clocks

The MSP430 microcontroller has quite a few clocks that can be used. There is, for example, the main system clock, MCLK; the sub-system clock, SMCLK; and the auxiliary clock, the ACLK. There are also clocks that are available solely for use with a particular peripheral. For example, the 10-bit A/D converters that are available in some of the MSP430 versions that we'll be using can be clocked by the ADC10OSC, approximately a 5 MHz clock. The complete description of these clocks is given in the TI publication SLAU144H.

Each of these clocks is generated by several possible sources. For example, MCLK can be generated from a crystal-controlled oscillator by connecting a crystal between the XIN and XOUT pins of the MSP430. Or MCLK can be generated externally and brought in through the XIN pin of the MSP430. There is also an internal low-frequency source, called VLO, which can serve as the MCLK source. Or MCLK can be generated from what is known as the digitally controlled oscillator (DCO), an internal resistor/capacitor type of oscillator.

And, for each clock and each peripheral, the clock can be "divided down" by some power of 2. So, for example, an 8 MHz clock could be divided down to 2 MHz.

Yikes! As with many powerful and flexible systems, the choices and setup can seem daunting. That's why, in the "Introducing the MSP430 Microcontroller" chapter, the following constraints on clock choice were made:

- The MCLK is always sourced from the DCOCLK
- The SMCLK is always sourced from the DCOCLK and will be equal to MCLK
- The ACLK is never used
- No peripheral-specific clocks will be used except for the ADC10OSC clock

It isn't that you'll never have a need for these other clock choices – as your MSP430 knowledge increases and your projects become more sophisticated, you almost certainly will have times when you'll want to explore these other choices. But it's important to start out with a workable subset of the device's capabilities so that the experience is not overwhelming.

The DCOCLK

Okay, so we're using the DCOCLK for just about all clocking on the MSP430 (with the possible exception of the A/D converter). The DCOCLK is an RC (resistor/capacitor) type of oscillator. What's that? It's a simple circuit that uses the periodic charge and discharge of a capacitor, along with the gain from an amplifier or comparator, to generate a square wave. It's worth understanding how such a circuit works, since you may have a need to create your own RC oscillator in the future. Figure 13.1 shows an example of such a circuit. This is not exactly how the RC oscillator circuit inside the MSP430 is implemented but it shows how such oscillators work in general.

Although the Figure 13.1 circuit uses an op amp, this is not a linear amplifier. The amplifier is actually behaving as a comparator. The output will flip between ground and its supply voltage, as we'll see shortly.

To help analyze this circuit, let's use a circuit simulator. Remember that? As in the earlier chapter, let's use the TINA circuit analyzer. The circuit is drawn in TINA as shown in Figure 13.2. Note that the voltage reference, *V*, should be about half the supply voltage (not shown in Figure 13.1) and that the op amp is assumed to have rail-to-rail outputs.

The op amp chosen in TINA is a TI TLV2772, which has rail-to-rail output. The output of this op amp can "slew" (that is, rise and fall) at a fairly quick 9 V/μsec. The op amp is powered from a 3.3 V supply, so the reference voltage (here split into two references, *V2* and *V3*, for clarity) is set to 1.6 V. Three voltage monitors, *VM1*, *VM2*, and *VM3*, are used.

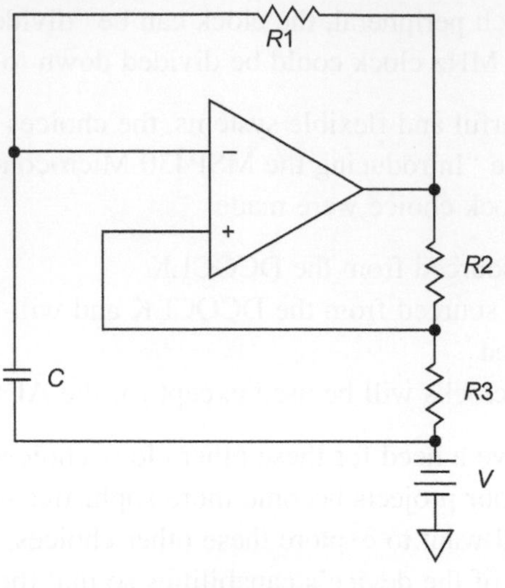

Figure 13.1
Example of an RC oscillator.

The TINA "Oscilloscope" feature is brought up and the "Trigger" is set to "Normal" mode, "Source" equal to *VM1*, and "Level" of 1000 mV. Figure 13.3 shows the virtual oscilloscope display.

The larger square wave is the op amp output, the smaller square wave is the attenuated version created by the resistor divider R2/R3, and the "triangle" waveform is the capacitor voltage. From Figures 13.2 and 13.3 we can see how the circuit works. When the op amp output goes low, the voltage at the noninverting input of the op amp (*VM3*) drops to ½(1.6 V) = 0.8 V. The capacitor, at that point, begins to discharge. When the capacitor voltage drops below the 0.8 V of the noninverting input of the op amp, the op amp output voltage goes to the supply rail (3.3 V), the noninverting op amp input voltage goes to ½(3.3 V + 1.6 V) = 2.45 V and the capacitor then begins to charge. When the capacitor voltage reaches 2.45 V, the op amp output voltage goes to the negative rail (0 V) and the cycle starts all over.

Frequency inaccuracy of the RC oscillator

Some oscillators, like crystal-controlled oscillators, have extremely accurate frequencies. For example, a "watch crystal" has a frequency of 32768 Hz. This frequency is determined by the crystal structure itself and even the least expensive such crystal oscillators can maintain the frequency to within 0.01%, with most much better than this.

RC oscillators, on the other hand, are dependent on the resistor and capacitor values to maintain frequency accuracy. Such components, especially when part of an integrated

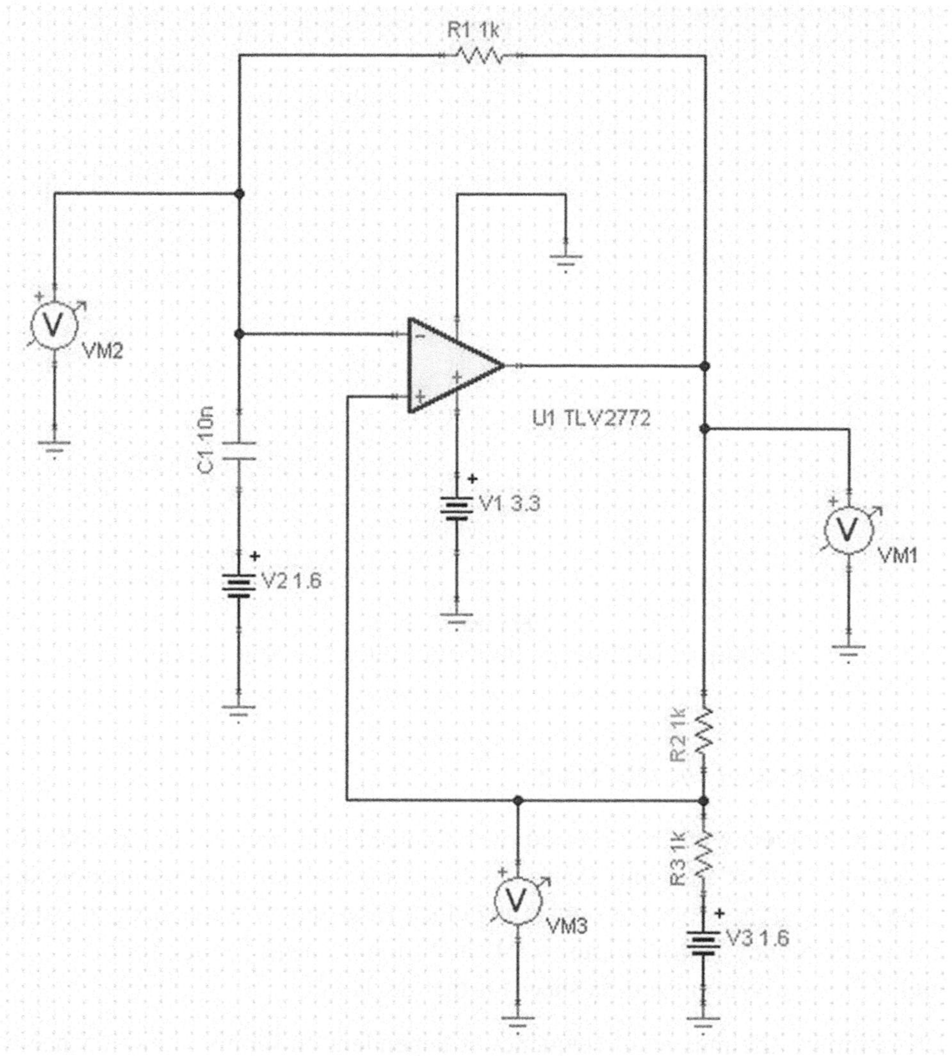

Figure 13.2
Circuit redrawn in TINA.

circuit, have component values that can be off by more than 1%. Capacitors and many resistors can have values that are in error by 5% or more. Worse, when these relatively imprecise components are used together, the total error can add, so that two 5% components used together in the same circuit produce an overall error that approaches 10%. Therefore, the DCOCLK is not the clock to use if you need extreme accuracy.

Fortunately, in our robot project, we don't need extreme frequency accuracy. Plus, the MSP430 DCOCLK can be programmed to use settings calibrated at the factory to eliminate some of this frequency error.

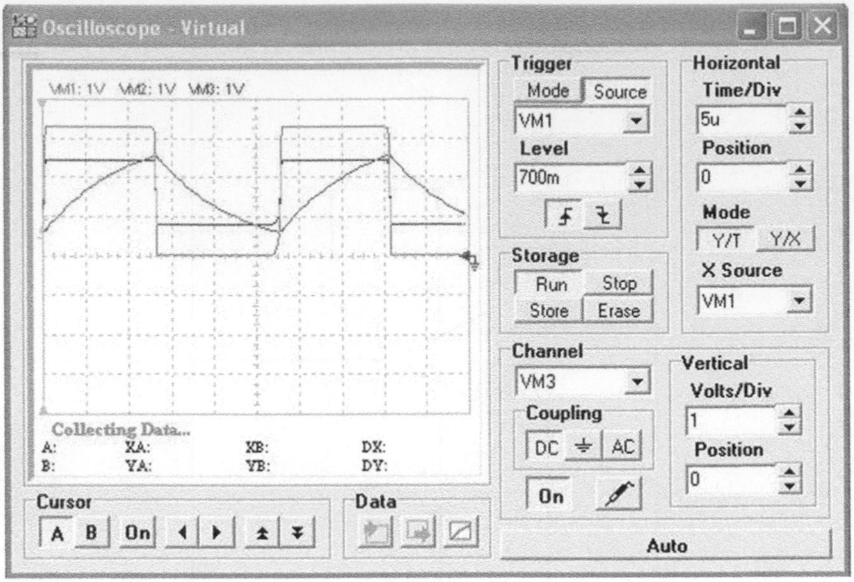

Figure 13.3
TINA virtual oscilloscope display for the RC oscillator circuit.

Setting the frequency of the DCOCLK

The DCOCLK frequency can be adjusted over a considerable range. Why would anyone want to do that? Well, it's obvious that, for a lot of applications where execution time needs to be as fast as possible, we'd want the clock frequency to be as high as can be made. But just what frequency is that? Figure 13.4 shows the frequency range for the MSP430G2452, as given in the data sheet for that device.

As the chart shows, the maximum frequency is dependent on supply voltage. Since we will be operating at 3.3 V or higher for the robot applications, 16 MHz is the maximum frequency. One note about this chart – there is mention here of flash memory programming. This is a feature that we won't need or use in this book, so ignore the different markings in the chart. Once again, this is a very useful feature for advanced applications, but something that we can and will do without.

Okay, so we can go as high as 16 MHz. Are there ever applications where we might want to have the clock frequency set lower? The answer is yes, quite often. It turns out that the power consumed by the microcontroller is highly dependent on the clock frequency. This is actually a quite common situation with CMOS circuits. Much of the power required to operate these devices is dissipated while making the integrated circuit transistors switch states. So the more often the transistors switch states, the higher the power consumption is.

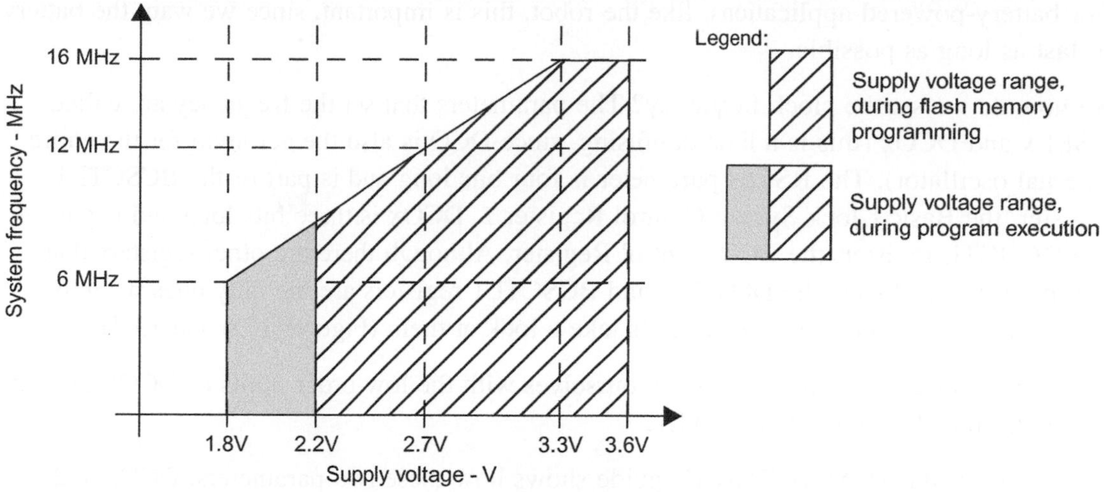

Figure 13.4
Allowable DCOCLK clock frequencies *(courtesy of Texas Instruments)*.

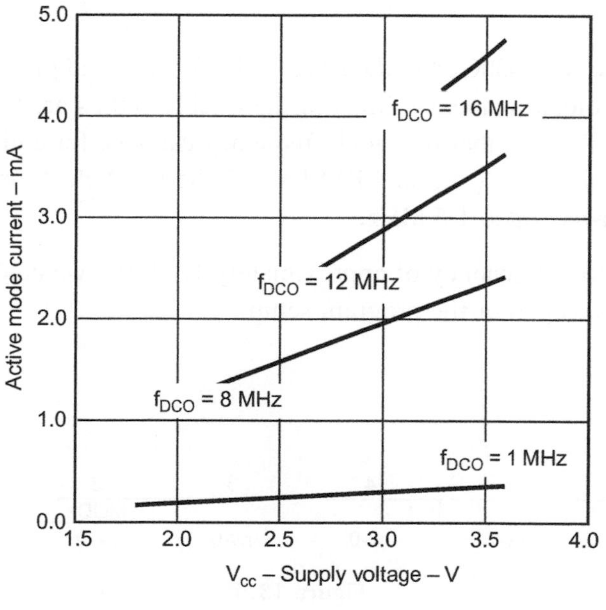

Figure 13.5
Variation of MSP430G2452 current with DCO clock frequency and voltage *(courtesy of Texas Instruments)*.

Figure 13.5 is also from the MSP430G2452 datasheet and shows the variation of microcontroller current with DCO clock frequency and with supply voltage. At 3.3 V, the MSP430G2452 uses more than 10 times as much current at 16 MHz as it does at 1 MHz.

For battery-powered applications, like the robot, this is important, since we want the battery to last as long as possible.

So how do we set the clock frequency? The parameters that set the frequency are called RSELx and DCOx (this is a little confusing, since DCO is also the acronym for the entire internal oscillator). The RSELx parameter is four bits long and is part of the BCSCTL1 register, the Basic Clock System Control Register 1. DCOx is three bits long and is part of the DCOCTL register, the DCO Control Register. Although there are other registers that can modify the clocks, the DCOCTL and BCSCTL1 registers are the only ones we will need for the projects in this book. Let's take a look at them (Figures 13.6 and 13.7).

For our purposes, we will not concern ourselves with the low-order 5 bits of DCOCTL and the high-order four bits of BCSCTL1.

A chart from the SLAU144H user's guide shows how these two parameters, DCOx and RSELx, vary the frequency (Figure 13.8). We can think of the RSELx parameter as the coarse clock frequency adjustment and the DCOx parameter as the fine adjustment, but we can see from Figure 13.8 that there is considerable overlap between these families of curves.

Table 13.1 is from the MSP430G2452 datasheet (SLAS722B) and gives us further insight into setting the DCO clock frequency. Most of these table entries are for DCOx = 3. Note the very wide range of values that the clock frequency can take for a given DCOx/RSELx value. For example, with DCOx = 3 and RSELx = 7, the DCO clock frequency can be anywhere from 0.80 MHz up to 1.5 MHz.

To produce a DCO clock frequency of approximately 1.1 MHz, we could include the following instructions as part of the program setup:

```
mov.b #060h,&DCOCTL
mov.b #7,&BCSCTL1
```

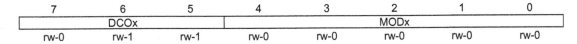

7	6	5	4	3	2	1	0
DCOx			MODx				
rw-0	rw-1	rw-1	rw-0	rw-0	rw-0	rw-0	rw-0

Figure 13.6
The DCOCTL register *(courtesy of Texas Instruments)*.

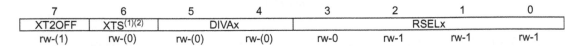

7	6	5	4	3	2	1	0
XT2OFF	XTS[(1)(2)]	DIVAx		RSELx			
rw-(1)	rw-(0)	rw-(0)	rw-(0)	rw-0	rw-1	rw-1	rw-1

Figure 13.7
The BCSCTL1 register *(courtesy of Texas Instruments)*.

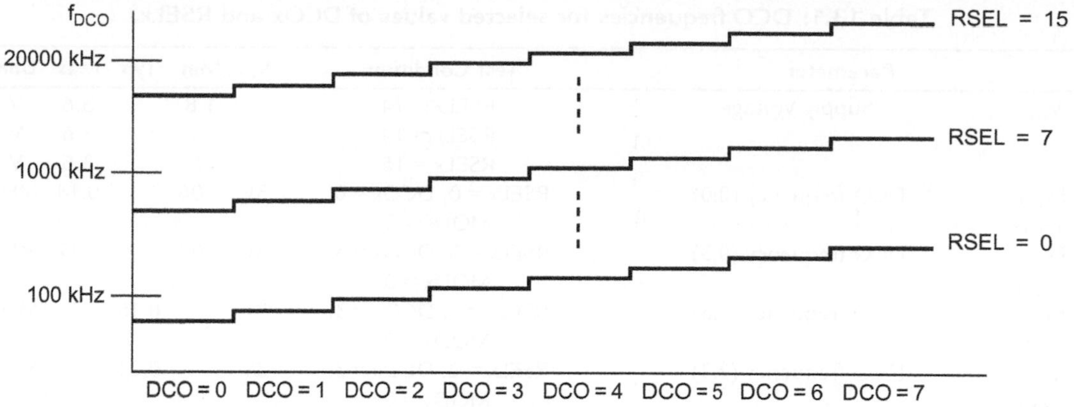

Figure 13.8
DCO clock frequency as function of DCOx and RSELx parameters *(courtesy of Texas Instruments)*.

Do you see how putting 60_{16} into the DCOCTL register causes DCOx = 3? And, of course, we have set RSELx = 7. Note that we don't have to worry about the memory location associated with these registers – the assembler (using information in the header file at the top of your program) figures that out for us. Note also that we need to use the "byte" version of the "mov" instruction, since these two registers, DCOCTL and BCSCTL1, are only 8 bits each.

Okay, so we have a nominal 1.1 MHz clock rate. But when we look at Table 13.1, we see that the clock can be anywhere from 0.8 MHz to 1.5 MHz for this setting. My own experience is that, at room temperature, this range is considerably reduced. Still, it would be nice to get a clock frequency that we can be confident is very near its nominal value. Is that possible?

Calibrated frequencies

Fortunately, the answer is yes. Texas Instruments has a set of calibrated frequencies for each MSP430 that the user can choose, and these clock frequencies will generally be within 1% or so of their stated frequency, if the MSP430 operates at room temperature and with a supply voltage at or near 3 V. The calibrated clock frequencies are given in the MSP430 datasheet for the specific version. For the MSP430G2452, the calibrated DCO clock frequencies are 1 MHz, 8 MHz, 12 MHz, and 16 MHz. To set the DCO clock to the calibrated 12 MHz, the following instructions would be used:

```
mov.b  #CALDCO_12MHZ,&DCOCTL
mov.b  #CALBC1_12MHZ,&BCSCTL1
```

Table 13.1: DCO frequencies for selected values of DCOx and RSELx.

Parameter		Test Conditions	V_{cc}	Min	Typ	Max	Unit
V_{CC}	Supply Voltage	RSELx < 14		1.8		3.6	V
		RSELx = 14		2.2		3.6	V
		RSELx = 15		3		3.6	V
$f_{DCO(0,0)}$	DCO frequency (0,0)	RSELx = 0, DCOx = 0, MODx = 0	3V	0.06		0.14	MHz
$f_{DCO(0,3)}$	DCO frequency (0,3)	RSELx = 0, DCOx = 3, MODx = 0	3V	0.07		0.17	MHz
$f_{DCO(1,3)}$	DCO frequency (1,3)	RSELx = 1, DCOx = 3, MODx = 0	3V		0.15		MHz
$f_{DCO(2,3)}$	DCO frequency (2,3)	RSELx = 2, DCOx = 3, MODx = 0	3V		0.21		MHz
$f_{DCO(3,3)}$	DCO frequency (3,3)	RSELx = 3, DCOx = 3, MODx = 0	3V		0.30		MHz
$f_{DCO(4,3)}$	DCO frequency (4,3)	RSELx = 4, DCOx = 3, MODx = 0	3V		0.41		MHz
$f_{DCO(5,3)}$	DCO frequency (5,3)	RSELx = 5, DCOx = 3, MODx = 0	3V		0.58		MHz
$f_{DCO(6,3)}$	DCO frequency (6,3)	RSELx = 6, DCOx = 3, MODx = 0	3V	0.54		1.06	MHz
$f_{DCO(7,3)}$	DCO frequency (7,3)	RSELx = 7, DCOx = 3, MODx = 0	3V	0.80		1.50	MHz
$f_{DCO(8,3)}$	DCO frequency (8,3)	RSELx = 8, DCOx = 3, MODx = 0	3V		1.6		MHz
$f_{DCO(9,3)}$	DCO frequency (9,3)	RSELx = 9, DCOx = 3, MODx = 0	3V		2.3		MHz
$f_{DCO(10,3)}$	RSELx = 10, DCOx = 3, MODx = 0	DCO frequency (10,3) 3V		3.4			MHz
$f_{DCO(11,3)}$	DCO frequency (11,3)	RSELx = 11, DCOx = 3, MODx = 0	3V		4.25		MHz
$f_{DCO(12,3)}$	RSELx = 12, DCOx = 3, MODx = 0	DCO frequency (12,3) 3V		4.30		7.30	MHz
$f_{DCO(13,3)}$	RSELx = 13, DCOx = 3, MODx = 0	DCO frequency (13,3) 3V		6.00		9.60	MHz
$f_{DCO(14,3)}$	RSELx = 14, DCOx = 3, MODx = 0	DCO frequency (14,3) 3V		8.60		13.9	MHz
$f_{DCO(15,3)}$	RSELx = 15, DCOx = 3, MODx = 0	DCO frequency (15,3) 3V		12.0		18.5	MHz
$f_{DCO(15,7)}$	RSELx = 15, DCOx = 7, MODx = 0	DCO frequency (15,7) 3V		16.0		26.0	MHz
S_{RSEL}	Frequency step between range RSEL and RSEL + 1	$S_{RSEL} = f_{DCO(RSEL+1,\ DCO)}/f_{DCO(RSEL,\ DCO)}$	3V		1.35		ratio
S_{DCO}	Frequency step between tap DCO and DCO + 1	$S_{DCO} = f_{DCO(RSEL,\ DCO+1)}/f_{DCO(RSEL,DCO)}$	3V		1.08		ratio
	Duty cycle	Measured at SMCLK output	3V		50		%

Over recommended ranges of supply voltage and operating free-air temperature unless otherwise noted.
Courtesy of Texas Instruments.

Such calibrated frequencies make it possible to have reasonably accurate DCO clock frequencies for most applications, eliminating the need for crystals or external oscillators.

Bibliography

[1] Computer History. <http://www.computersciencelab.com/ComputerHistory/HistoryPt4.htm>.

[2] MSP430x2xx Family User's Guide. Texas Instruments Literature Number: SLAU144H, December 2004; revised April 2011. <http://www.ti.com/lit/ug/slau144i/slau144i.pdf>.

[3] MSP430G2x52/MSP430G2x12 MIXED SIGNAL MICROCONTROLLER. Texas Instruments Literature Number: SLAS722B, December 2010; revised March 2011. <http://www.ti.com/lit/ds/slas722e/slas722e.pdf>.

CHAPTER 14

Parallel and Serial Input/Output Ports

Chapter Outline

As we talked about in the chapter on the MSP430 Introduction, the thing that makes microcontrollers different from, say, just a central processing unit (CPU) is the fact that there are peripheral circuits included on the chip. In the case of the MSP430, those peripherals can include A/D converters, timers, and parallel and serial I/O. In this chapter, we will take a look at the parallel and serial I/O.

Parallel I/O

Let's review the flip-flop (Figure 14.1). The value of the data bit, which is either a logic 0 or logic 1, is "captured" on the rising edge of the clock pulse and the output maintains that captured state until the next clock pulse rising edge (Figure 14.2).

The input port

As discussed earlier, a register is a collection of such flip-flops. An 8-bit-wide register is shown in Figure 14.3. As drawn, this is part of a microcontroller and is referred to as an input *port*, to avoid confusion with the microcontroller's general-purpose registers and to emphasize the fact that it is a construct for bringing information in and out (in this case, in).

MSP430-based Robot Applications.
DOI: http://dx.doi.org/10.1016/B978-0-12-397012-1.00014-X

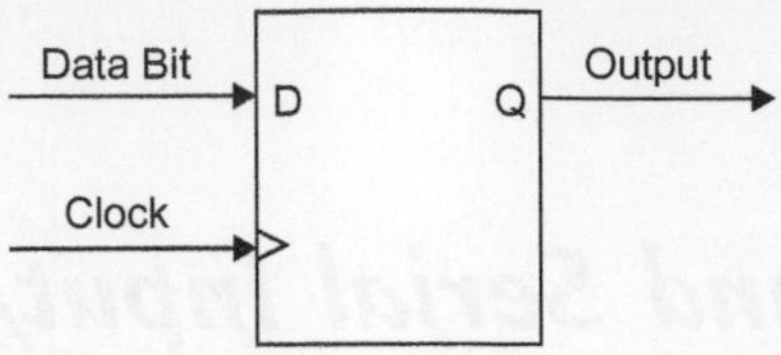

Figure 14.1
The D flip-flop.

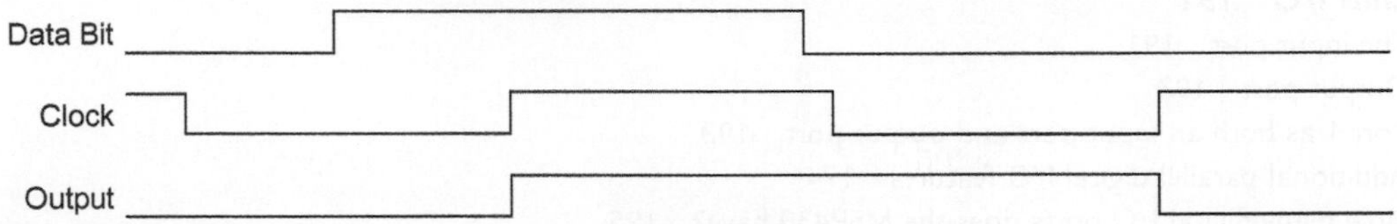

Figure 14.2
D flip-flop timing.

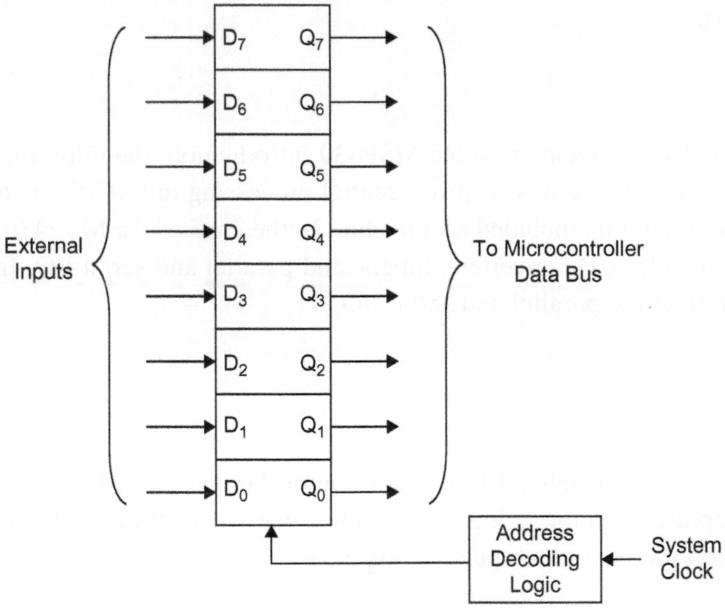

Figure 14.3
Eight-bit-wide input port of a microcontroller.

Note the clock input at the bottom. This actually goes to all eight of the flip-flops making up the port, even though, to reduce clutter, it's only shown going to the bottom flip-flop.

Note also the box labeled "Address Decoding Logic". We only want to clock the flip-flops (capture the external data) when the port is addressed. In the MSP430, the command to do this might be something like this:

```
mov.b       &P1IN,R7
```

When this instruction is executed, the address decoding logic gates the system clock so that the external data at the port inputs are captured. The data thus captured is then transferred to the low-order byte of the MSP430 general-purpose register, R7.

One more thing to note about the above assembly language instruction – the Port 1 Input, P1IN, is actually memory location 20_{16} in the MSP430. However, there's no need to remember that or use that. When you reference P1IN in the above instruction, the assembler, using information provided to it by the header file that you include (for example, "MSP430G2452.h"), knows the absolute address, 20_{16}, and substitutes that address in the actual machine code.

Output port

Let's say that we wish to perform the opposite operation from an input – we wish to generate output bits that will control things outside the microcontroller. In that case, the Data Bit of Figure 14.1 would be generated within the microcontroller and, when the microcontroller issues a clock pulse to capture this Data Bit input, that value, 1 or 0, appears on the output. In fact, we could generate an entire 8 bits of output with an output port, in a manner very similar to the input port of Figure 14.3 (see Figure 14.4).

For this output type of port, the assembly language instruction:

```
mov.b        R7,&P1OUT
```

moves the contents of the low-order byte of general-purpose register, R7, to the Port 1 outputs, Q_0 through Q_7. Again, even though the Port 1 output memory location is 21_{16}, there is no need for us to remember or use that. The assembler knows, from the included header file, that 21_{16} is the Port 1 output memory location and makes the substitution in the machine code.

Port 1 as both an input port and output port

Just a second – the two assembly language instructions given so far refer to Port 1 for both input operations and output operations. Are these the same physical port? How can that be?

It turns out that each bit of Port 1 can be set up as either an input or an output. Figure 14.5 shows a single bit of a port that can handle either input or output. If both switches are set to their high positions, the top flip-flop will be used and this port pin is then acting as an

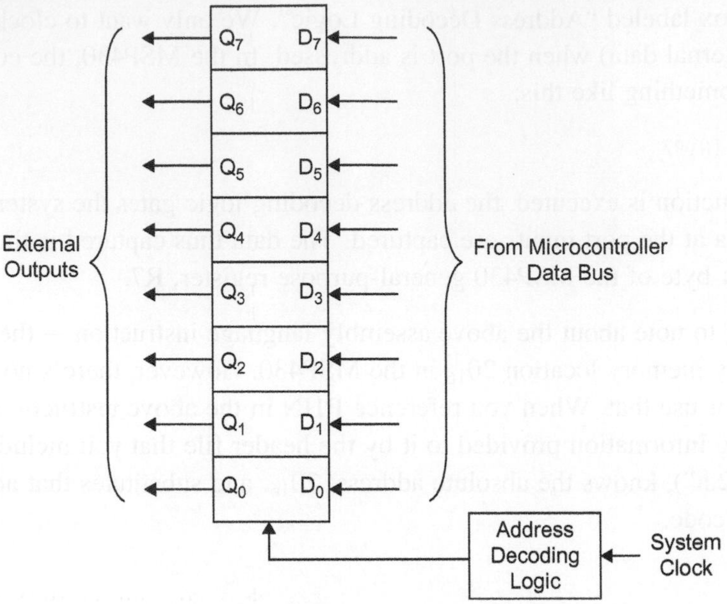

Figure 14.4
Eight-bit-wide output port of a microcontroller.

output. If both switches are set to their low positions, the bottom flip-flop is used, and the port pin is acting as an input.

In the diagram, the switch positions are set by a signal called Direction. In the MSP430, the Port 1 direction for each of the 8 bits of the port is set by a control register called P1DIR. A logic 1 in a particular position of this control register makes the Port 1 bit an output, while a logic 0 makes it an input. If we wanted to, say, make the high-order four bits of the port be outputs and the low-order four bits be input, we would use the instruction:

```
mov.b       #0F0h,&P1DIR
```

Note that the input clock is generated when an instruction references P1IN, whereas the output clock is generated when an instruction references P1OUT.

Additional parallel digital I/O features

The internal electronics of a port pin are actually more complex than what is shown in Figure 14.5. There is, for example, a resistor at each pin, the other end of which can be connected, through control register settings, to the supply voltage (referred to as a "pull-up" resistor) or to ground (a "pull-down" resistor). This can be a useful feature but, like other powerful, but (possibly) overwhelming aspects of the MSP430, we're going to leave this one out in the robot applications.

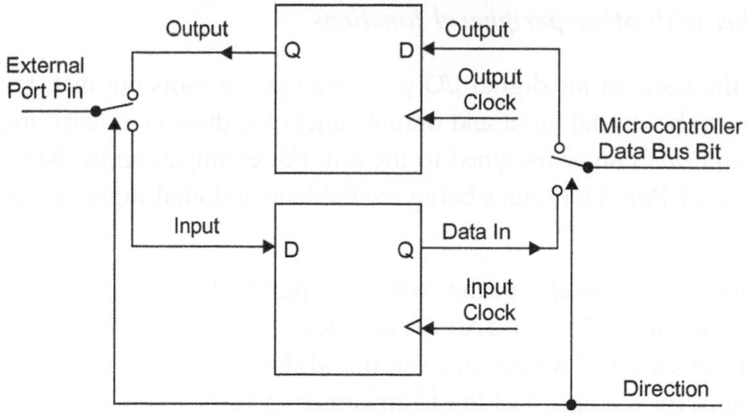

Figure 14.5
Single port I/O bit.

An especially powerful feature of the digital I/O pins is the ability to generate interrupts based on a specific port input going from 0 to 1 or from 1 to 0. This can be very handy when the program we write needs to respond quickly to the presence of a change in the input state. But we'll avoid this feature also, in the interest of simplifying things.

Still another powerful feature of some MSP430 port pins is the ability to create, without the need for external components, an oscillation. This is useful for applications like capacitive touch, where the human body capacitance from a user touching the port pin (usually through a connection to a physical panel) changes the oscillation frequency in a manner that the MSP430 can sense. But like these other features – well, you guessed it – we won't need it for the robot so we leave that as an advanced feature that you can consider in the future.

How many digital I/O ports does the MSP430 have?

So far, we've only talked about the MSP430 Port 1. All DIP versions of the MSP430 (that is, the MSP430 versions that work with the LaunchPad product) have at least this 8-bit port. The 20-pin versions of the MSP430 DIP microcontrollers have a second 8-bit port, Port 2, available for digital input and output. The 14-pin DIP versions of the MSP430 have only the two most significant bits of Port 2, bits P2.6 and P2.7, available.

Some of the larger MSP430 packages include quite a few digital I/O ports. However, these packages use the high-density pin spacing that makes use by hobbyists difficult or impossible.

Sharing I/O pins with other peripheral functions

We can't leave the issue of the digital I/O pins quite yet. It turns out that, for most I/O port pins, in addition to the digital input and output functions, there is at least one additional peripheral function that can be assigned to the pin. For example, on the MSP430G2452, P1.0 (that is, bit 0 of Port 1), besides being available as a digital input or output, can be used as

- an analog input for channel 0 of the A/D converter (A0)
- the clock for the internal timer TA0 (TA0CLK)
- the Auxiliary clock (ACLK) output from the MSP430
- an analog input for channel 0 of the Comparator_A (CA0).

Of these, we will make use of only the first in this book. In fact, in general the only peripheral function besides digital input and output that we'll use in these robot projects will be the analog-to-digital converter functions.

How do we select these alternate peripheral functions? There are select bits within the I/O port control register that select which peripheral we wish to use. However, since the only non-digital-I/O functions that we will use are the A/D functions, we won't have to worry about those select bits. Instead we set up pins that we wish to have as A/D functions within the A/D control registers. We'll discuss the A/D setup in detail in a subsequent chapter devoted to data acquisition.

Serial I/O

Although we won't use serial I/O in this book's robot project, it's worth mentioning the subject at this point, since it is something that most designers will use sooner or later. It's often necessary to transfer one or more bytes of information between the microcontroller and some external circuit, such as another microcontroller, an external control device, etc.

Of course, we could make this transfer using the digital I/O bits, but that would require all or most of the I/O bits on the microcontroller. It would also result in a lot of wires or printed circuit board traces. To get around this problem, designers today often transmit the data serially. The use of serial data transmission includes everything from electronic door locks on cars, which often use what is called a local interconnect network (LIN), to fiber optic gigabit Ethernet networks capable of transmitting data at speeds of 1 gigabit per second and higher.

Some of these serial data protocols include a means for the transmitter and receiver to synchronize the data in such a way that no separate clock signal is required. However, the two most common types of serial transmission between MSP430 microcontrollers, the serial

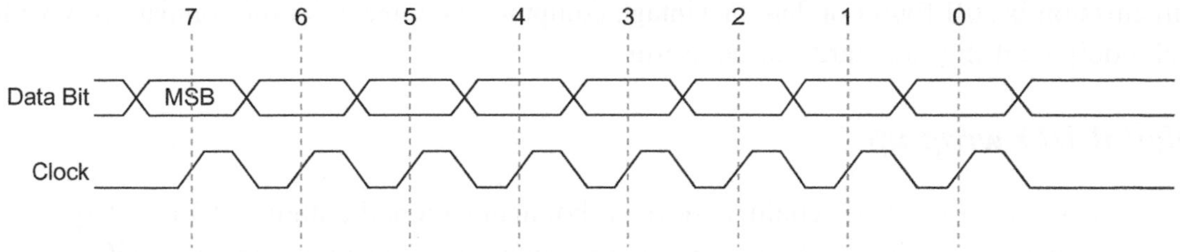

Figure 14.6
Typical serial peripheral interface (SPI) waveforms.

peripheral interface (SPI) and the inter-integrated circuit (I^2C), have a separate clock signal sent along with the data to let the data receiver know when to clock in the data. Figure 14.6 shows a typical SPI data transmission.

Whereas parallel data transmission requires a separate *wire* for each bit, serial data transmission requires a separate *time interval* for each bit. In both the SPI and I^2C protocols, and in many other serial data protocols as well, there is a single master unit (the unit usually being referred to as a "node") and one or more slave units (nodes). The master unit creates the clock for all data transmission. Note that these protocols are intended to accommodate *networks*, in which there may be many devices communicating with one another.

The nice thing about the MSP430 is that the internal byte-to-serial bit conversion and the synchronization of the clock signal to these serial data bits are taken care of automatically by the microcontroller. Once the serial interface is set up properly, the program simply loads the byte into the serial interface buffer and the transmission of the serial data is handled transparently by the microcontroller. Likewise, if the microcontroller is to receive a byte, the entire transmission is handled by microcontroller hardware and does not require program intervention to succeed. The program simply reads the byte that it has in its buffer. Data rates for an SPI network can be as high as 20 MHz (although the data transmission speed for the microcontrollers we're considering won't be higher than 16 MHz due to the MSP430 clock speed limit).

Universal asynchronous receiver/transmitter (UART)

A third type of serial data transmitter/receiver that is available on some, but not all, MSP430 microcontrollers is the UART. The serial I/O on some personal computers is based on this protocol. It is sometimes referred to as RS-232, although true RS-232 has large, bipolar voltage swings, not the 0 to 3 V swings found with an MSP430. This form of serial data transmission was developed in the early 1960s. It is intended for low-speed data exchange between just two nodes. In spite of these limitations, this type of data transmission is still found on lots of vintage computer equipment, so for compatibility with such equipment this standard can be useful.

Digital I/O wrap-up

For most applications, and certainly for the robot application, the digital I/O is a very important part of the design. Most modern microcontrollers, including the MSP430, can support individually set input or output functions on each bit of the I/O ports.

Serial data transmission, while important for many microcontroller applications and supported by the MSP430, will not be used in this project.

Bibliography

[1] J. Axelson, Serial Port Complete, Lakeview Research, LLC, Madison, WI, 2007.

[2] J. Davies, MSP430 Microcontroller Basics, Elsevier-Newnes, 2008.

[3] MSP430G2452 Datasheet. Literature Number: SLAS722B, December 2010; revised March 2011. <http://www.ti.com/lit/ds/symlink/msp430g2152.pdf>.

[4] MSP430x2xx Family User's Guide. Literature Number: SLAU144H. <http://www.ti.com/general/docs/litabsmultiplefilelist.tsp?literatureNumber = slau144h>.

CHAPTER 15

Timers and Counters

Chapter Outline

Timers are among the most-used of microcontroller peripherals. Timers are basically counters with associated comparators and registers for recording the time of occurrence for a particular microcontroller event or for switching a function on or off at a particular time. We can, in fact, divide all of the timer applications into three categories: (1) those that detect some occurrence of a specified event (for example, an input port pin changing state) and record the time of the occurrence; (2) those that compare the timer counter contents to some particular number for the purpose of making some decision at that time; and (3) those that simply use the timer contents to make software timing decisions.

The first two categories are referred to as "capture" and "compare". Capture refers to the act of storing the time that a particular event occurred and compare refers to comparing the timer counter contents to some preset number, usually for the purpose of performing some function at that preset time, such as changing the state of an output bit or executing some set of instructions in the program.

The robot project will make use of the simplest of these three categories, which is the third choice. We will periodically reset the timer, then let it free-run and check its contents to see when certain time intervals have been exceeded. This is similar to the compare mode except that we don't use any of the timer's compare hardware, making the compare entirely in software. Although we will not use either the capture or compare hardware, we will give a brief overview here, since these will likely be timer modes you will want to use in the future.

Capture – timestamping events

The word "timestamp" originally referred to a rubber stamp used in offices that recorded, with ink, the date and time that a document was received. A postmark from the post office

MSP430-based Robot Applications.
DOI: http://dx.doi.org/10.1016/B978-0-12-397012-1.00015-1

is an example of this original type of timestamp (although the post office usually records just the date, not the time of day).

In modern times, timestamp refers to electronically storing the time of an event. Such an event is usually the changing of a digital input pin from 1 to 0 or from 0 to 1. The time is usually just the local system time. For example, if the timer counter is being incremented by a 1 MHz clock, then system time is measured in microseconds. The timer doesn't necessarily know the absolute time of day. It's just incrementing every microsecond. Of course, the counter has finite length (that is, there are a finite number of bits to the counter) so eventually the counter "rolls over". For example, a 16-bit counter counts from 0 to 65535, then "rolls over" to 0. In this case, the timer can only keep track of intervals up to a little over 65 milliseconds without some additional record-keeping within the microcontroller's program.

The capture mode of the timer can be extremely useful for measuring the interval of time between two events. When an event occurs, the contents of the timer counter are stored in a register within the MSP430 dedicated to this purpose. The MSP430 can be set up to also allow interrupts, so that upon a prescribed event occurring, program control is immediately transferred to some interrupt service routine. The interrupt service routine might take some special action such as storing the timer's capture time in some memory location for later retrieval.

In spite of the potential usefulness of the timer capture mode, there will not be a need for it in the robot project, so this information is just included to make you aware of the capability.

Timer compare mode

The timer compare mode provides hardware that continuously compares the timer counter to some previously set number and sets a bit within the timer control registers or otherwise changes the state of some part of the microcontroller hardware so that the program can sense this compare result. Figure 15.1 shows a diagram of the MSP430 timer function. It is simplified considerably from the timer block diagram shown in the Texas Instruments *MSP430x2xx Family User's Guide*. Among other things, it:

- shows only the compare aspects of timer hardware, not capture
- shows only the sub main clock (SMCLK) as a clocking option for the timer counter
- Shows only the Compare 0 and Compare 1 functions, even though many MSP430 versions have additional comparator options

However, it depicts the timer functions as needed to implement pulse-width modulation, the next topic in this section.

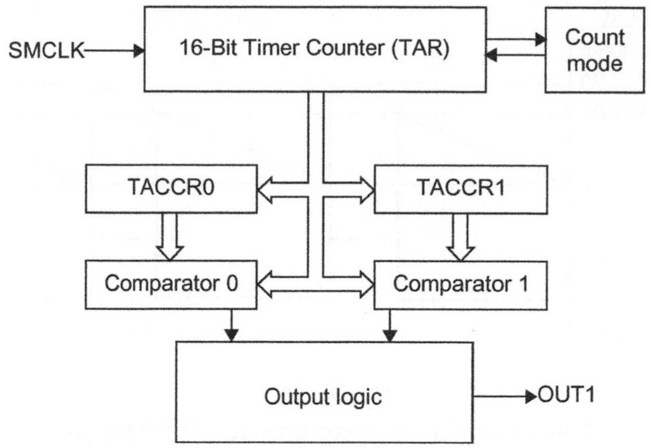

Figure 15.1
Simplified diagram of the timer compare functions.

Pulse-width modulation

An example will help illustrate the use of the timer comparator function. This is not something that you will need for the robot project that we will get to soon in the book, but it is something that may be helpful to you in the future in understanding how to create precisely timed pulses.

An often-used timing technique is pulse-width modulation (PWM). PWM consists of repetitively setting a bit high for some time, t_H, then setting it low for some interval, t_L. This periodic rectangular waveform has a period:

$$T = t_H + t_L. \tag{15.1}$$

We discussed such a waveform earlier with respect to adjusting motor drive (although the waveform there will be created with software, rather than timer hardware, due to the very low resolution needed). Such waveforms are also used to implement Class D switching-type audio amplifiers.

The MSP340 is able to implement PWM by using the following functions within the timer control section of the microcontroller:

- The comparators TACCR0 and TACCR1
- The continuous UP mode, which starts counting at 0 and counts up to the number stored in TACCR0, then resets to 0
- The OUT1 output signal, which can be connected to an external pin of the MSP430
- The output mode Set/Reset, which sets OUT1 to 1 when the timer reaches the number in TACCR1 and resets OUT1 to 0 when the timer reaches the number in TACCR0

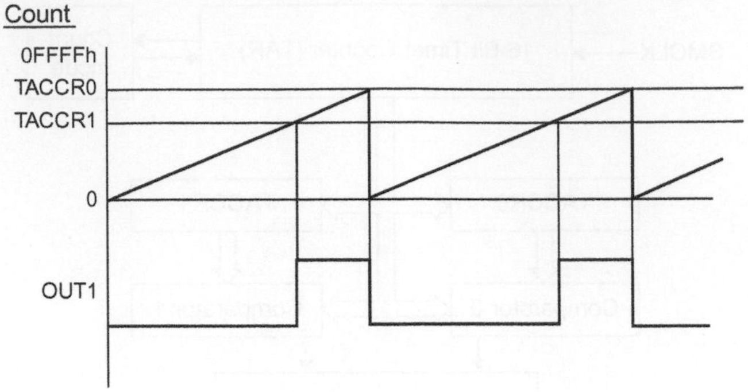

Figure 15.2
Timer counter set up to produce PWM signal.

Figure 15.2 illustrates the sequence of events that will produce the PWM repetitive waveform. The timer is set up to continuously count to the number that we place in the timer control register TACCR0. We place some smaller number in TACCR1. When the timer counter reaches the number in TACCR1, the PWM output (OUT1) goes high. When the timer counter later reaches the number stored in TACCR0 the output goes low and the whole sequence starts again. If we designate the number in TACCR0 as n_0 and the number in TACCR1 as n_1, then we can see that the PWM signal is low for counts 0 through $(n_1 - 1)$ and is high for counts n_1 through n_0.

Let's try this on the MSP430 LaunchPad with a program. We'll use a 1 MHz clock for the timer and produce a 200 microsecond pulse that repeats every 1 millisecond. Since we want to repeat every millisecond, we'll need TACCR0 to contain the number 1000_{10}. Since we want the pulse to start at 800 microseconds, we'll place the number 800_{10} in TACCR1. We'll do this on an MSP430G2211, which happens to be one of the MSP430 variants sent with this LaunchPad. Any of the DIP versions of the MSP430 should work fine for this example.

Before writing the program, let's look at the control registers for the MSP430 Timer_A function. TACTL is the Timer_A Control Register, and it is organized as in Figure 15.3, which is copied from the Timer_A section of the *MSP430x2xx Family User's Guide* (SLAU144H).

As the Guide mentions, these bits are all reset to 0 during the microcontroller's power-on reset (POR). However, there are some bits that we will want to set. Specifically, we want to choose SMCLK as the timer clock and we want to use the UP mode. So we want the two TASSEL bits to be 10_2 and the two MC bits to be 01_2. To do this, use the command:

```
mov     #TASSEL_2+MC_1,&TACTL
```

15	14	13	12	11	10	9	8
Unused						TASSELx	
rw-(0)	rw-(0)	rw-(0)	rw-(0)	rw-(0)	rw-(0)	rw-(0)	rw-(0)

7	6	5	4	3	2	1	0
IDx		MCx		Unused	TACLR	TAIE	TAIFG
rw-(0)	rw-(0)	rw-(0)	rw-(0)	rw-(0)	rw-(0)	rw-(0)	rw-(0)

Unused	Bits 15-10	Unused	
TASSELx	Bits 9-8	Timer_A clock source select	
		00	TACLK
		01	ACLK
		10	SMCLK
		11	INCLK (INCLK is device-specific and is often assigned to the inverted TBCLK) (see the device-specific data sheet)
IDx	Bits 7-6	Input divider. These bits select the divider for the input clock.	
		00	/1
		01	/2
		10	/4
		11	/8
MCx	Bits 5-4	Mode control. Setting MCx = 00h when Timer_A is not in use conserves power.	
		00	Stop mode: the timer is halted.
		01	Up mode: the timer counts up to TACCR0.
		10	Continuous mode: the timer counts up to 0FFFFh.
		11	Up/down mode: the timer counts up to TACCR0 then down to 0000h.
Unused	Bit 3	Unused	
TACLR	Bit 2	Timer_A clear. Setting this bit resets TAR, the clock divider, and the count direction. The TACLR bit is automatically reset and is always read as zero.	
TAIE	Bit 1	Timer_A interrupt enable. This bit enables the TAIFG interrupt request.	
		0	Interrupt disabled
		1	Interrupt enabled
TAIFG	Bit 0	Timer_A interrupt flag	
		0	No interrupt pending
		1	Interrupt pending

Figure 15.3

The TACTL Timer_A Control Register *(courtesy of Texas Instruments)*.

How do we know that "TASSEL_2" is the symbol that represents $0000\ 0010\ 0000\ 0000_2$? Why not "TASSEL2" or TASS_2? And how do we know that "MC_1" is the symbol for $0000\ 0000\ 0001\ 0000_2$? It turns out that all of this information is available for us to see in the same way that it's available for the MSP430 Assembler to see – it's in the header file, "msp430g2211.h". If you're not sure where the header file is stored, just do a search on your computer for that name.

Okay, so what other Timer_A control registers need to be set? Well, TACCR0 needs to be loaded with the number 1000, and TACCR1 needs to be loaded with the number 800. We will also need to set the output mode for compare control register 1 to Set/Reset:

```
mov        #OUTMOD_3,&TTACTL1
```

This statement causes the output bit to set when the Timer_A counter (register TAR) reaches the number in the TACCR1 register and resets when TAR reaches the number in the TACCR0 register.

TERMINAL NAME	NO. 14 N, PW	NO. 16 RSA	I/O	DESCRIPTION
P1.0/ TA0CLK/ ACLK/ CA0	2	1	I/O	General-purpose digital I/O pin Timer0_A, clock signal TACLK input ACLK signal output Comparator_A+, CA0 input(1)
P1.1/ TA0.0/ CA1	3	2	I/O	General-purpose digital I/O pin Timer0_A, capture: CCI0A input, compare: Out0 output Comparator_A+, CA1 input(1)
P1.2/ TA0.1/ CA2	4	3	I/O	General-purpose digital I/O pin Timer0_A, capture: CCI1A input, compare: Out1 output Comparator_A+, CA2 input(1)
P1.3/ CA3/ CAOUT	5	4	I/O	General-purpose digital I/O pin Comparator_A+, CA3 input(1) Comparator_A+, output(1)
P1.4/ SMCLK/ CA4/ TCK	6	5	I/O	General-purpose digital I/O pin SMCLK signal output Comparator_A+, CA4 input(1) JTAG test clock, input terminal for device programming and test

Figure 15.4

Excerpt from the MSP430G2211 datasheet showing the Out1 output on P1.2 *(courtesy of Texas Instruments).*

But just where is this OUT1 bit on the pins of the MSP430? To determine this, take a look at the datasheet for the MSP430G2211 (TI publication SLAS695G). Figure 15.4 shows an excerpt of Table 2 of that document. As you can see, bit 2 of Port 1 (P1.2) is assigned the OUT1 function.

However, as we know from the chapter on serial and parallel I/O ports and from looking at the datasheet table above, there are multiple signals that can be assigned to P1.2. How do we make sure that OUT1 is the one? To answer this, take a look at another table from the datasheet (Figure 15.5). Here we see that bit 2 of the P1DIR control register and bit 2 of the P1SEL control register need to be set in order to select TA0.1 (the OUT1 function).

Consequently, the two statements

```
bis.b       #4,&P1DIR
bis.b       #4,&P1SEL
```

need to be included. Note that the ".b" version of the BIS command needs to be used because P1DIR and P1SEL are only a single byte, not a 16-bit word. Putting a constant 4 in these registers to set bit 2 of each of these registers may seem a little confusing. Remember, for each of these registers, we need to OR the bit 2 position with a 1 so that it is set (that's actually what the BIS command does – BIS is the MSP430 version of an OR). Figure 15.6 shows what each of these registers should look like after the BIS operation.

PIN NAME (P1.x)	x	FUNCTION	CONTROL BITS/SIGNALS	
			P1DIR.x	P1SEL.x
P1.0/	0	P1.x (I/O)	I: 0; O: 1	0
TA0CLK/		TA0CLK	0	1
ACLK		ACLK	1	1
P1.1/	1	P1.x (I/O)	I: 0; O: 1	0
TA0.0		TA0.CCI0A	0	1
		TA0.0	1	1
P1.2/	2	P1.x (I/O)	I: 0; O: 1	0
TA0.1		TA0.CCI1A	0	1
		TA0.1	1	1
P1.3	3	P1.x (I/O)	I: 0; O: 1	0

Figure 15.5
Excerpt from the MSP430G2211 datasheet *(courtesy of Texas Instruments).*

7	6	5	4	3	2	1	0
X	X	X	X	X	1	X	X

Figure 15.6
Setting P1SEL bit 2 and P1DIR bit 2 to 1.

One more thing about the program – a calibrated 1 MHz DCOCLK clock source is used in order to produce a fairly precise clock frequency. The program is shown in Figure 15.7.

Note that once the clock and timer are set up, the program is finished as far as any serious business goes. The program simply loops continually to HangAround. The PWM signal is produced entirely independently of the software, compliments of the Timer_A hardware that is internal to the MSP430. Figure 15.8 shows the PWM waveform produced by the MSP430 at pin P1.2.

Timers and counters wrap-up

Timers are extremely important in modern microcontroller applications. They are used quite often in real-time operating systems, where a microcontroller needs to go from one task to another at specified times within a continuous execution loop. Quite often, the timer is used in conjunction with interrupts to accomplish such task management.

Although we will use the idea of scheduling for this robot project and will make use of the Timer_A function in the MSP430 to maintain proper timing, we will not be using interrupts. Be aware, however, that development systems exist that can help set up such sophisticated task management.

```
; This program continuously generates 200 microsecond pulses at 1 millisecond
;   intervals, using the Timer_A function of the MSP430

#include "msp430G2211.h"

;-------------------------------------------------------------------------------
            ORG     0F800h                     ; Program beginning location
;-------------------------------------------------------------------------------

RESET       mov.w   #0280h,SP                  ; Initialize stackpointer
StopWDT     mov.w   #WDTPW+WDTHOLD,&WDTCTL     ; Stop watchdog timer
            call    #PortSetup
            call    #ClockSetup
            call    #TimerSetup
HangAround
            jmp     HangAround

;-------------------------------------------------------------------------------
PortSetup
            ; Make P1DIR.2 and P1SEL.2 equal to 1
            bis.b   #4,&P1DIR
            bis.b   #4,&P1SEL
            ret
;-------------------------------------------------------------------------------
ClockSetup
            ; Set DCOCLK for 1 MHz calibrated
            ; Select DCOCLK for MCLK and SMCLK
            clr.b   &DCOCTL ; Select lowest DCOx and MODx settings
            mov.b   &CALBC1_1MHZ,&BCSCTL1 ; Set range
            mov.b   &CALDCO_1MHZ,&DCOCTL ; Set DCO step + modulation
            ret
;-------------------------------------------------------------------------------
TimerSetup
            ; Use the SMCLK as the source for the timer
            ; UP mode
            mov     #TASSEL_2+ID_0+MC_1,&TACTL  ; Use SMCLK for timer, UP mode
            mov     #1000,&TACCR0    ; Just an arbitrarily big number to start
            mov     #800,&TACCR1     ; OUT high from 800 to 1000 each cycle
            mov     #OUTMOD_3,&TACCTL1
            ret

;-------------------------------------------------------------------------------
            ORG     0FFFEh
            DW      RESET
            END
```

Figure 15.7
PWM Example Program.

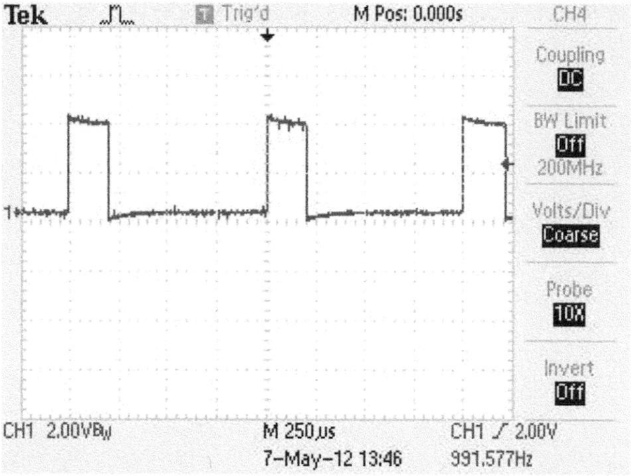

Figure 15.8
Pulse-width-modulated signal produced at P1.2 of MSP430.

Bibliography

[1] J. Davies, MSP430 Microcontroller Basics, Elsevier-Newnes, 2008.

[2] MSP430G2452 Datasheet. Literature Number: SLAS722B. <http://www.ti.com/lit/ds/symlink/msp430g2152.pdf>, December 2010; revised March 2011.

[3] MSP430x2xx Family User's Guide. Literature Number: SLAU144H. <http://www.ti.com/general/docs/litabsmultiplefilelist.tsp?literatureNumber = slau144h>.

CHAPTER 16

Data Acquisition

Chapter Outline

One of the puzzles circuit designers had to solve in the early days of computers was how to convert analog voltages from real-world sensors into digital numbers that computers could understand. The companion problem was how to convert these digital numbers back into analog voltages that external electronic equipment needed.

The solutions to these puzzles are called analog-to-digital (A/D) converters and digital-to-analog (D/A) converters, respectively, and are the topic of this chapter. Because we will

MSP430-based Robot Applications.
DOI: http://dx.doi.org/10.1016/B978-0-12-397012-1.00016-3

need to understand D/A converters in order to understand A/D converters, we'll look at D/A converters first.

Digital-to-analog converters

This chapter will look at some ways of implementing such a D/A function.

One common way of implementing such a D/A is a resistor network like the one shown in Figure 16.1. Here the individual bits, V_i, are outputs from the MSP430 microcontroller. Although the example shown is for a 4-bit D/A, the idea can be extended to any number of bits.

The Radio Shack Color Computer, popular with computer hobbyists in the 1980s before PCs were around, used a 6-bit version of this circuit to inexpensively achieve moderate digital-to-analog output capability.

Using superposition to analyze D/A circuits

Analyzing the Figure 16.1 circuit is made easier if we know about something called the superposition principle. This is a theorem that can be useful any time there are multiple sources in a resistive network. The theorem basically states that, for a circuit with multiple resistors and multiple voltage sources, the response of the circuit can be determined by determining the response from each voltage source separately, then summing all such responses (the theorem is actually more general than this and includes current sources, in addition to voltage sources, but this abbreviated version of the theorem will suffice for what we want to do). During the determination of the partial response from each individual voltage, the other voltage sources are replaced with short circuits. Let's try this out with the Figure 16.1 circuit. Figure 16.2 redraws this circuit showing each bit as the voltage source that it is.

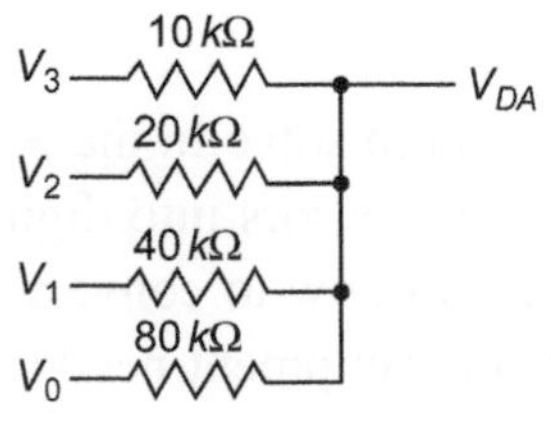

Figure 16.1
Simple D/A based on resistor network.

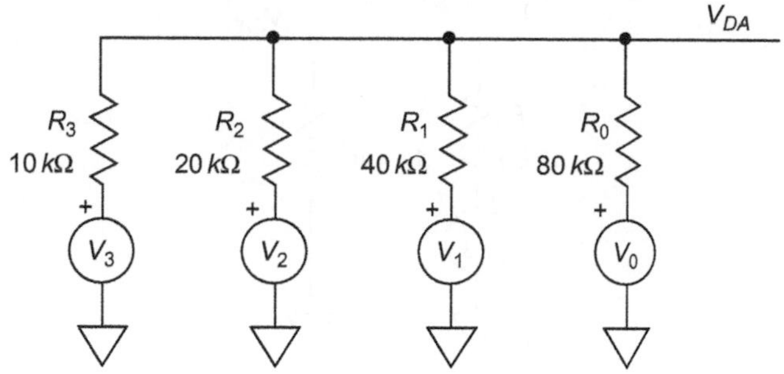

Figure 16.2
Figure 16.1 redrawn to show the bits as voltage sources.

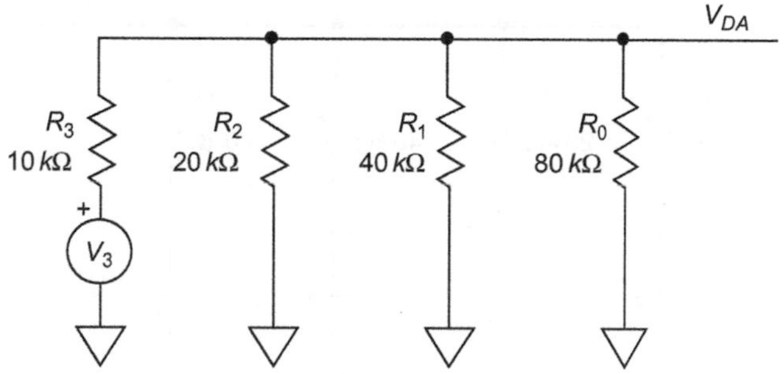

Figure 16.3
First step in using the superposition principle response from V_3.

Now, let's apply the superposition principle. Let's find the response due to just V_3. To do that, we replace each of the other three voltage sources with a short circuit, as shown in Figure 16.3.

It is straightforward to compute the parallel combination of R_0, R_1, and R_2:

$$R_0||R_1||R_2 = (R_0||R_1)||R_2 = (80k||40k)||20k = \left(\frac{(80k)(40k)}{80k + 40k}\right)||20k = 26.67k||20k = 11.4k \tag{16.1}$$

Thus, the contribution from V_3 is given by the circuit of Figure 16.4.

A resistor divider like this has a response given by:

$$V_{DA,3} = \frac{11.4k}{(11.4k + 10k)} V_3 = 0.533 \cdot V_3 \tag{16.2}$$

The extra subscript 3 on the V_{DA} variable in Eq. 16.2 indicates that this is just a partial response due to the contribution of V_3 only.

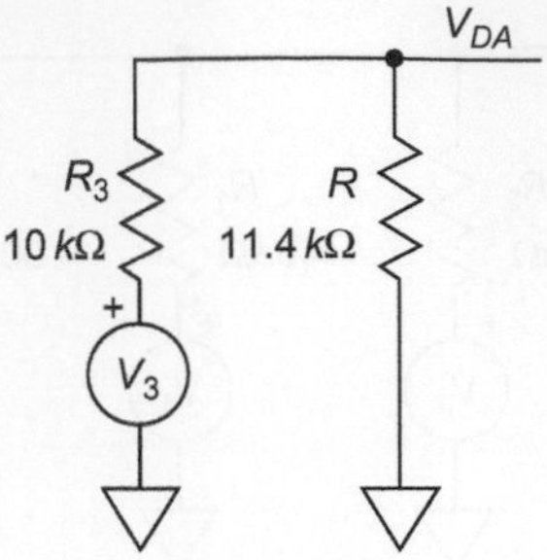

Figure 16.4
Equivalent circuit to the Figure 16.3 circuit.

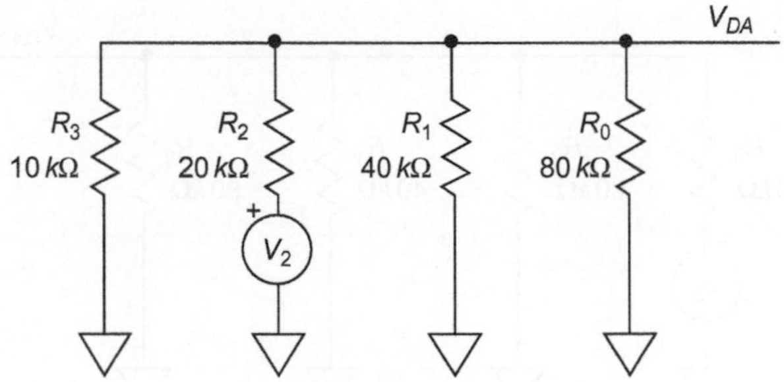

Figure 16.5
D/A contribution from V_2.

Similarly, we can analyze the response due to just V_2 (Figures 16.5 and 16.6). This resistor divider has the response:

$$V_{DA,2} = \frac{7.3k}{(7.3k + 20k)} V_2 = 0.267 \cdot V_2. \tag{16.3}$$

In a similar manner, we can show that the partial response due to V_1 is $0.133 \cdot V_1$ and the partial response due to V_0 is $0.0667 \cdot V_0$. Thus the total response to the four bits of input is given by the sum of these partial responses:

$$V_{DA} = 0.533 \cdot V_3 + 0.267 \cdot V_2 + 0.133 \cdot V_1 + 0.0667 \cdot V_0. \tag{16.4}$$

Table 16.1 and Figure 16.7 show the response of the D/A converter to the 16 possible input combinations.

Of course, the D/A inputs will typically not have a 1 volt amplitude, as was assumed in Table 16.1. However, the particular voltage amplitude of the digital inputs simply scales the output. Quite often a precise reference voltage is used for the high amplitude.

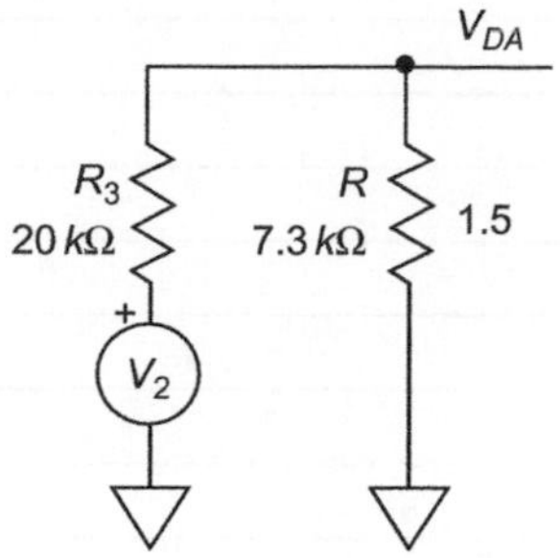

Figure 16.6
Equivalent circuit to Figure 16.5 circuit.

Table 16.1: D/A response to digital inputs.

V_3	V_2	V_1	V_0	V_{DA}
0	0	0	0	0.000
0	0	0	1	0.067
0	0	1	0	0.133
0	0	1	1	0.200
0	1	0	0	0.267
0	1	0	1	0.333
0	1	1	0	0.400
0	1	1	1	0.467
1	0	0	0	0.533
1	0	0	1	0.600
1	0	1	0	0.666
1	0	1	1	0.733
1	1	0	0	0.800
1	1	0	1	0.866
1	1	1	0	0.933
1	1	1	1	1.000

Note that, while we've just described the use of an array of resistors, in which each successive resistor is (ideally) twice the value of its more significant neighbor, we can also implement the function using arrays of capacitors. Note also that there are other resistor arrangements besides the power-of-two array just described for implementing D/A converters.

PWM D/A functions

The passive component (resistor or capacitor array) types of D/A circuits just described are common ways to implement D/As in standalone converter ICs, yet microcontrollers seldom implement such functions. Instead, most microcontrollers, including most versions of the

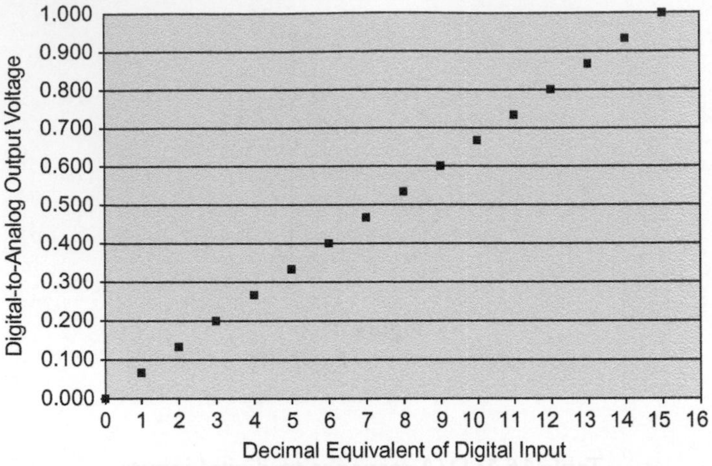

Figure 16.7
Graph of the Table 16.1 response.

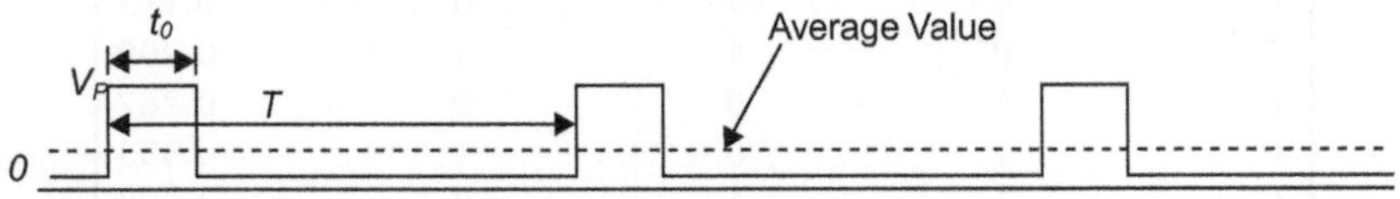

Figure 16.8
A PWM sequence for creating a pulsed D/A.

MSP430 microcontrollers, use a timing method known as pulse-width modulation (PWM) to implement the digital-to-analog conversion. We talked about PWM briefly in the motors chapter of the book and in the timers chapter as well. Here we'll look at using such a method for generating analog voltages.

With PWM D/A conversion, a repeating sequence of pulses is generated (Figure 16.8). The *average* value of this sequence is given by:

$$v_{avg} = \frac{t_0}{T} V_P. \tag{16.5}$$

where V_P is the pulse amplitude, T is the repetition period, and t_0 is the pulse width. By digitally controlling the width of this pulse, the average value changes.

The beauty of this approach is its simplicity. We've discussed the fact that MSP430s (and most microcontrollers) contain timers that allow the automatic generation of pulses as prescribed intervals. Once set up, the timer pulse width and pulse repetition frequency ($1/T$) will be maintained by the MSP430 hardware without further intervention from the program. Moreover, the analog averaging can be achieved with just a lowpass filter. Figure 16.9

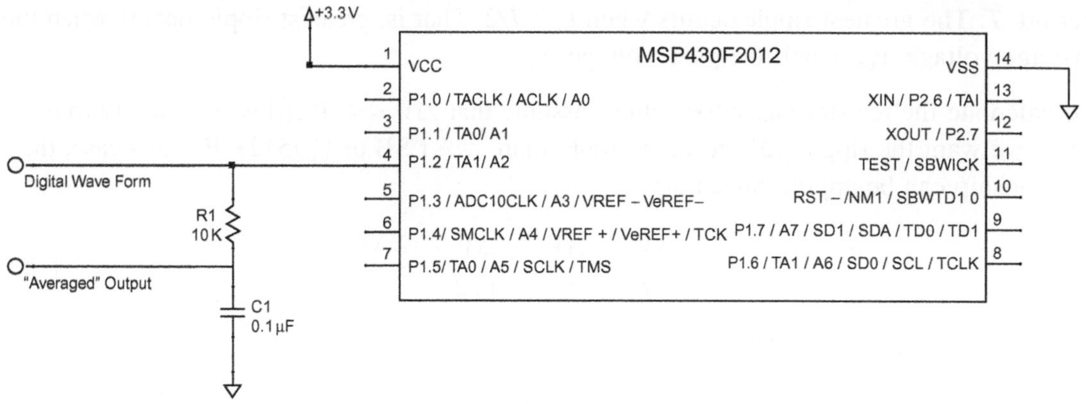

Figure 16.9
The MSP430F2012 circuit used to produce the PWM D/A output.

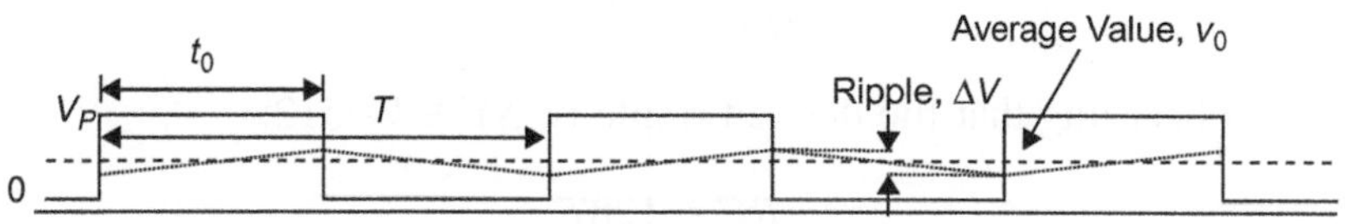

Figure 16.10
Ripple of a PWM-generated D/A voltage.

shows an MSP430F2012 set up for such a D/A output. Note that the only additional components required are a single resistor and capacitor to create the lowpass filter. The averaged waveform is shown in Figure 16.10.

The primary weakness of the PWM D/A is that it produces a ripple voltage – in this case, an approximately triangular deviation from the desired analog output. While choosing larger values of resistor and capacitor for the lowpass filter reduces this ripple voltage, it also slows down the rate at which the D/A can change its output, so a trade-off must be made.

Knowing that we need to minimize this ripple but not overdo it, what is a good limit to set for the ripple and what resistor/capacitor combination will guarantee that the ripple remains below this limit? We can calculate this fairly easily.

Assume that we seek a D/A output with 8 bits of accuracy. So the least significant bit (LSB) should have 1/256 the value of the maximum value, V_p. With regard to ripple, ± ½ LSB of ripple is a reasonable limit for most applications. Let's say that the output is a pulse train with amplitude $V_p = 3.3$ V, so that the filtered (averaged) amplitude varies from 0 V to 3.3 V, depending on how long the pulse interval, t_O, is, with respect to the repetition period, T. The greatest ripple occurs when $t_O = T/2$. That is, greatest ripple occurs when the average voltage, v_O, is half the peak voltage, V_p.

To calculate the resistor/capacitor values, assume that $\Delta V << V_p$ (this is certainly true, since we want the ripple, ΔV, to be no more than $\pm$ ½ LSB or $V_p/512$). For this case, the relationship can be approximated as:

$$\frac{i}{C} \approx \frac{\Delta V}{\Delta T} = \frac{\Delta V}{t_0} = \frac{\Delta V}{T/2} = \frac{2\Delta V}{T}. \tag{16.6}$$

Since, for this case, $v_O = V_P/2$:

$$i = \frac{V_P - v_0}{R} = \frac{V_P}{2R}, \tag{16.7}$$

Eq. 16.6 can be written as:

$$\frac{V_P}{2RC} = \frac{2\Delta V}{T}. \tag{16.8}$$

Rearranging, and observing that the desired ripple is $\Delta V \leq V_P/512$:

$$RC = 128T \tag{16.9}$$

With the MSP430 clock set to 8 MHz and with a period, T, set to 256 clock cycles, that is, $T = 32$ µsec, the resistor/capacitor combination becomes 128 $T = 4$ msec. For $C = 0.1$ µF, the resistor, R, is around 40 kΩ. In order to stick with standard values, a resistor of 39 kΩ is used.

Note in the Figure 16.9 schematic that the output is generated at P1.2 of the MSP430F2012. This is the output associated with the timer compare register in the MSP430.

The complete assembly language program is very similar to the one written for the timers chapter and is shown in Figure 16.11. The actual code is shown in black, while comments are shown in green. The first two instructions starting at location RESET simply (1) define where in the MSP430 random access memory (the kind that you can read *and* write to) the stack is supposed to begin, and (2) turn off the Watchdog Timer.

Next, several *subroutines* are called. The first of these, P1Setup, establishes P1.2 as the PWM output. The next one, SetupClock, initializes the peripheral registers associated with the clock function so that an internal clock of approximately 8 MHz is generated. The third subroutine, SetupTimer, sets up the timer to repeatedly count from 0 to 256.

Next, register R5 is initialized to 128 (half the full-scale output of 256). This number is then transferred to the timer register, TACCR1. The MSP430 sets the P1.2 (referred to as OUT1 when it performs the timer function) whenever the timer counts from 256 to 0 and it

```
;*********************************************************************************
; MSP430F2012 Project
;
; This program uses the Pulse-Width Modulated output of the MSP430, along with
;  an external RC filter, to produce an analog voltage from a digital number
;  stored in register R5
;
#include "msp430x20x2.h"
;-------------------------------------------------------------------------------
        ORG     0FA00h              ; Progam Start (1K Flash device)
;-------------------------------------------------------------------------------
RESET      mov.w   #0280h,SP                ; Set stackpointer (128B RAM device)
StopWDT    mov.w   #WDTPW+WDTHOLD,&WDTCTL ; Stop watchdog timer
        call   #P1Setup
        call   #SetupClock
        call   #SetupTimer
        ; The number in R5, divided by 256, is the duty cycle of the PWM output
        mov    #128,R5
        mov    R5,&TACCR1
Loop
        jmp    Loop

;-------------------------------------------------------------------------------
SetupClock
        ; This subroutine sets up the clock parameters:
        ;  MCLK and SMCLK generated by the digital clock oscillator (DCO)
        ;  Nominal 8 MHz frequency
        ;  DCOx=3, MODx=0, RSELx=13
        mov.b  #DCO1+DCO0,&DCOCTL
        mov.b  #RSEL3+RSEL2+RSEL0,&BCSCTL1
        ret

;-------------------------------------------------------------------------------
P1Setup
        ; P1.2 is the OUT1 pin
        ; All I/O pins except P1.0 and P1.2 will be left in their power-up
        ; default state
        bis.b  #05h,&P1DIR
        mov.b  #04h,&P1SEL
        ret


;-------------------------------------------------------------------------------
SetupTimer
        ; Choose SMCLK for the timer clock source
        ;   SMCLK = 8 MHz
        ; Up mode with TACCR0 setting the repetition period
        ;   TACCR0 = 256 (that is, repetition frequency = 8MHz/256 = 32kHz)
        mov    #TASSEL_2+MC_1,&TACTL
        mov    #OUTMOD_6,&TACCTL1
        mov    #256,&TACCR0
        ret

;-------------------------------------------------------------------------------

        ORG    0FFFEh               ; MSP430 RESET Vector
        DW     RESET                ;
        END
```

Figure 16.11

Program generating the PWM D/A output.

resets when the timer counter reaches the value in TACCR1, which is 128. In this way, a 50% duty cycle square wave is produced.

To try this program, an MSP430 LaunchPad microcontroller was used. Figure 16.12 shows the LaunchPad board, along with a vector board that serves as a convenient "piggyback" board, sometimes referred to in the electronics industry as a mezzanine card or a

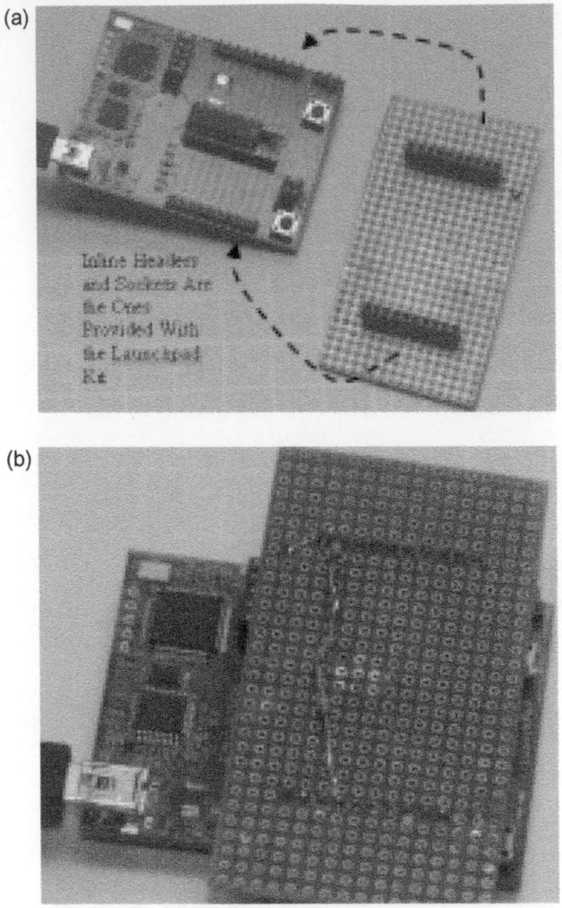

Figure 16.12
(a) The LaunchPad board (left) along with the piggyback board (bottom side shown).
(b) The LaunchPad board with piggyback board installed.

daughterboard. The piggyback board uses the single inline connectors that come with the LaunchPad kit. For this particular experimental setup, the piggyback board has only the 39 kΩ resistor and 0.1 μF capacitor for averaging.

Figure 16.13a shows the actual waveforms sampled at the points in Figure 16.9 labeled "Digital Waveform" and "Averaged Waveform". As can be seen there, the digital output of the microcontroller is a square wave with 50% duty cycle and average value that is half of the peak voltage of 3.3 V.

Figure 16.13b is the waveform with the oscilloscope set to AC-coupled input and with the channel sensitivity set to 20 mV/division. This allows measurement of the actual ripple, which can be seen to be approximately ±5 mV.

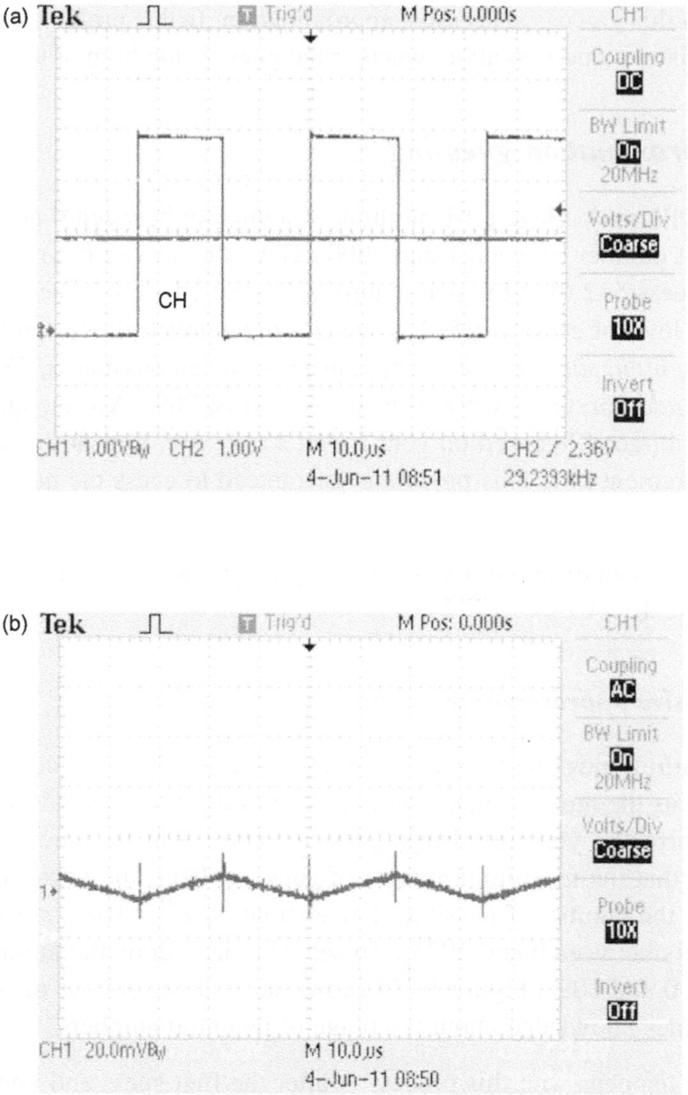

Figure 16.13
(a) Oscilloscope trace of the PWM pulse and filtered D/A output. (b) Filtered output at 20 mV/div showing ripple.

Analog-to-digital conversion

Many A/D converters and most of the MSP430 microcontroller A/D converters use a method of conversion referred to as successive approximation. To get an idea of how this works let's try a little exercise that shows how, without any prior knowledge, we can guess a number, between 1 and 1000, that someone is thinking, and do so with just 10 guesses. The process that we will use in this game is successive approximation. In this process, each guess will be refined on the basis of a question about whether the guess is too high or too low.

Successive approximation guessing

Here's how this will work. Have a friend think of a number between 0 and 1023 (or just ask them to pick a number between 1 and 1000). Have them write it down (but don't look). Make your first guess 512 (half the maximum). Ask your friend whether your guess of 512 is too high or too low (of course, if 512 is actually *the* number, they need to tell you that). If this guess is *too high*, *subtract* 256 from your previous guess (that is, $512 - 256 = 256$). If 512 is *too low*, *add* 256 to your previous guess (that is, 768). Ask the question again and then add or subtract 128 based on your friend's response. Repeat this process until the increment/decrement is 1. This process is guaranteed to guess the number in 10 guesses or less!

Let's try this out with an example. Let's say that your friend writes down 411. The process is charted in Figure 16.14.

How does successive approximation work?

Of course, it is entirely possible to guess the number in less than 10 guesses (had your friend picked 512 as the number, you would have guessed it in one guess) but, if the process is done correctly, you will always guess the number in no more than 10 guesses. It is no coincidence that the maximum number of guesses (10) is the power to which 2 must be raised to equal the number of possible choices (that is, $2^{10} = 1024$, where 1024 is the number of numbers between 0 and 1023). To see why, let's draw the number line for all the numbers between 0 and 1023. Figure 16.15 shows successive versions of the number line as each iteration refines knowledge about the range of potential numbers.

Think about what happens with this process – after the first guess and your friend's response to it, you can eliminate *half* of all possible numbers as the potential correct answer. So when you guess 512 and your friend says the number is lower than that, you can automatically eliminate all the numbers between 512 and 1023 as possible answers. This is shown in the top line of the figure as a grayed-out region.

When you next guess 256 and your friend says that this is too *low*, you can eliminate all the numbers between 0 and 256 as possible answers – you've again eliminated half of all the numbers that remained after the first iteration. After the second iteration, there are only $\frac{1}{4}(1024) = [1/(2^2)](1024) = 256$ numbers remaining as possible guesses. Each iteration reduces the range of potential guesses by half. After 10 guesses, there will be only $[1/(2^{10})](1024) = (1/1024)(1024) = 1$ number remaining – that is, that's your final answer.

Your Guess	Friend's Response	Next Increment/Decrement	Action
512	Too high	256	Subtract 256
256	Too low	128	Add 128
384	Too low	64	Add 64
448	Too high	32	Subtract 32
416	Too high	16	Subtract 16
400	Too low	8	Add 8
408	Too low	4	Add 4
412	Too high	2	Subtract 2
410	Too low	1	Add 1
411	Correct Guess!		

Figure 16.14
Successive approximation guessing game for the number 411.

How can we apply this technique to analog-to-digital conversion? Let's say that we need to convert voltages in the range 0 V to 10.23 V to a digital representation (typically, the range would be 0 V to 10 V, or some other integer number, but 10.23 V makes the arithmetic a little simpler in this example). If our goal is to have 1024 uniformly distributed possible guesses about what the analog voltage is, then these guesses will take on the values 0.00 V, 0.01 V, 0.02 V, . . ., 10.22 V, 10.23 V. Ideally, our analog-to-digital converter will identify the digital number whose associated analog voltage most closely matches the actual analog input to the converter.

Note that one difference between this example and the previous one is that, in the guessing game, there were 1024 discrete (in that case, integer) possible numbers and the object was

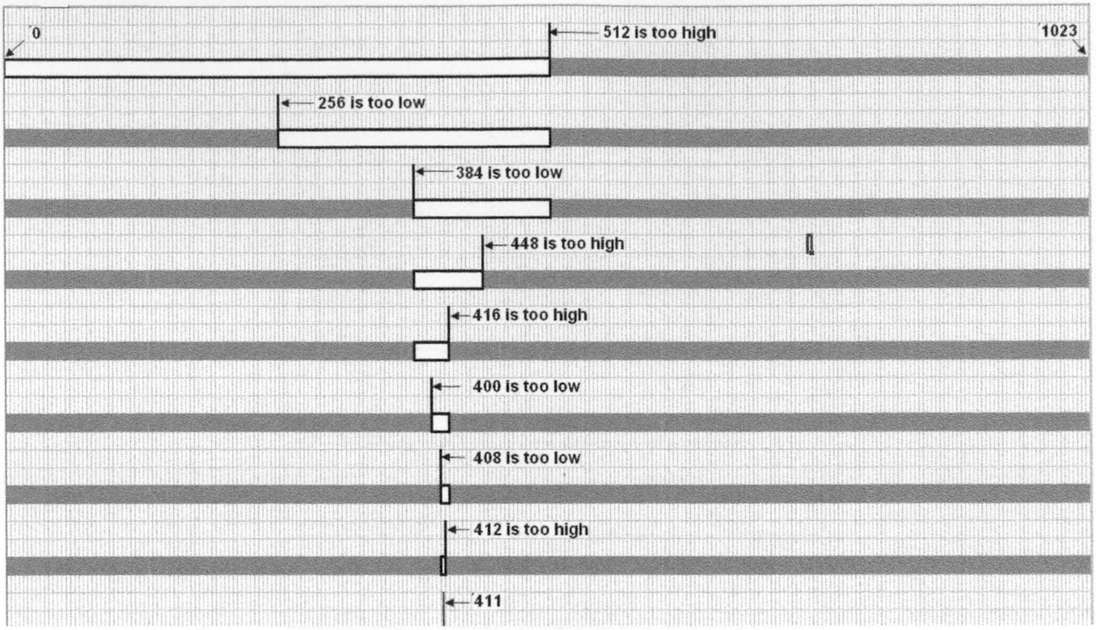

Figure 16.15
Successive approximation example showing range reduction for each successive iteration.

to identify the *exact* number. With this analog-to-digital converter example, the analog voltage to be converted is part of a *continuous range* of numbers between 0.00 and 10.23 and the object is to identify the closest match to that number. The exact voltage might be 2.11346... V, for example, but the best that our 10-bit converter can do is to identify the number 211 (corresponding, in this case, to 2.11 V) as the correct result.

Figure 16.16 shows a block diagram for a successive approximation analog-to-digital converter. The block diagram includes successive approximation logic. This implements the algorithm that we used earlier in our guessing game, using 512 (for a 10-bit converter) as the initial guess and then adding or subtracting progressively smaller powers of 2 until all 10 guesses have been made.

As we saw earlier in the D/A section of this chapter, a 10-bit D/A is a straightforward circuit with the exception that the resistors making up that circuit must have high precision.

When the successive approximation logic determines the next guess, this is clocked into a register by the clock signal, A/D Clock, which we will talk more about shortly. The data is then presented to the D/A, which generates an analog voltage corresponding to the digital input. Note that the 10 bits flowing between the successive approximation logic and the register and between the register and the D/A are represented by a single block arrow, with a "10" inside to indicate that this represents 10 bits of data.

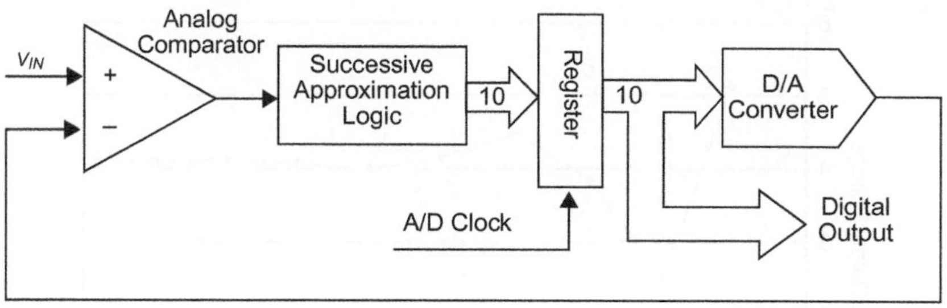

Figure 16.16
The A/D block diagram.

The comparator's role is simple – it compares the input voltage, v_{IN}, to the D/A-generated voltage (the guess) and outputs a binary signal. As drawn, the comparator creates a binary 1 output if the guess voltage is too low and a binary 0 if the guess voltage is too high. Thus, the comparator serves the same function as your friend in the previous example, letting the decision logic know whether its last guess was too high or too low.

In this A/D example, the first voltage that the successive approximation logic/register/ digital-to-analog converter generates is the mid-level voltage, 5.12 V. Let's say that for this example, the input voltage is 4.11 V. In that case, the A/D will proceed through the same set of digital numbers as our guessing game example did. The D/A voltages generated at each iteration of the successive approximation algorithm are shown in Figure 16.17.

Note the reason for the register and clock in the A/D – all of the components within the A/D take some amount of time to settle to their final value. When the A/D Clock signal goes high, the register itself takes some time to change its outputs to the new state, the D/A converter takes some time, the comparator takes more time and the successive approximation logic takes still more time. Thus, the A/D Clock must be run at a rate slow enough to guarantee that, upon receiving a clock edge, all of these processes settle to their final value before the next clock edge occurs.

Non-ideal sampling

The idea of an analog-to-digital converter is that we convert the voltage of an input waveform at some *instant in time*. But the A/D converter takes time for its various stages to settle and 10 iterations of such guess/settle steps are required for a 10-bit A/D converter. For some applications this makes little difference. In a later chapter, we'll develop a "high-voltage" converter circuit, in which the microcontroller samples the high voltage to see if it's too high or too low. For that simple, very-low-frequency application, the A/D conversion time is of little concern.

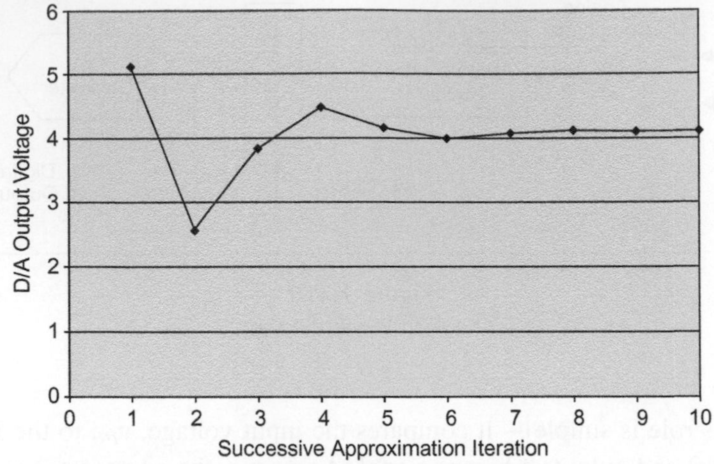

Figure 16.17
Successive approximation D/A output for a 4.11 V input to 10-bit A/D.

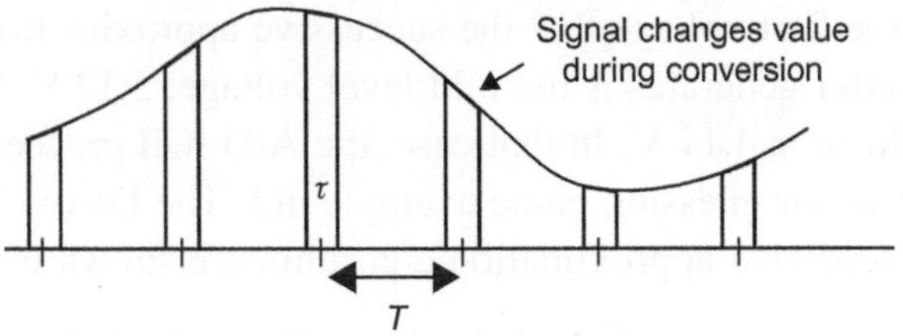

Figure 16.18
Continuously sampled waveform showing A/D conversion time, τ.

For other applications, the conversion time is a problem. One problem that may occur as the result of the A/D converter's non-instantaneous conversion time is an apparent timing or amplitude jitter when trying to capture a series of samples at some uniform sampling rate. Figure 16.18 illustrates the problem.

In this figure, the waveform is sampled at a uniform rate of $1/T$. During the A/D conversion time, τ, the signal continues to change value. Also during this time, the A/D will be trying successive guesses (Figure 16.17). The changing input waveform can cause the A/D to arrive at a digital output that corresponds to the waveform at some point within the conversion time, but without a precise "instant" relative to the start of conversion. The result is that the digitized signal will exhibit this uncertainty as timing or amplitude jitter (random timing jitter produces random amplitude jitter). Is this jitter a big deal? Your application will determine your project's sensitivity to this phenomenon. Just be aware that such a potential problem exists.

The sample-and-hold amplifier

Fortunately there's an easy solution to this timing uncertainty problem and the solution is built into MSP430 A/D converters. Preceding the MSP430 A/D converter is a sample-and-hold amplifier. Such a circuit might be as simple as the one shown in Figure 16.19. Here the input charges the capacitor to its voltage, v_{IN}, when the switch closes, which happens when the sample clock signal goes high. The sample clock pulse is very narrow and when it goes back low, causing the switch to open, the capacitor holds the voltage, v_{IN}. With a sufficiently narrow pulse, the sampled signal, v_{SAMPLE}, will appear like the sampled waveform in Figure 16.20.

We can see from Figure 16.20 that the sample-and-hold amplifier does capture the waveform at some instant in time and then holds that voltage until the next sample command. Or at least it comes pretty close to doing that. As with most things in electronic design, the reality can be a little less simple than we'd like or than we expected. In fact, the sample-and-hold amplifier:

- Doesn't immediately change to the new input voltage when it receives the sample clock but, instead, slews at some fixed rate, due to
  - Finite charging current for the sample-and-hold capacitor
  - Op amp's limited slew rate
- The sample pulse (and therefore the switch "on" time) is non-zero
  - This is known as acquisition time or acquisition aperture
  - This results in some conversion jitter (although usually much better than without the sample-and-hold)
- The amplifier experiences voltage "droop" due to leakage current from the op amp's non-inverting input

In very critical applications such as high-fidelity digital audio, the above bulleted limitations of the sample-and-hold amplifier can make its design as challenging as that of the A/D converter itself. For any robot applications in this book and for most applications

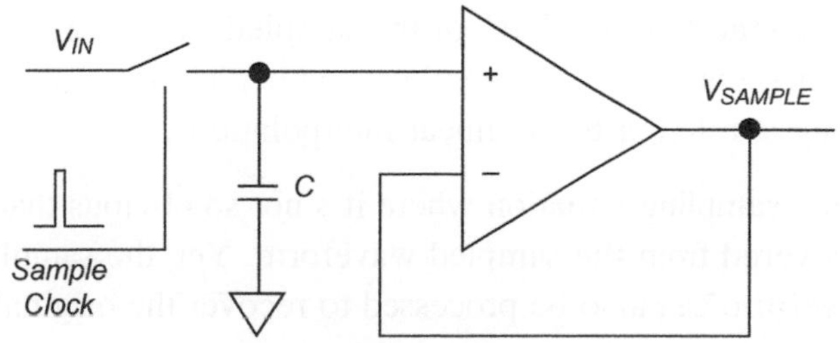

Figure 16.19
Simple sample-and-hold amplifier.

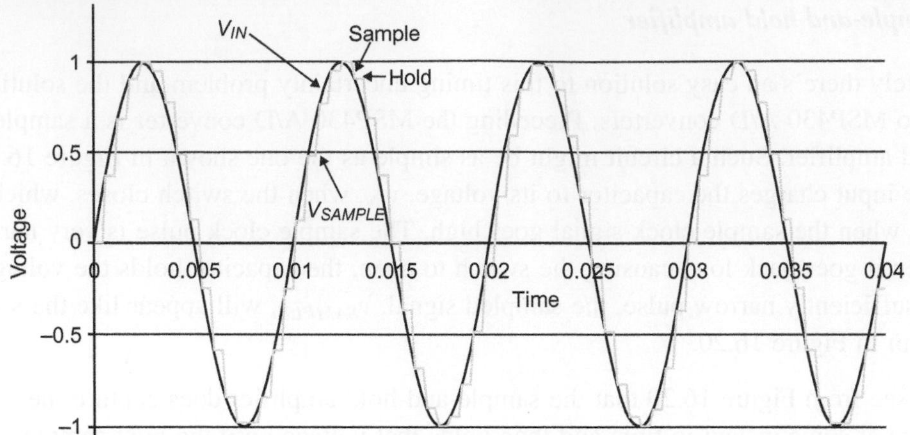

Figure 16.20
Sample-and-hold waveforms for a 100 Hz sine wave input.

that you will encounter for the MSP430 10-bit A/D converters, you can be confident that the internal sample-and-hold amplifier will eliminate significant jitter as the result of A/D conversion timing and we'll leave things at that.

Nyquist-Shannon sampling theorem

Claude Shannon has already been introduced in an earlier chapter of this book on computer logic. Shannon, Harry Nyquist, and a number of other individuals in the first half of the twentieth century all arrived at the conclusion that a waveform, if sampled at some minimum rate (usually referred to, these days, as the Nyquist rate), can be exactly reconstructed.

Looking at Figure 16.20, it's easy to see how we could get at least close to the original waveform by simply "connecting the dots" of the sampled waveform – that is, drawing a straight line from the beginning of each sample to the beginning of the next sample. This straight-line connecting is referred to as "linear interpolation".

Figure 16.21 presents a sampling situation where it's not so obvious that the original waveform can be recovered from the sampled waveform. Yet, the sampling theorem states that this sampled waveform can also be processed to recover the original waveform.

So just what are the Nyquist-Shannon requirements for signal reconstruction? The main requirement is that, for a signal sampled at uniform intervals of $T_S = 1/(2W)$, the sampled

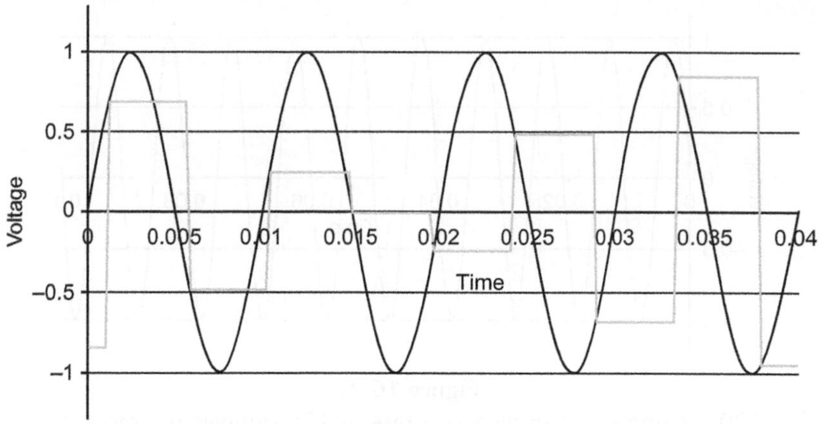

Figure 16.21
100 Hz sinewave sampled 217 times per second.

signal must be bandlimited to W. In other words, in the Figure 16.21 example, where $T_S = 1/(217\text{ Hz})$, the original signal must have a bandwidth no greater than 108.5 Hz, a limit which the 100 Hz sinewave just meets.

In addition, perfect reconstruction (or something close to perfect reconstruction) requires interpolation techniques more sophisticated than the piecewise-linear interpolation method. The complexity of the interpolation method increases as the sampling rate approaches the Nyquist rate, which is why designers try to convert at rates considerably higher than the Nyquist rate, if possible.

Aliasing

So what happens if we make the mistake of sampling at too low a rate? Figure 16.22 shows an example, where the 100 Hz sinewave is now sampled at 125 samples/second, clearly in violation of the Sampling Theorem.

Note that the sampled waveform repeats every 40 milliseconds. In fact, if we sampled a 25 Hz sinewave at 125 samples per second, we would obtain exactly this same pattern (see Figure 16.23).

It is not a coincidence that the 25 Hz sinewave created by undersampling a 100 Hz sinewave at a 125 sample/second rate turns out to have, as its frequency, the difference of these two original frequencies (the 125 Hz sampling rate minus the 100 Hz signal frequency). This creation of a new waveform at the difference frequency is an example of a phenomenon known as *aliasing*.

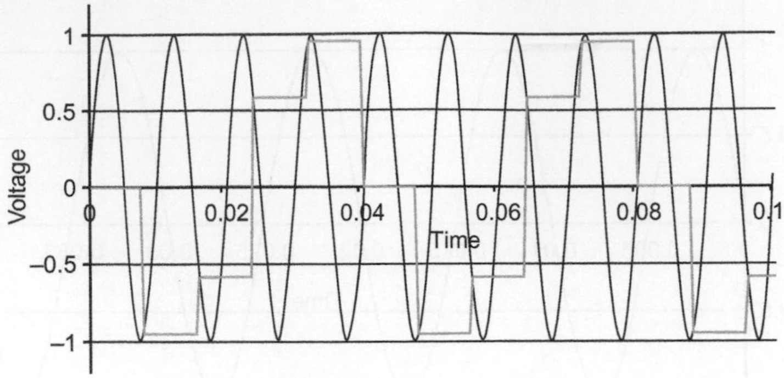

Figure 16.22
100 Hz sinewave sampled at a rate of 125 samples per second.

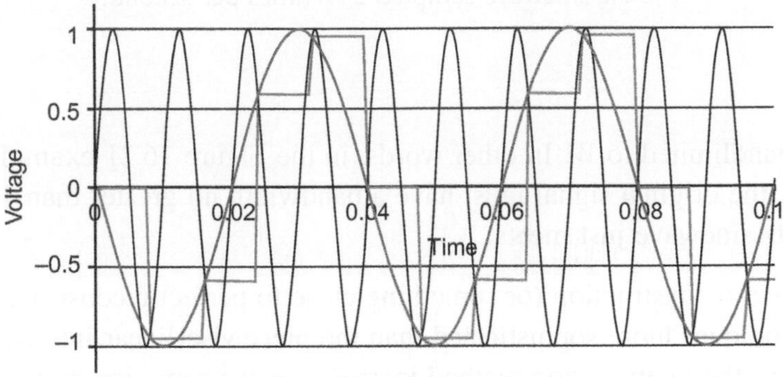

Figure 16.23
25 Hz sinewave intersecting sample points of the undersampled waveform.

To avoid aliasing, lowpass filters are usually added ahead of the sample-and-hold amplifier and A/D converter. A simple RC filter, for example, eliminates most of a signal's high-frequency content beyond frequencies of $1/(2\pi RC)$. Because a simple, first-order lowpass filter like this still allows some higher frequencies to pass through (although with attenuation) and because reconstruction of a waveform sampled at a rate near the Nyquist rate is difficult, such filters are usually designed to have a cut-off well below the theoretical limit predicted by the sampling theorem.

Other types of A/D converters

For many control and other applications, a relatively modest A/D converter resolution – for example, 10 bits – is sufficient. However, there are lots of instrumentation, audio, and other applications that require higher resolutions. Most audio-quality A/D converters used today employ 20 or more bits. This means that the least significant bit of the A/D will have

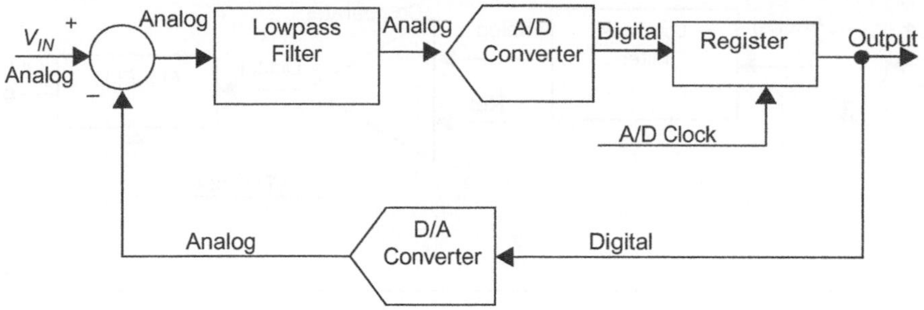

Figure 16.24
The ΣΔ A/D converter.

a value that is only $2^{-20} = 0.00001\%$ of the most significant bit. The problem with trying to achieve this kind of high resolution with the successive approximation A/D converter is that the D/A used inside the successive approximation A/D requires precise components, and matching components to these extremely low tolerances is difficult. As we will see shortly in the section of this chapter on D/A converters, it is relatively easy to achieve 6 or 7 bits of resolution in a D/A and it is only moderately difficult to achieve as much as 12 bits of resolution in a D/A. However, with each additional bit of resolution that the D/A designer tries to introduce, all of the D/A components must be made more accurately by about a factor of two.

The sigma-delta converter

With the advent of digital audio, a different type of architecture, called sigma-delta (sometimes called delta-sigma) came into vogue to handle this problem. Quite often this is simply represented by its Greek symbols, ΣΔ.

The basic building blocks of the ΣΔ converter are shown in Figure 16.24. At first glance, it doesn't seem that this is anything particularly exciting or particularly different from the successive approximation A/D. However, what makes the ΣΔ converter useful is that we can make the A/D and D/A converters just single-bit devices. A single-bit A/D is the same as a comparator. A single-bit D/A might be nothing more than just the output of the register. And, of course, the register itself is just a single flip-flop. A very simple ΣΔ converter is shown in Figure 16.25, consisting of just a summing amplifier, a lowpass filter, a comparator and a flip-flop.

But how can we obtain, say, 20-bit resolution from a single-bit system? The key is in oversampling. The A/D Clock in a ΣΔ converter runs at a very high rate – rates as high as 100× the Nyquist rate are not uncommon with these converters. The result is an output bitstream that looks like a random sequence of very fast 1 s and 0 s. However, if the oversampling rate is high enough, the *average* of that bitstream will closely follow the input

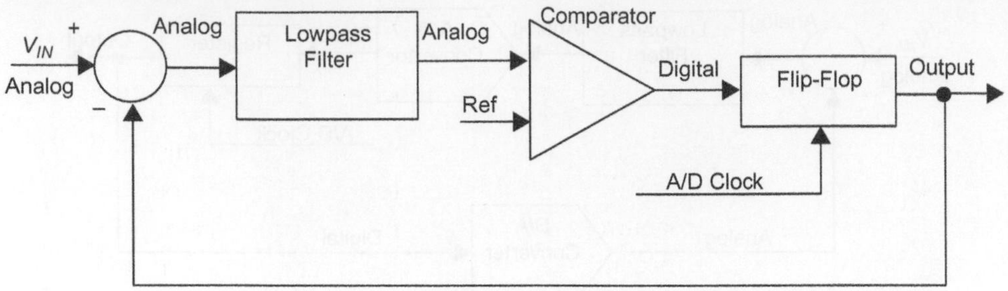

Figure 16.25
A very simple ΣΔ A/D converter.

voltage, v_{IN}. If we placed a lowpass filter at the output of this converter, we could then recover an accurate replica of the input voltage, since the lowpass filter performs the averaging function. In fact, in the most common use of these converters, a digital lowpass filter takes the output of this circuit and creates such an average.

ΣΔ A/D converters in MSP430 microcontrollers

There are some MSP430 converters, for example, the MSP430F2013, which have ΣΔ A/D converters. However, most of the MSP430 versions that are accommodated by the LaunchPad development board have 10-bit successive approximation A/D converters inside.

Is resolution the same as accuracy?

One question that comes up a lot about A/D converters is whether resolution and accuracy are the same. When a manufacturer offers a 10-bit A/D, they're talking about the resolution. That is, the device produces a result with 10 bits. Let's say that this A/D converts analog signal signals in the range 0 V to 3 V. In that case, the least significant of the 10 bits has a weight of just slightly less than 3 mV (3 V/1024). But if the A/D produces a result of, say, 512, can we be certain that the actual analog voltage is 1.5 V $\pm$ 1.5 mV?

The answer, of course, is no. Few 10-bit A/D converters have true 10-bit accuracy. Think about some of the errors that the A/D can make:

1. The A/D will have some voltage offset at its input
2. The A/D will add noise to the analog signal
3. Mismatch of D/A components will distort the A/D results.

A/D noise

A/D noise (beyond the quantization "noise" that even an ideal converter produces due to its finite number of levels) can arise from a number of sources. The analog components of the converter can exhibit thermal noise, which exists in any analog design and is the result of thermal agitation of electrons within a resistance. In relatively low-resolution converters (for example, 10 bits) this type of noise can probably be ignored.

In microcontroller A/D converters, a significant source of noise can occur due to power supply noise. Keep in mind that, while the A/D converter of such a microcontroller is making its conversion, the microcontroller's clocks and, therefore, its internal logic are rapidly changing their states. This can cause the power supply voltage to fluctuate as these state changes mean that the electrical current demand is rapidly changing. A voltage variation of just a few tens of millivolts, if it shows up in the A/D circuitry, can cause the A/D's least significant bits to vary.

A/D distortion due to component mismatch

Remember the nice, perfect-looking D/A response of Figure 16.7? That illustration assumed all the resistors were their exact calculated value. In fact, as we've already talked about with such D/As, the components can only be guaranteed to be within some range of values. This is referred to as the resistor *tolerance*. Let's say that our 10 kΩ resistor had a 1% tolerance. Then the true value could be anywhere in the range 9.9 kΩ to 10.1 kΩ.

If we construct a 4-bit D/A, like the one in that example, such errors are probably tolerable. However, for a 10-bit D/A, where the least significant bit has a weight only 0.1% of the most significant bit, such precision would produce some very noticeable and, probably, objectionable results.

A/D resolution conclusions

The important thing to note about A/D resolution is that it is definitely *not* the same as accuracy. If you have an application that truly requires the A/D to operate with accuracy near its resolution, be sure to read the A/D datasheet specifications and consider how the parameters listed in the datasheet will affect your application.

The MSP430 10-bit A/D

Not all the versions of MSP430 microcontrollers compatible with the TI LaunchPad have A/D converters. Those that do use a 10-bit successive approximation converter, which is what we'll concentrate on in the remainder of this chapter. These A/D converters are preceded by an analog multiplexer. So, just what is an analog multiplexer?

The analog multiplexer

The analog multiplexer is just a collection of transistor circuits, in which each transistor circuit is designed either to pass the input signal with (ideally) no attenuation or to block it

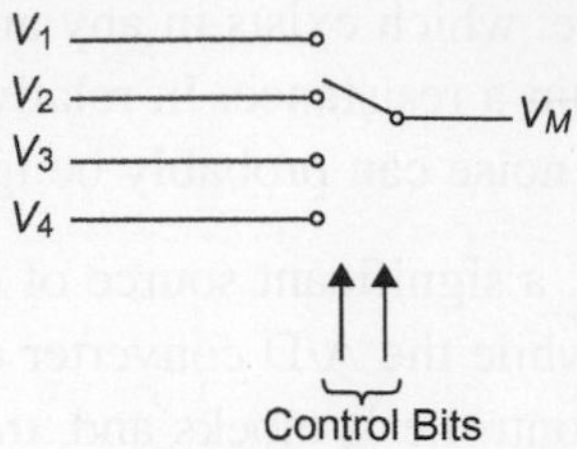

Figure 16.26
Four-input analog multiplexer.

totally. The idea is to have just one of these analog switch circuits pass its signal at any given time. Control bits determine which "1-of-*N*" switches is the one that passes its signal. Figure 16.26 shows a 1-of-4 analog multiplexer.

A typical MSP430 microcontroller will have eight external pins that can be configured as an analog input to the A/D. There are also some internal signals, such as ground, $V_{CC}/2$, and temperature that can be sampled with some of these microcontrollers (the individual datasheets will provide this information). The multiplexer connects to the sample-and-hold amplifier, which feeds the A/D converter.

A/D example

An example will help show how the MSP430 A/D works. Let's first take a look at the registers for the A/D. The associated registers are:

- ADC10AE0, the input enable register
- ADC10CTL0, the control register 0
- ADC10CTL1, the control register 1
- ADC10MEM, which contains the result of the last A/D conversion

There are also three registers (ADC10DTC0, ADC10DTC1, and ADC10SA) that are used to perform continuous transfers of A/D results directly into the microcontroller's memory without any intervention by the CPU – a very cool feature, but an advanced one that we will not use in this book. In addition, there is a second input enable register, ADC10AE1, which controls inputs when there are more than eight analog inputs, but we will not have a need for these.

ADC10AE0 register

Let's look first at the input enable register. The ADC10AE0 register has 8 bits, each bit corresponding to a different analog input channel (Figure 16.27).

ADC10AE0x							
rw-(0)	rw-(0)	rw-(0)	rw-(0)	rw-(0)	rw-(0)	rw-(0)	rw-(0)

ADC10AE0x	Bits 7-0	ADC10 analog enable. These bits enable the corresponding pin for analog input. BIT0 corresponds to A0, BIT1 corresponds to A1, etc. 0 Analog input disabled 1 Analog input enabled

Figure 16.27
ADC10AE0 register *(courtesy of Texas Instruments).*

If we have a design in which analog inputs on channels 1, 3, and 7 (labeled A1, A3, and A7 on datasheet drawings) are to be converted, we would move 86_{16} (that is, $1000\ 1010_2$) into this register:

```
mov.b       #086h,&ADC10AE0
```

Remember how, in the chapter on I/O ports, we talked about the primary function of each I/O pin on the microcontroller, which is digital, and the secondary, or peripheral, function for each pin? Selection between the two sets of I/O (primary and secondary) can be accomplished through the use of a select bit corresponding to that particular pin. It turns out that, in the case of analog inputs for the A/D, the ADC10AE0 register takes care of this for us. Setting the bit in that register automatically sets the corresponding I/O pin for the analog input function and no further configuration is required for that particular pin.

ADC10CTL0 register

ADC10CTL0 is a 16-bit register and its bit definition is shown in Figure 16.28. Note that the bits that are grayed-out can only be changed when the ENC bit is 0. *Program instructions that try to change these bits when ENC = 1 will be ignored.* On the other hand, A/D conversions will not take place *unless* ENC = 1. So make sure you pay attention to the state of this bit in your program. Note also that all bits of this register are cleared when the microcontroller is first powered up.

In understanding the functions of these bits, let's first identify the bits that can be ignored for purposes of the robot that we have set out to build. The MSC bit is a bit that enables the type of automatic transfer of A/D readings directly to memory, as we talked about a little earlier – cool, but we won't use it (it's an advanced feature). The ADC10IE bit enables interrupts when the A/D conversion is complete – also nice, but we've sworn off interrupts for now. Finally, there is the ability to make the MSP430's internal voltage reference available on one of the microcontroller's pins. This is done by using REFOUT = 1, but we won't have a need for this capability, so we'll leave REFOUT = 0.

(a)

15	14	13	12	11	10	9	8
SREFx			ADC10SHTx		ADC10SR	REFOUT	REFBURST
rw-(0)	rw-(0)	rw-(0)	rw-(0)	rw-(0)	rw-(0)	rw-(0)	rw-(0)

7	6	5	4	3	2	1	0
MSC	REF2_5V	REFON	ADC10ON	ADC10IE	ADC10IFG	ENC	ADC10SC
rw-(0)	rw-(0)	rw-(0)	rw-(0)	rw-(0)	rw-(0)	rw-(0)	rw-(0)

(shaded) Can be modified only when ENC = 0

SREFx	Bits 15-13	Select reference
		000 $V_{R+} = V_{CC}$ and $V_{R-} = V_{SS}$
		001 $V_{R+} = V_{REF+}$ and $V_{R-} = V_{SS}$
		010 $V_{R+} = Ve_{REF+}$ and $V_{R-} = V_{SS}$
		011 $V_{R+} =$ Buffered Ve_{REF+} and $V_{R-} = V_{SS}$
		100 $V_{R+} = V_{CC}$ and $V_{R-} = V_{REF-}/V_{eREF-}$
		101 $V_{R+} = V_{REF+}$ and $V_{R-} = V_{REF-}/V_{eREF-}$
		110 $V_{R+} = Ve_{REF+}$ and $V_{R-} = V_{REF-}/V_{eREF-}$
		111 $V_{R+} =$ Buffered Ve_{REF+} and $V_{R-} = V_{REF-}/V_{eREF-}$
ADC10SHTx	Bits 12-11	ADC10 sample-and-hold time
		00 4 × ADC10CLKs
		01 8 × ADC10CLKs
		10 16 × ADC10CLKs
		11 64 × ADC10CLKs
ADC10SR	Bit 10	ADC10 sampling rate. This bit selects the reference buffer drive capability for the maximum sampling rate. Setting ADC10SR reduces the current consumption of the reference buffer.
		0 Reference buffer supports up to ~200 ksps
		1 Reference buffer supports up to ~50 ksps
REFOUT	Bit 9	Reference output
		0 Reference output off
		1 Reference output on
REFBURST	Bit 8	Reference burst.
		0 Reference buffer on continuously
		1 Reference buffer on only during sample-and-conversion

(b)

MSC	Bit 7	Multiple sample and conversion. Valid only for sequence or repeated modes.
		0 The sampling requires a rising edge of the SHI signal to trigger each sample-and-conversion.
		1 The first rising edge of the SHI signal triggers the sampling timer, but further sample-and-conversions are performed automatically as soon as the prior conversion is completed
REF2_5V	Bit 6	Reference-generator voltage. REFON must also be set.
		0 1.5 V
		1 2.5 V
REFON	Bit 5	Reference generator on
		0 Reference off
		1 Reference on
ADC10ON	Bit 4	ADC10 on
		0 ADC10 off
		1 ADC10 on
ADC10IE	Bit 3	ADC10 interrupt enable
		0 Interrupt disabled
		1 Interrupt enabled
ADC10IFG	Bit 2	ADC10 interrupt flag. This bit is set if ADC10MEM is loaded with a conversion result. It is automatically reset when the interrupt request is accepted, or it may be reset by software. When using the DTC this flag is set when a block of transfers is completed.
		0 No interrupt pending
		1 Interrupt pending
ENC	Bit 1	Enable conversion
		0 ADC10 disabled
		1 ADC10 enabled
ADC10SC	Bit 0	Start conversion. Software-controlled sample-and-conversion start. ADC10SC and ENC may be set together with one instruction. ADC10SC is reset automatically.
		0 No sample-and-conversion start
		1 Start sample-and-conversion

Figure 16.28

(a) Register ADC10CTL0 definition (*courtesy of Texas Instruments*). (b) ADC10CTL0 register definition, continued (*courtesy of Texas Instruments*).

Of the remaining bits, the SREF bits determine the reference for the A/D and, therefore the range. Although the MSP430 allows external reference voltages to be used, most applications can be simplified by using an internal reference. Therefore, SREF will either be set equal to 000_2 or to 001_2. When SREF = 000_2, the reference is the microcontroller's supply voltage. This can be handy, particularly when the analog circuits feeding the analog channel run from the same voltage supply and therefore have the same 0 to V_{CC} signal range. When SREF = 001_2, the microcontroller's internal reference of either 1.5 V or 2.5 V will be used as the upper limit of the range. These are particularly useful when you are looking for a more precise reference.

If you decide to use the internal reference (1.5 V or 2.5 V) then make sure REFON = 1. On the other hand, if you use V_{CC} as the reference, you can leave this bit at 0 and save a little current. Also, if you choose the internal reference, the REF2_5 V bit chooses between the 1.5 V and 2.5 V (REF2_5 V = 0 chooses 1.5 V, else 2.5 V).

The digital result produced by the A/D converter is the number given by:

$$N_{ADC} = 1023 \cdot \frac{V_{IN} - V_{R-}}{V_{R+} - V_{R-}}. \tag{16.10}$$

When SREF = 0 or 1, as it will in our applications, $V_{R-} = 0$ V, so the equation can be simplified to:

$$N_{ADC} = 1023 \cdot \frac{V_{IN}}{V_{R+}}. \tag{16.11}$$

V_{R+} will be either 1.5 V, 2.5 V, or V_{CC}, depending on how you've configured the A/D reference.

The ADC10SHT bits determine the length of time that the sample-and-hold amplifier is left in the Sample mode, before starting the A/D conversion. This can be important since the sample-and-hold takes some time to charge to its final value. The sample time is measured in ADC10 clock intervals. So, if your ADC10 is being clocked by the ADC10OSC clock (~5 MHz) then, if you chose 64 clock cycles as the sample time (ADC10SHT = 3), the sample time will be $64/(5 \times 10^6) = 12.8$ μsec.

The ADC10SR and REFBURST are associated with the reference buffer, which is an internal amplifier that provides the extra drive capability needed to deliver the internal reference voltage to the A/D. For most applications, ADC10SR = 1 and REFBURST = 1 will give adequate performance and will reduce microcontroller current consumption.

ADC10ON turns the power on to the microcontroller A/D. Why would you ever want to have it off? Well, the A/D consumes current, so turning the A/D off when it is not in use can save power. Your application, and just how power-sensitive it is, will determine whether you take this extra step.

The ADC10IFG is the interrupt flag. It is set to 1 when the A/D conversion result is available. Even though its usual function is to initiate an interrupt, it can be useful as a flag that can be read by software (we will use it this way in the example program at the end of this chapter). If you read it from software, be sure to reset the flag to 0 before the next A/D conversion.

As already noted, ENC must be 0 when the program makes changes to those bits in registers ADC10CTL0 and ADC10CTL1 that are shown in gray. And ENC must be 1 when the program is performing an A/D conversion.

ADC10SC is the bit that actually starts the conversion. Setting it to 1 is what causes the conversion to start. It automatically resets to zero.

Finally, note that you do not need to remember where these bits are in the register. Note, in the example given at the end of the chapter, how these bits can be easily set up using the constants whose labels are defined in the header file for the particular version of the MSP430 device that you're using.

ADC10CTL1 register

The second control register associated with the MSP430 A/D function is the ADC10CTL1. Its bit definitions are shown in Figure 16.29.

The four INCH bits are the control bits for the multiplexer (see Figure 16.26), so they select which one signal is presented to the A/D converter. The distinction between these input channel bits and the bits in the ADC10AE0 register can be a little confusing. Setting the bit in the ADC10AE0 register says that you're intending to use the associated analog channel as an input to the A/D. However, it does not cause that signal to then be connected to the A/D. That function (switching the intended channel to the A/D) is performed by the INCH bits of the ADC10CTL1 register. Thus, we could have, say, three different channels, for example A1, A3, and A7, that are selected as analog channels to be used with the A/D. We do that with the ADC10AE0 register. As far as which channel is the one actually presented to the A/D for conversion at any given time – that is determined by the state of the INCH bits. Note that the INCH bits can only be changed when the ENC bit of the ADC10CTL0 register is 0.

For simplicity, the remaining bits of the ADC10CTL1 can be left in their power-on state of 0. Among other things, this means that the result will be in "straight binary", by which is meant that 000_{16} corresponds to 0 V and $3FF_{16}$ corresponds to V_{R+} (1.5 V, 2.5 V, or V_{CC}, depending on how things have been configured).

(a)

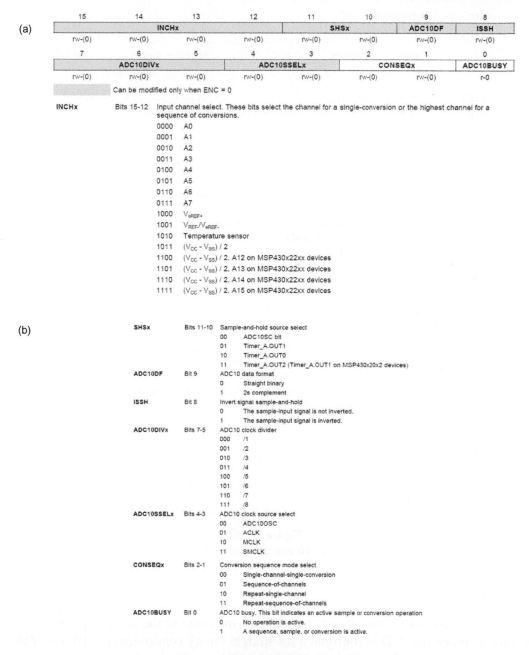

15	14	13	12	11	10	9	8
INCHx				SHSx		ADC10DF	ISSH
rw-(0)	rw-(0)	rw-(0)	rw-(0)	rw-(0)	rw-(0)	rw-(0)	rw-(0)

7	6	5	4	3	2	1	0
ADC10DIVx			ADC10SSELx		CONSEQx		ADC10BUSY
rw-(0)	rw-(0)	rw-(0)	rw-(0)	rw-(0)	rw-(0)	rw-(0)	r-0

Can be modified only when ENC = 0

INCHx Bits 15-12 Input channel select. These bits select the channel for a single-conversion or the highest channel for a sequence of conversions.

0000	A0
0001	A1
0010	A2
0011	A3
0100	A4
0101	A5
0110	A6
0111	A7
1000	V_{eREF+}
1001	V_{REF-}/V_{eREF-}
1010	Temperature sensor
1011	$(V_{CC} - V_{SS}) / 2$
1100	$(V_{CC} - V_{SS}) / 2$, A12 on MSP430x22xx devices
1101	$(V_{CC} - V_{SS}) / 2$, A13 on MSP430x22xx devices
1110	$(V_{CC} - V_{SS}) / 2$, A14 on MSP430x22xx devices
1111	$(V_{CC} - V_{SS}) / 2$, A15 on MSP430x22xx devices

(b)

SHSx	Bits 11-10	Sample-and-hold source select	
		00	ADC10SC bit
		01	Timer_A.OUT1
		10	Timer_A.OUT0
		11	Timer_A.OUT2 (Timer_A.OUT1 on MSP430x20x2 devices)
ADC10DF	Bit 9	ADC10 data format	
		0	Straight binary
		1	2s complement
ISSH	Bit 8	Invert signal sample-and-hold	
		0	The sample-input signal is not inverted.
		1	The sample-input signal is inverted.
ADC10DIVx	Bits 7-5	ADC10 clock divider	
		000	/1
		001	/2
		010	/3
		011	/4
		100	/5
		101	/6
		110	/7
		111	/8
ADC10SSELx	Bits 4-3	ADC10 clock source select	
		00	ADC10OSC
		01	ACLK
		10	MCLK
		11	SMCLK
CONSEQx	Bits 2-1	Conversion sequence mode select	
		00	Single-channel-single-conversion
		01	Sequence-of-channels
		10	Repeat-single-channel
		11	Repeat-sequence-of-channels
ADC10BUSY	Bit 0	ADC10 busy. This bit indicates an active sample or conversion operation	
		0	No operation is active.
		1	A sequence, sample, or conversion is active.

Figure 16.29

(a) The ADC10CTL register (*courtesy of Texas Instruments*). (b) ADC10CTL1 register definition, continued (*courtesy of Texas Instruments*).

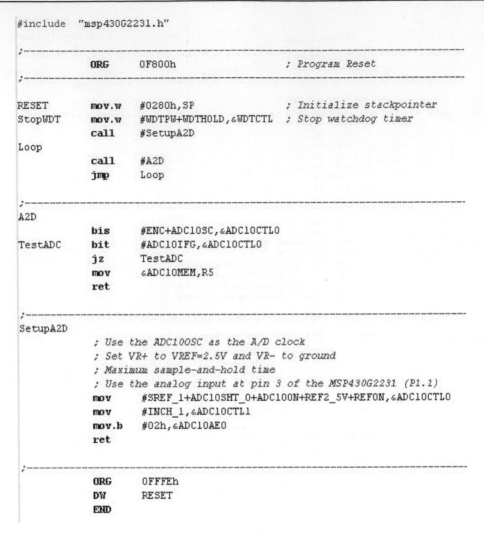

```
#include  "msp430G2231.h"

;-------------------------------------------------------------------------------
            ORG     0F800h                  ; Program Reset
;-------------------------------------------------------------------------------

RESET       mov.w   #0280h,SP               ; Initialize stackpointer
StopWDT     mov.w   #WDTPW+WDTHOLD,&WDTCTL  ; Stop watchdog timer
            call    #SetupA2D
Loop
            call    #A2D
            jmp     Loop

;-------------------------------------------------------------------------------
A2D
            bis     #ENC+ADC10SC,&ADC10CTL0
TestADC     bit     #ADC10IFG,&ADC10CTL0
            jz      TestADC
            mov     &ADC10MEM,R5
            ret

;-------------------------------------------------------------------------------
SetupA2D
            ; Use the ADC10OSC as the A/D clock
            ; Set VR+ to VREF=2.5V and VR- to ground
            ; Maximum sample-and-hold time
            ; Use the analog input at pin 3 of the MSP430G2231 (P1.1)
            mov     #SREF_1+ADC10SHT_0+ADC10ON+REF2_5V+REFON,&ADC10CTL0
            mov     #INCH_1,&ADC10CTL1
            mov.b   #02h,&ADC10AE0
            ret

;-------------------------------------------------------------------------------
            ORG     0FFFEh
            DW      RESET
            END
```

Figure 16.30
ADC10 example program.

ADC10MEM

The ADC10MEM is a 16-bit register that contains the result of the last completed A/D conversion. When the A/D is configured for straight binary conversion, the 10 bits of the conversion will be in the least significant 10 bits of the register.

ADC10 example

The program in Figure 16.30 is a simple program that continually loops around an analog-to-digital conversion on the analog channel A1 (P1.1) of an MSP430G2231 installed on a

LaunchPad. In order to test it, the positive end of an AAA battery was connected to P1.1 of the LaunchPad board and the negative end was connected to the board's ground. Using a voltmeter, the battery voltage was measured to be 1.830 V (fresh battery).

Note that the A/D is configured to use the 2.5 V internal reference (REFON = 1, REF2_5 V = 1, and SREF = 1). Using Eq. 16.11, the reading from the A/D should be:

$$N_{ADC} = 1023 \cdot \frac{1.830}{2.5} = 749_{10} = 2ED_{16}.$$

When the program was run, then execution stopped (break), the ADC10MEM register read $2EA_{16}$ – an error of about 7 mV.

Bibliography

[1] J. Candy, T. Gabor, Oversampling Delta-Sigma Converters, IEEE Press, 1992.

[2] J. Davies, MSP430 Microcontroller Basics, Elsevier-Newnes, 2008.

[3] MSP430G2452 Datasheet. Literature Number: SLAS722B. <http://www.ti.com/lit/ds/symlink/msp430g2152.pdf>, December 2010; revised March 2011.

[4] MSP430x2xx Family User's Guide. Literature Number: SLAU144H. <http://www.ti.com/general/docs/litabsmultiplefilelist.tsp?literatureNumber = slau144h>.

[5] Ott, H. Noise Reduction Techniques in Electronic Systems, second ed. John Wiley and Sons, 1988.

[6] Pohlmann, K., Principles of Digital Audio, second ed. Prentice Hall Computer Publishing, 1989.

[7] J. Proakis, M. Salehi, Communication Systems Engineering, Prentice Hall, 1994.

CHAPTER 17

Circuit Building

Chapter Outline

So far we've talked a lot about circuits. That's good – we need to understand the theory behind how these things work. The question is: can we actually build such circuits make them work? To do that, we'll need to: (1) build circuits with enough quality to eliminate, or at least keep to some very low level, any construction errors, and (2) be able to perform troubleshooting that uncovers either construction or design errors. This chapter covers the building part of this requirement. In a later chapter we'll discuss troubleshooting.

Types of circuit construction

Building a circuit assembly from an existing printed circuit board

The easiest way to build an electronic assembly is to start out with an existing printed circuit board (PCB) provided through a kit or as a bare board. The robot that is the focus of this book is available this way. Figure 17.1 is the bare board for this robot kit. This is a double-sided printed circuit board, meaning that there is metal on both sides of the board. Very simple designs can be built using a single-sided printed circuit board while more complex designs, such as the ones inside your home computer, may have many layers sandwiched in between the two outside layers. Those boards are referred to as multilayer boards.

MSP430-based Robot Applications.
DOI: http://dx.doi.org/10.1016/B978-0-12-397012-1.00017-5

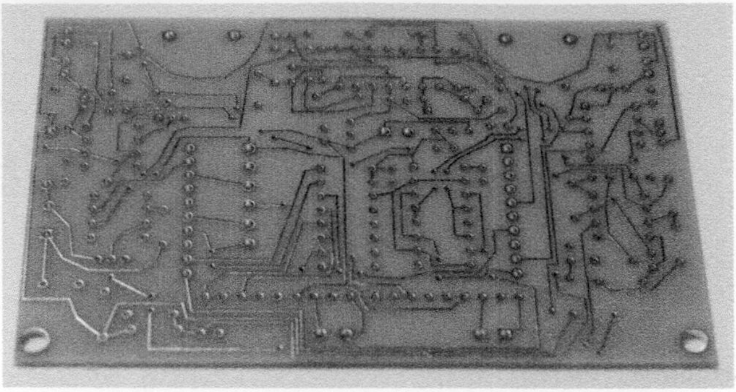

Figure 17.1
Printed circuit board for the robot project described in this book (from Bitstream Technology).

One of the nice things about building an existing printed circuit board is that the design is (or, at least, should be) already worked out. The design bugs should therefore be nil. And there's not a lot of work involved in building the assembly. You simply place the components where the instructions say to and solder the parts in place.

Of course, you still need to know *how* to solder, but that's not a hard skill to learn. In fact, let's talk about what that requires.

How to solder

Good soldering requires some practice. If you haven't soldered before, you might want to purchase one of the little practice kits that are available from a number of online electronics kit websites.

The following tips are taken from a book by Ward Silver, *Circuitbuilding for Dummies*:

- Start with a clean printed circuit board, free of grease, oxide, liquids. Metal surfaces should be bright and shiny.
- Apply the soldering iron tip to both the printed circuit board pad and the lead protruding through that pad.
- Apply the solder at the junction of the soldering iron tip, pad, and lead.
- The solder should melt immediately. When it melts, feed a small amount onto the joint, then pull the solder away.
- The iron should be on the joint for just a second or two. The solder should flow around the lead, filling the joint between the pad and the lead.
- The finished solder joint should be shiny. Check that the solder has not flowed to adjacent leads or pads.

One additional note: don't use too large a soldering tip. Soldering tips come in lots of sizes and for the devices that you'll be soldering, usually the smaller the better. Large soldering tips will make it far too easy to wind up with solder flowing to adjacent traces and pads.

- Pros of building an electronics assembly from a kit:
  - Dependable, proven design
  - Fast construction
  - Minimal construction errors
- Cons of building an electronics assembly from a kit:
  - Requires soldering
  - Very little design flexibility

Making your own printed circuit boards

The last bullet above is particularly important. When you build an assembly from a kit, you're building someone else's design. That's good or bad, depending on how you look at it. It's good in that the design should be free of bugs. But it takes away your ability to try things.

One thing that you can do to retain the advantages of a printed circuit board while still having the ability to create your own design is to lay out and build, or have built for you, your own printed circuit board.

You can make your own printed circuit boards, using schematic capture and PCB layout software, then transferring the PCB layout (that is, the traces and pads), using a transfer printed using a laser printer, to a copper-clad board. Copper-clad board is then chemically etched, with the copper etched away except for the intended traces and pads. The holes for the component leads are then drilled out on a drill press. The details of how to do this are available at a number of websites. The one I've included in the references is "Making your own PCBs at Home" at http://www.qsl.net/k5lxp/projects/PCBFab/PCBFab.html.

Also note that the schematic capture and PCB layout software is available free or at little cost from a number of companies. The article calls out Eagle software from CadSoft.

There are a number of reasons why this is not an option that's often exercised by hobbyists. Among the disadvantages:

- Cost of materials
- Time and effort required
- Design may have bugs, in which case process will likely have to be repeated
- Only single-sided boards are usually attempted

This last bullet has to do with the fact that, if you have a two-sided board, you must "maintain registration". By that is meant that the component pads on one side of the board

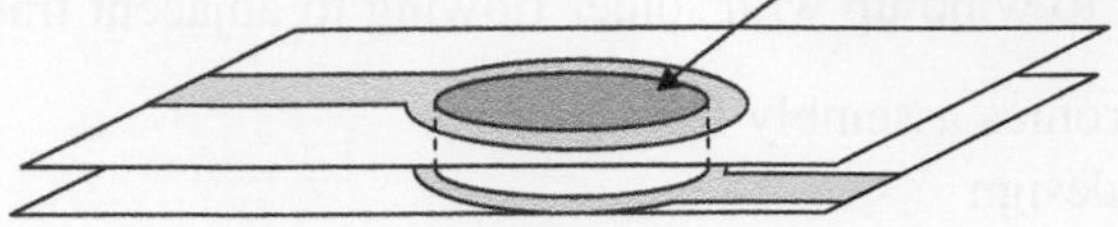
Copper plating on sides of hole
makes conduction between top and
bottom layers possible

Figure 17.2
Pad on professionally made double-sided printed circuit board has plated-through holes.

must precisely line up with the pads on the other side of the board. To illustrate the problem, Figure 17.2 shows such a pad. Note that the pads must line up so that when the hole is drilled, it is in the center of both the top and bottom pads.

Figure 17.2 also illustrates another problem with trying to make your own double-sided PCBs. Whereas professionally made PCBs usually have "plated-through holes" – that is, copper covers the sides of the hole, as well as the top and bottom pads – homemade PCBs usually do not have plated-through holes. This means that every component lead will have to be soldered at both the top and bottom pads in order to guarantee conduction of current from one layer to the other.

Outsourcing printed circuit board fabrication

If you have a design that you want to try and would like to have a professionally made printed circuit board fabricated, there are a number of companies that can do that for you. I have had considerable experience with ExpressPCB, which will make three copies of the *same* circuit design for around $60. The company provides you with proprietary software for schematic capture and layout that is excellent. The downside of this is, of course, the expense. And keep in mind that any big design boo-boos will cost you for another fab iteration.

Building a circuit assembly on a solderless board

A very popular choice for circuit building among college labs and some hobbyists is the solderless board, also called protoboards (Figure 17.3). Such boards have the big advantage that you don't have to solder any parts. Unfortunately, they have a lot of disadvantages, including:

- Wires are easily undone – board must be handled with extreme care
- Tedious for large circuits
- Prone to wiring error – each wire has to be individually routed by hand
- Hard to keep lead lengths short
- Hard to troubleshoot

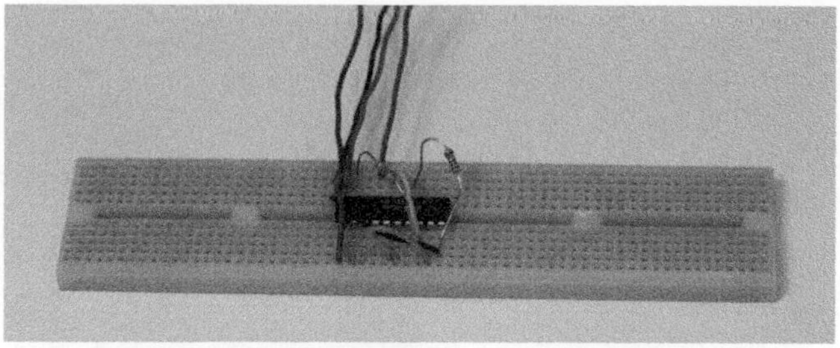

Figure 17.3
Simple circuit on a solderless board.

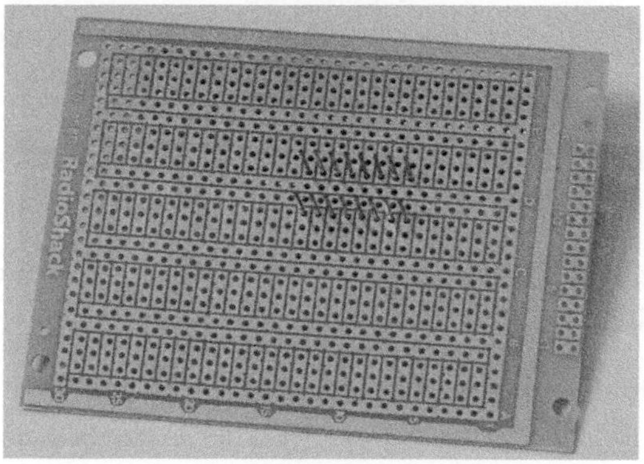

Figure 17.4
"Perf" board bottom side showing copper traces (16-pin DIP socket leads protruding through).

- Wire connectors internal to the boards lose their spring tension over time
- Board is heavy and bulky, making portable use (such as on a robot) less than optimal
- Accommodates leaded parts only – can't handle surface mounted parts

Because of these many drawbacks, solderless boards are rarely found in professional electronics labs.

Building a circuit assembly on vector board, using solder

A popular method for electronics hobbyists and others building a circuit is to use what is referred to as "perf board". This is a "board" – really a piece of composite material – that is pre-drilled with holes at regular intervals, usually with 0.1" spacing. These boards can be plain or can be copper-plated, like the one shown in Figure 17.4. The copper-plated board is nice because it includes two copper "bus" strips that run up and down the board. These can

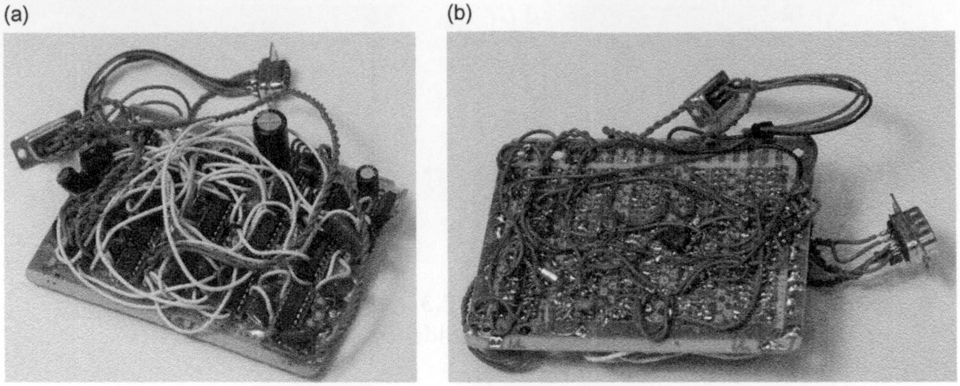

Figure 17.5
(a) Circuit built using perf board and point-to-point soldered wires (topside). (b) Bottom side of the panel (a) circuit.

be used as a supply voltage "bus" and a ground "bus". In electronics jargon, the term "bus" means a power strip.

Note also that, where the DIP socket pins protrude through, there are common copper strips that allow up to two additional solder connections to be made.

Figure 17.5 shows an actual circuit built using such a perf board with copper bus patterns. This board is not a robot board and has about twice as many ICs as the printed circuit board in Figure 17.1 accommodates.

These photos illustrate one of the problems with this type of construction – the assembly creates a rat's nest of wires which makes troubleshooting difficult.

To summarize the pros and cons:

- Pros
  - Accommodates new circuit design
- Cons
  - Tedious for large circuits
  - Prone to wiring error – each wire has to be individually routed by hand
  - Hard to keep lead lengths short
  - Hard to troubleshoot
  - Can't handle surface-mounted parts without adapters

Building a circuit assembly on vector board, using wirewrap

Another circuit construction method employing perf board, but without the need for soldering, is the wirewrap board. These boards are generally constructed with components

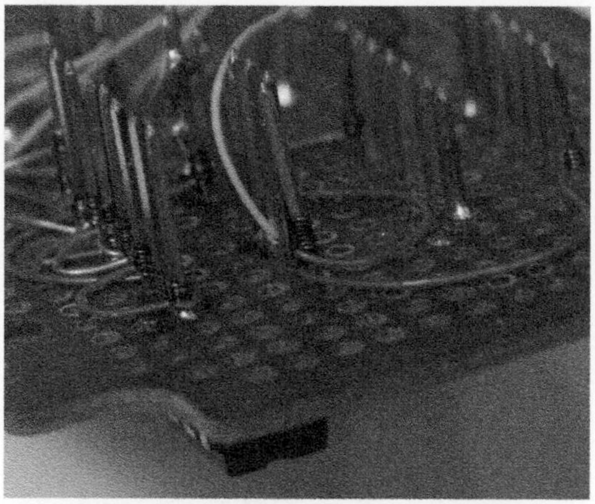

Figure 17.6
Wirewrap method of building boards.

mounted in sockets that have very long pins (Figure 17.6). The wire is wrapped around these pins using a wirewrapping tool (Figure 17.7). The tool actually has two holes at its end – one that slides over the socket pin and one right next to it into which is inserted the wire.

As you can see from Figure 17.6, the socket pins are long enough to accommodate at least three different wires wrapped around them. The pros and cons for this construction method are similar to the ones for the perf board with soldering:

- Pros
  - Accommodates new circuit design
  - No soldering
- Cons
  - Tedious for large circuits
  - Prone to wiring error – each wire has to be individually routed by hand
  - Hard to keep lead lengths short
  - Can't handle surface mounted parts without adapters

Deadbugging

One other construction method, listed here only for completeness, so that you'll be familiar with it if you hear the term, is "deadbugging". In this construction method, the integrated circuits and other components are glued to a board (often a copper-clad board). The components are glued to the board upside down, so that their leads are up (hence, the name "deadbugging" – an 8-pin DIP mounted like this looks like a dead spider lying on its back). Wires are then soldered directly to the component leads.

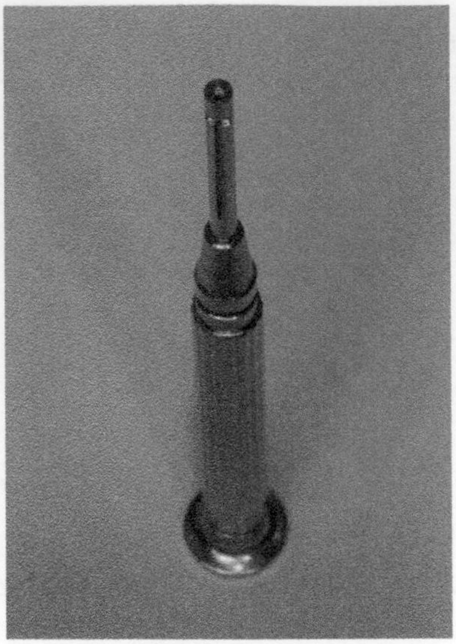

Figure 17.7
Wirewrapping tool.

This is not a construction method for beginners (actually, it's not a construction method that I would recommend for anyone, but especially not for beginners). It is unforgiving in the sense that the components, once glued down, may be difficult to remove and the wires connecting components, once soldered, may be difficult to remove if there are multiple wires going to a lead.

Where to buy parts

There are a number of companies that you can buy parts from. RadioShack still sells some components, although the selection of ICs, resistors, and capacitors is quite limited. Several companies sell parts over the Internet, among them DigiKey (http://www.digikey.com) and Mouser (http://www.mouser.com). Both of these companies have an extremely wide selection of parts, have very reasonable prices, and will sell parts in low quantity.

Bibliography

[1] Making your own PCBs at home. <http://www.qsl.net/k5lxp/projects/PCBFab/PCBFab.html>.
[2] H.W. Silver, Circuitbuilding for Dummies, Wiley, 2008.

CHAPTER 18

Using Sensors to Avoid Collisions

Chapter Outline

Robots, like humans, need both the ability to move and the ability to sense their environment. The human body's ability to sense is truly amazing. Think about the simple task of standing. The human body contains a vast network of sensors that relay information about balance, feeling (for example, nerve endings telling the brain that there is excess pressure on a particular part of the body), visual cues that aid in stability, etc. All of this information is processed by the brain, which then issues commands to the body's network of muscles to tighten or loosen in a way that maintains the person's vertical position.

In the case of our robot, the sensors will be much simpler and their main function will be to avoid colliding with things in their environment. The sensors used for collision avoidance in our robot (and in most robots) use the same concept as an active radar – energy is projected out from the robot and some of that energy is returned to the robot as a result of reflections from nearby objects. Sensing the magnitude and the delay of this return signal

MSP430-based Robot Applications.
DOI: http://dx.doi.org/10.1016/B978-0-12-397012-1.00018-7

provides information that the robot's microcontroller can then use in making decisions about its direction and speed.

The two types of energy that we will consider using for the robot's collision avoidance are acoustical and optical. The acoustic transducers (that is, transmitters and receivers) are ultrasonic. Ultrasonic transducers are really no different from the speakers and microphones used for audio applications except that whereas speakers and microphones are designed to work in the range of human hearing – for example, 50 Hz to 20 kHz – ultrasonic transducers work in the range above human hearing. So an ultrasonic transmitter can generate lots of energy without you hearing it (although it may drive your dog crazy).

The optical transmitters that we will use are light-emitting diodes (LEDs). These are very similar to the LEDs in a TV or other remote control unit. Like those remotes, our robot's transmitters will use LEDs that transmit in the infrared region, so we won't be able to see this light with our eyes. The optical detectors that we will use are called photodiodes.

Ultrasonic generation and detection

Ultrasonic generation

Most ultrasonic transducers are made from piezoelectric material. Such materials change dimensions when an electric field is applied across them. A thin slice of such a material will therefore expand and contract with a changing voltage across the two sides. If you think about it, that sounds a lot like a speaker.

In the case of the piezoelectric transducer, significant output energy is produced when the frequency is such that the transducer is at *resonance*. This resonant frequency occurs for a precise set of dimensions for the piezoelectric device. Before we discuss this further we need to introduce the concept of wavelength as it relates to a waveform's frequency.

Frequency and *period* are concepts that we've mentioned in previous chapters and which you're likely familiar with. A waveform, typically a sinewave, travels periodically through some voltage – for example, its positive peak – and we refer to the time interval between such points as the *period* (Figure 18.1). The reciprocal of the period is the *frequency*, f, given by:

$$f = \frac{1}{T}, \tag{18.1}$$

where T is the period. Typically frequency is measured in cycles per second or, more commonly, the equivalent term of *Hz*.

The wavelength, λ, is the distance that the waveform travels in one period (cycle). How can we know what that is? Well, if we know the waveform's frequency, f, then we know the waveform's period, T. Knowing the waveform's period and knowing the speed at which the

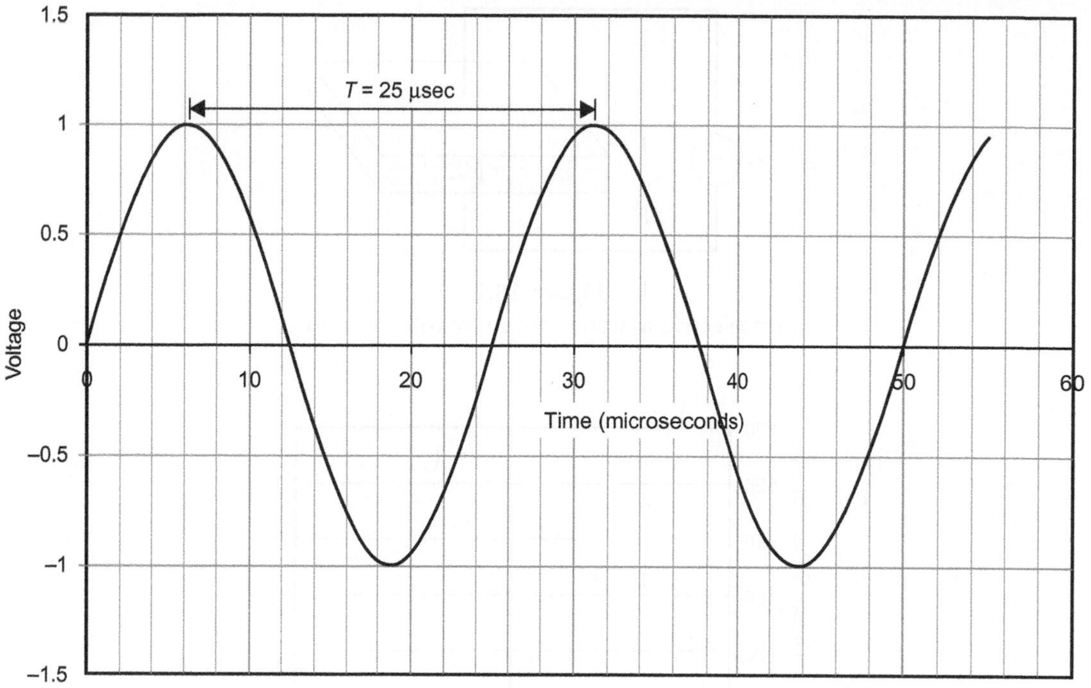

Figure 18.1
Period of a 40 kHz sinewave.

waveform propagates through the air, we then know the waveform's one-cycle distance, which is the wavelength. Let's write that down as an equation:

$$\lambda = vT = \frac{v}{f} \tag{18.2}$$

where v is the velocity of wave propagation.

Let's try an example. Let's say that we have an ultrasonic waveform with frequency 40 kHz. In dry air at 68°F the speed of sound is 1126 feet per second. What is the wavelength?

$$\lambda = \frac{v}{f} = \frac{1126}{40 \times 10^3} = 0.02815 \text{ ft} = 0.34 \text{ in.} \tag{18.3}$$

The speed of sound is slower through the piezoelectric ceramic material, so a wavelength within the material will be somewhat smaller. At any rate, when the thickness of the piezoelectric material is half a wavelength (Figure 18.2), the wavefront created at one face of the material is exactly in synch (that is, *in phase*) with the movement of the other face of the material, and the circuit is in resonance.

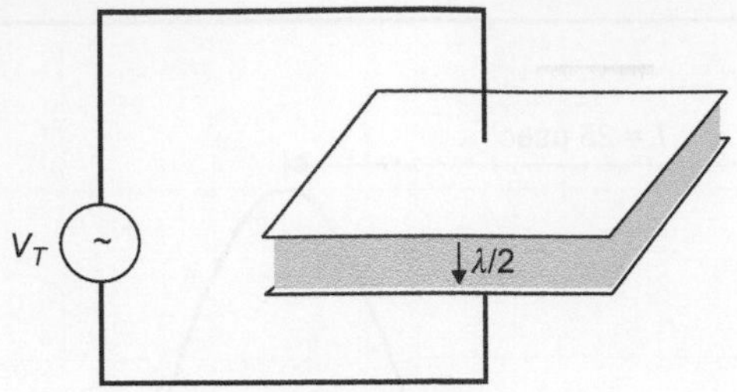

Figure 18.2
Piezoelectric material driven by voltage source.

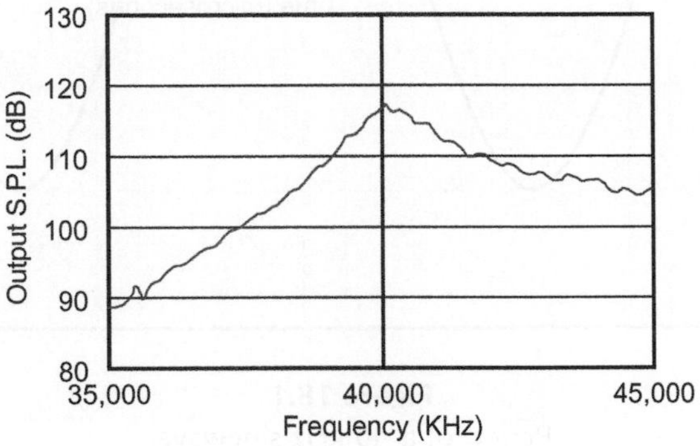

Figure 18.3
Sound pressure level for 40T-10AW ultrasonic transducer *(courtesy of APC International, Limited)*.

The sound pressure level produced by an APC International 40T-10AW transmitter driven with a 10 Vrms signal is shown in Figure 18.3. Note the sharp resonance at 40 kHz.

Advantages of using ultrasound

One of the primary advantages of using ultrasound as the energy for the robot to sense is the fact that the energy propagates relatively slowly – at the speed of sound rather than the speed of light. Let's look at a typical example of an object in the robot's path. Let's say it's a chair leg that is 2 feet away, as in Figure 18.4.

In this case, the ultrasonic pulses travel the 2 feet to the chair leg, reflect off the chair leg, and then travel the 2 feet back to the ultrasonic sensor on the robot, for a round-trip path of 4 feet. With a speed of sound of 1126 feet per second, then the transit time is given by:

$$t_{transit} = \frac{\text{distance}}{\text{speed}} = \frac{4\text{ ft}}{1126\text{ ft/sec}} = 3.6\text{ msec.} \tag{18.4}$$

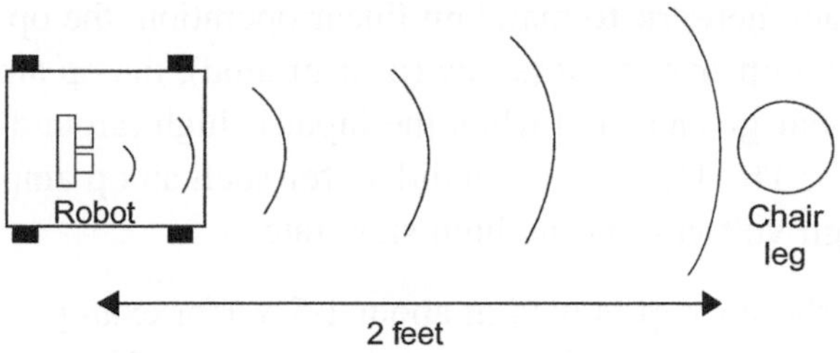

Figure 18.4
Chair leg in the path of ultrasonic transmission.

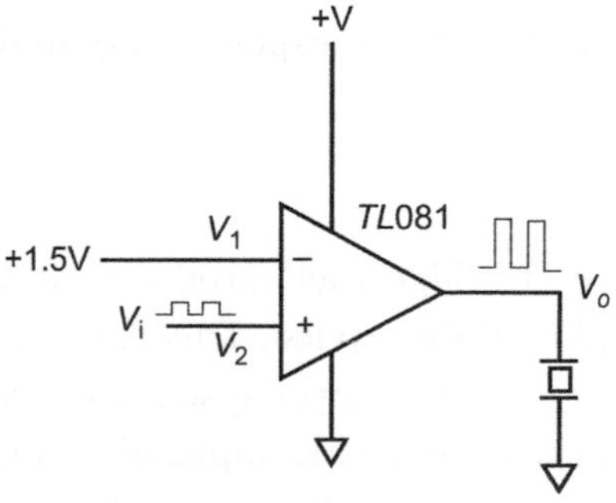

Figure 18.5
Example of single-ended ultrasonic driver.

By measuring the time that it takes for the ultrasonic pulse to return to the ultrasonic sensor, we can accurately determine the distance to the object. This type of measurement is referred to as time-of-flight (TOF). It works well with a microcontroller because 3.6 msec is a reasonably long time. No high-speed electronics are required to make the measurement.

Driving the ultrasonic transmitter

Electrically, what do we need to produce sound from such a transducer? We know that the drive signal (for the Figure 18.3 device) will need to be at or near 40 kHz frequency. Fortunately, we can use a 40 kHz square wave rather than a 40 kHz sine wave, simplifying waveform generation. From the device's datasheet, we know that the device can handle voltages of the order of 30 V.

Single-ended drive

One way to drive the piczoelectric ultrasonic transducer is to ground one side of it and drive the other side with the sine-wave or square-wave signal. A very simple way to do this is shown in Figure 18.5, where an op amp is used as a comparator – rather than setting the op amp inside a feedback network to maintain linear operation, the op amp is run open-loop (no feedback). When the op amp input is low (near ground), the op amp output will be at its minimum (that is, near ground) and when the input is high (around +3 V), the op amp output will be near V_{CC}. The TL081 is a candidate for such an op amp because of its ability to handle relatively high voltages and its high slew rate.

The inverting input of the op amp is held at about 1.5 V (for example, from a resistor divider) so that the op amp output swings from about 0 to +3 V (the TL081 op amp does not have rail-to-rail output, so the low voltage is a little higher than ground and the high voltage is less than +3 V).

The waveform will look like Figure 18.6a. The waveform can actually be broken down into its DC and AC components. The transducer responds only to the AC component of the signal, shown in Figure 18.6b.

Bridge-tied drive

In one of the earlier chapters the H-bridge was introduced, in which we drive a device (for example, a motor) from both sides of the device. This is sometimes referred to as a bridge-tied load, and is sometimes used in audio power systems to produce high output power from a relatively low voltage. In this ultrasonic application, the two drivers could be two op amps, for example, the op amps of a TL082 IC (Figure 18.7), which is the dual version of the TL081 used in the single-ended example.

The two op amps will be inverted from one another thanks to the fact that the top op amp has the microcontroller-generated waveform coming into its non-inverting input, while the bottom op amp has that signal coming into its inverting input. The two outputs will therefore occur as in Figure 18.8.

The transducer simply responds to the voltage *across* it – that is, $V_O^+ - V_O^-$. As Figure 18.8 shows, this difference of the two waveforms is approximately the same waveform as the AC component of the single-ended driver, but with a supply voltage only half that needed for the single-ended driver. Instead of having to produce a 30 V supply, the bridge-tied driver produces approximately the same result with a 15 V supply voltage.

MOSFET IC drive

The 4000 series digital MOSFET integrated circuits can operate from voltages as high as 20 V (check the datasheet to determine the maximum voltage). One device in particular, the 40109, is well suited to this application. The 40109 is a "level translator". That is, it takes as its input a signal from one power supply voltage system, say 3 V, and, at its output, produces a higher-voltage digital waveform. The 40109 is a quad device. One way to use it is to set it up as a bridge-tied driver, with two channels connected in parallel for one side and two channels connected in parallel for the other side, as shown in Figure 18.9. In this

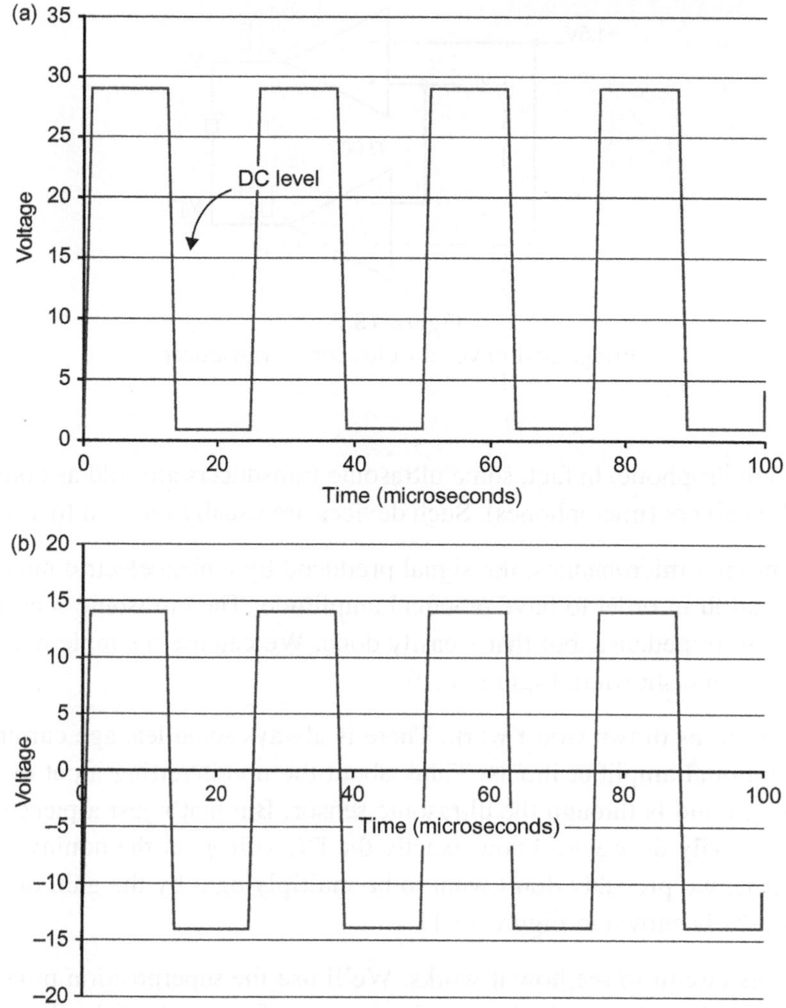

Figure 18.6
(a) Single-ended drive with 15 V DC level. (b) Single-ended drive, AC component only.

application, it performs the same function as the TL082 used in the above example (although it needs an inverter that the TL082 application didn't need).

Ultrasonic detection

Just as applying electrical energy to a piezoelectric material can cause it to change its dimensions, changing the dimensions of the material (by applying force) will cause a piezoelectric device to generate electrical energy. It's not hard to see how this effect can be

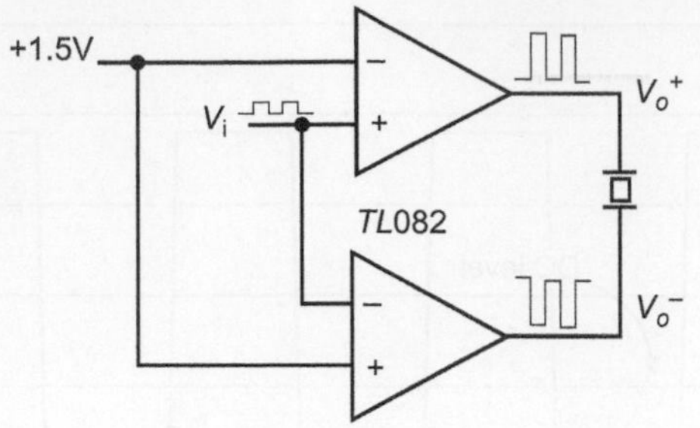

Figure 18.7
Bridge-tied driver for ultrasonic transducer.

used to create a microphone. In fact, some ultrasonic transducers are sold as both transmitters (speakers) and receivers (microphones). Such devices are usually referred to as *transceivers.*

As with conventional microphones, the signal produced by a piezoelectric microphone will require amplification in order to have practical amplitude. The ultrasonic sensor needs to interface to a high impedance, but that's easily done. We can use a simple non-inverting amplifier, like the one shown in Figure 18.10.

Actually, the circuit as drawn won't work. There is always some leakage current from each of the two operational amplifier inputs. Think about the non-inverting input of the op amp. Its only path to ground is through the ultrasonic sensor. But that's just a piece of ceramic material. And not only do we not know exactly the DC voltage at the noninverting input but, whatever it is, we probably don't want to be multiplying it by the gain of 10. A circuit that actually works is shown in Figure 18.11.

Let's analyze this circuit to see how it works. We'll use the superposition principle introduced in an earlier chapter, looking at the response from each of the three voltages (V_U, $V_{CC}/2$, and $V_{CC}/2$ – we'll treat the two bias voltages as separate voltages, V_1 and V_2). We'll ignore the fact that the individual output responses may produce outputs outside the 0-to-V_{CC} range of the op amp – what's important is that the combined result is inside this range. Figure 18.12 shows the circuit.

First let's look at the partial response due to V_U (Figure 18.13). Here the capacitor blocks DC and low-frequency signals but passes higher-frequency signals with a gain of 10:

$$V_{O,VU} \approx \left(1 + \frac{R_2}{R_1}\right) V_U \quad = 10V_U, \; f >> \frac{1}{2\pi R_3 C}$$
$$\approx 0, \; f << \frac{1}{2\pi R_3 C}. \qquad (18.5)$$

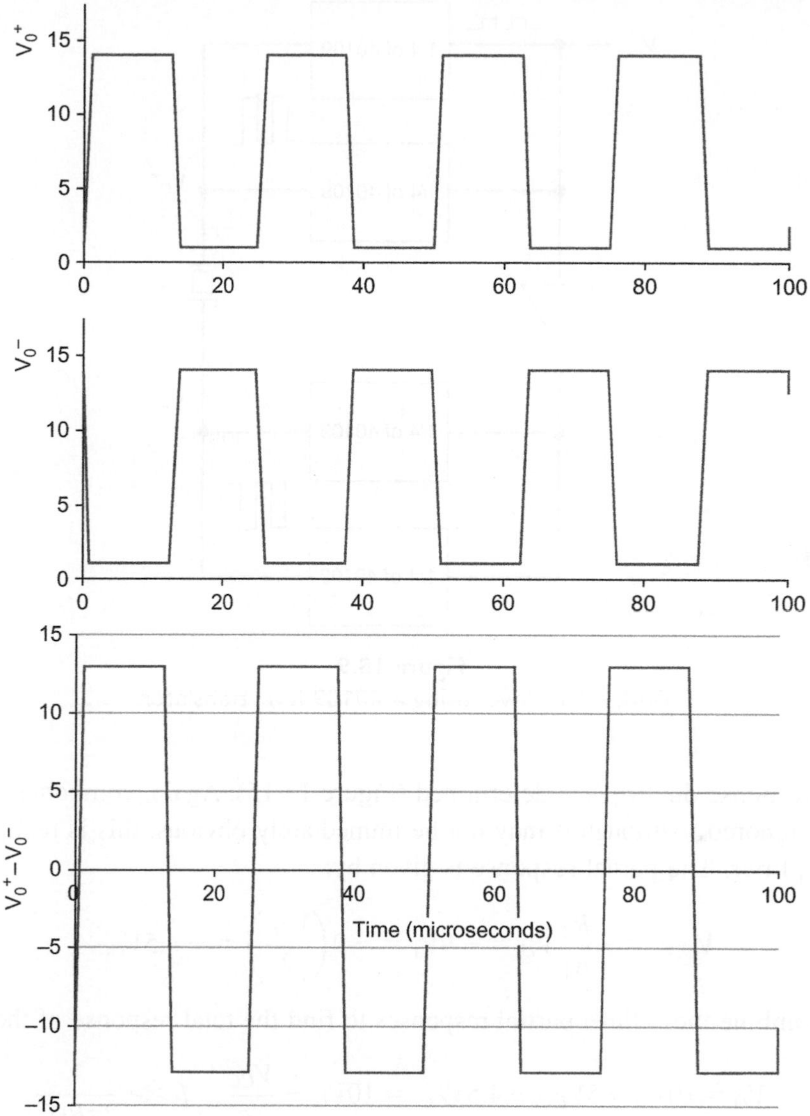

Figure 18.8
Bridge-tied driver waveforms.

Next, we find the partial response to V_1, using the circuit of Figure 18.14. Since V_1 is just a DC signal, and since the capacitor, C, completely blocks DC, we can ignore the components to the left of R_3. The partial response is given by:

$$V_{O,V_1} = \left(1 + \frac{R_2}{R_1}\right) V_U = 10V_1 = 5V_{CC}. \tag{18.6}$$

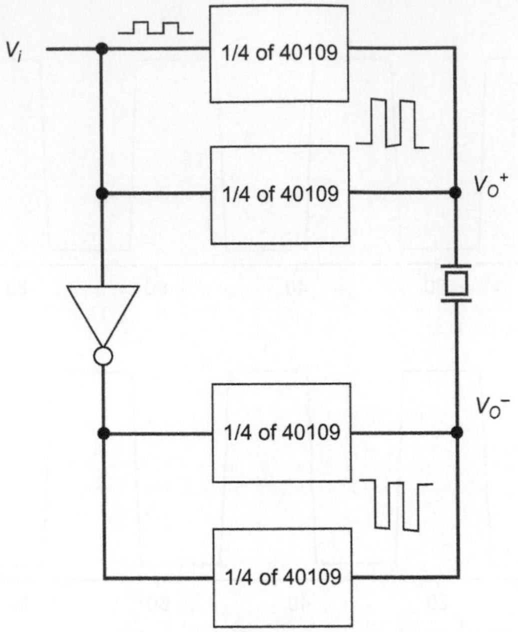

Figure 18.9
Bridge-tied driver using a 40109 level translator.

Finally, the response due to V_2 is determined (Figure 18.15). Again, components to the left of R_3 can be ignored. Although it may not be immediately obvious, this is really just an inverting amplifier. The partial response is given by:

$$V_{O,V_2} = -\frac{R_2}{R_1}V_U = -9V_1 = -9\left(\frac{V_{CC}}{2}\right) = -4.5V_{CC}. \tag{18.7}$$

Now, let's combine these three partial responses to find the total response of the circuit:

$$\begin{aligned} V_O = 10V_U + 5V_{CC} - 4.5V_{CC} &= 10V_U + \frac{V_{CC}}{2}, \quad f >> \frac{1}{2\pi R_3 C} \\ &\approx \frac{V_{CC}}{2}, \quad f << \frac{1}{2\pi R_3 C}. \end{aligned} \tag{18.8}$$

Thus, we have the desired gain, 10, for the ultrasonic sensor voltage, V_U, and the output (like the op amp input) has a DC level right in the middle of the voltage range, $V_{cc}/2$.

In actual applications, the gain will be set higher – 100 is a typical gain. There is no reason why the amplifier shown in these figures can't be scaled, but be careful that the gain bandwidth of the op amp is up to the job. The frequency of the signal being

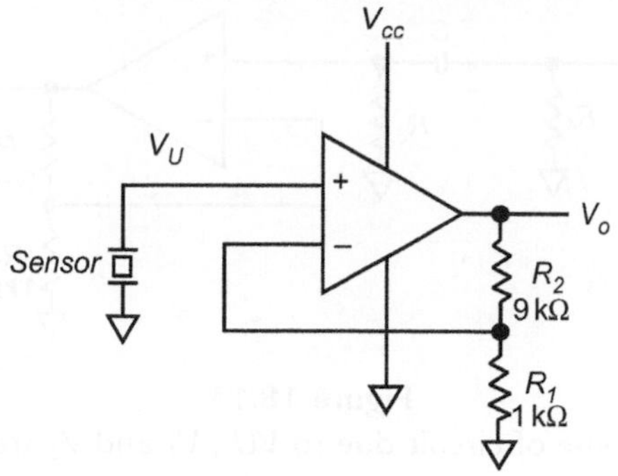

Figure 18.10
Ultrasonic sensor amplifier — *incomplete circuit.*

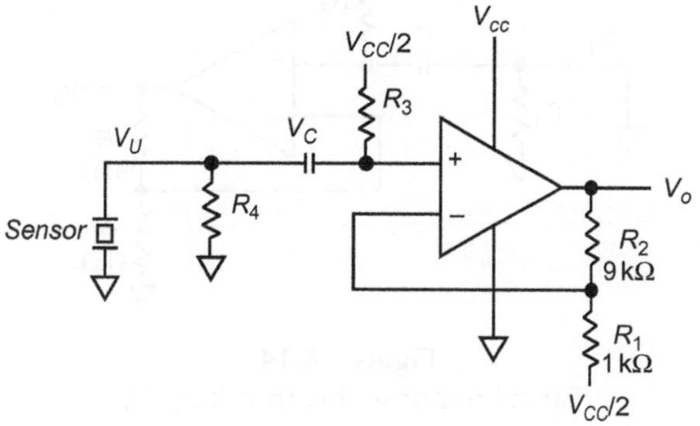

Figure 18.11
Ultrasonic detection circuit with AC coupling.

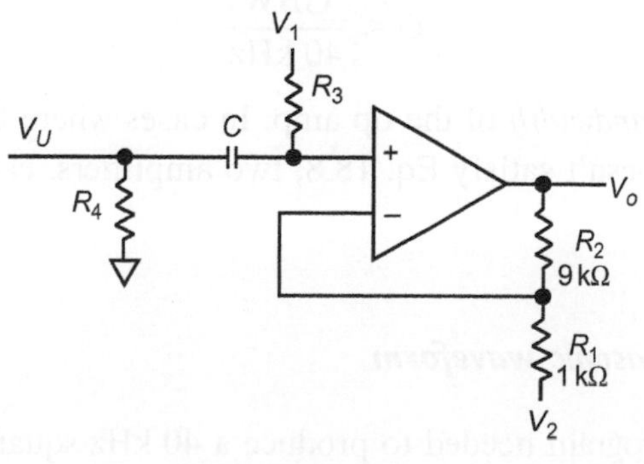

Figure 18.12
Ultrasonic sensor amplifier for superposition principle analysis.

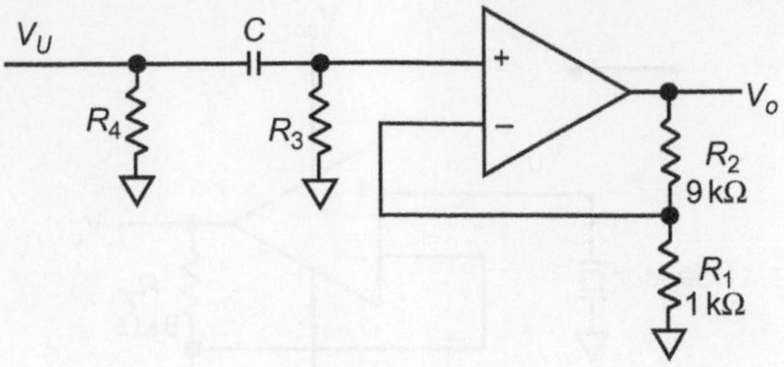

Figure 18.13
Partial response of circuit due to *VU* ; V_1 and V_2 are grounded.

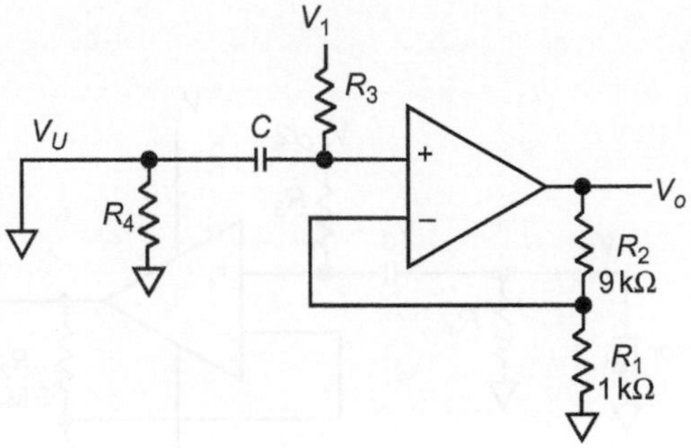

Figure 18.14
Partial response due to voltage V_1.

amplified is of the order of 40 kHz, and the gain of the circuit, *G*, must satisfy the relationship:

$$G < \frac{\text{GBW}}{40\ \text{kHz}} \tag{18.9}$$

where GBW is the *gain bandwidth* of the op amp. In cases where the gain needs to be high but the op amp's GBW doesn't satisfy Eq. 18.8, two amplifiers, cascaded, can be used to provide the overall gain.

Program to produce ultrasonic waveform

Figure 18.16 shows the program needed to produce a 40 kHz squarewave for use in driving an ultrasonic transmitter. Note that the main loop of this program consists of just the five instructions between Mainloop and the jump back to Mainloop.

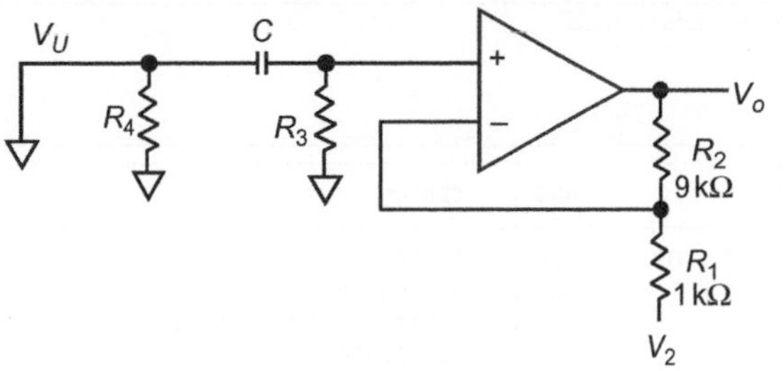

Figure 18.15
Partial response due to V_2.

The program produces 18 square-wave cycles at a frequency of 40,000 cycles per second. Note that "nop" instructions are used to "pad out" the timing so that the square-wave high amplitude and low amplitude have the right timing (12.5 μsec at each of the two amplitudes). Also, the repetition rate is 60 Hz – the 18-cycle "burst" occurs every 1/60 seconds.

Envelope detection

Figure 18.17 shows one side of a bridge-tied drive signal to an ultrasonic transducer, along with the amplified response of the sensor. There is some clutter in front of the transducer and sensor (small tools) but beyond that is a simulated chair leg (an 18-inch-long wooden rod, about 2 inches in diameter). This is located a little more than 2 feet from the transducer/sensor and produces the large sensor waveform at about 4 milliseconds.

Note that what we're really looking for is the onset of reflections. So what we really want to detect is when the envelope of the received waveform goes above some threshold. For that function we can use a rectifier-filter.

The AM receiver

Early amplitude-modulated (AM) radio relied on a type of detector that we can use for detecting the ultrasonic pulse returns. Conventional AM radio is really just a sinusoidal waveform operating at a single radio frequency (referred to as the *carrier*). The amplitude of this waveform is modulated (multiplied) by the voice or music waveform that is to be broadcast:

$$u(t) = A_c[1 + m(t)]\cos(2\pi f_c t) \qquad (18.10)$$

where A_c is the amplitude of the carrier, $m(t)$ is the message (voice or music), and f_c is the carrier frequency. Figure 18.18 shows a simple example where the message,

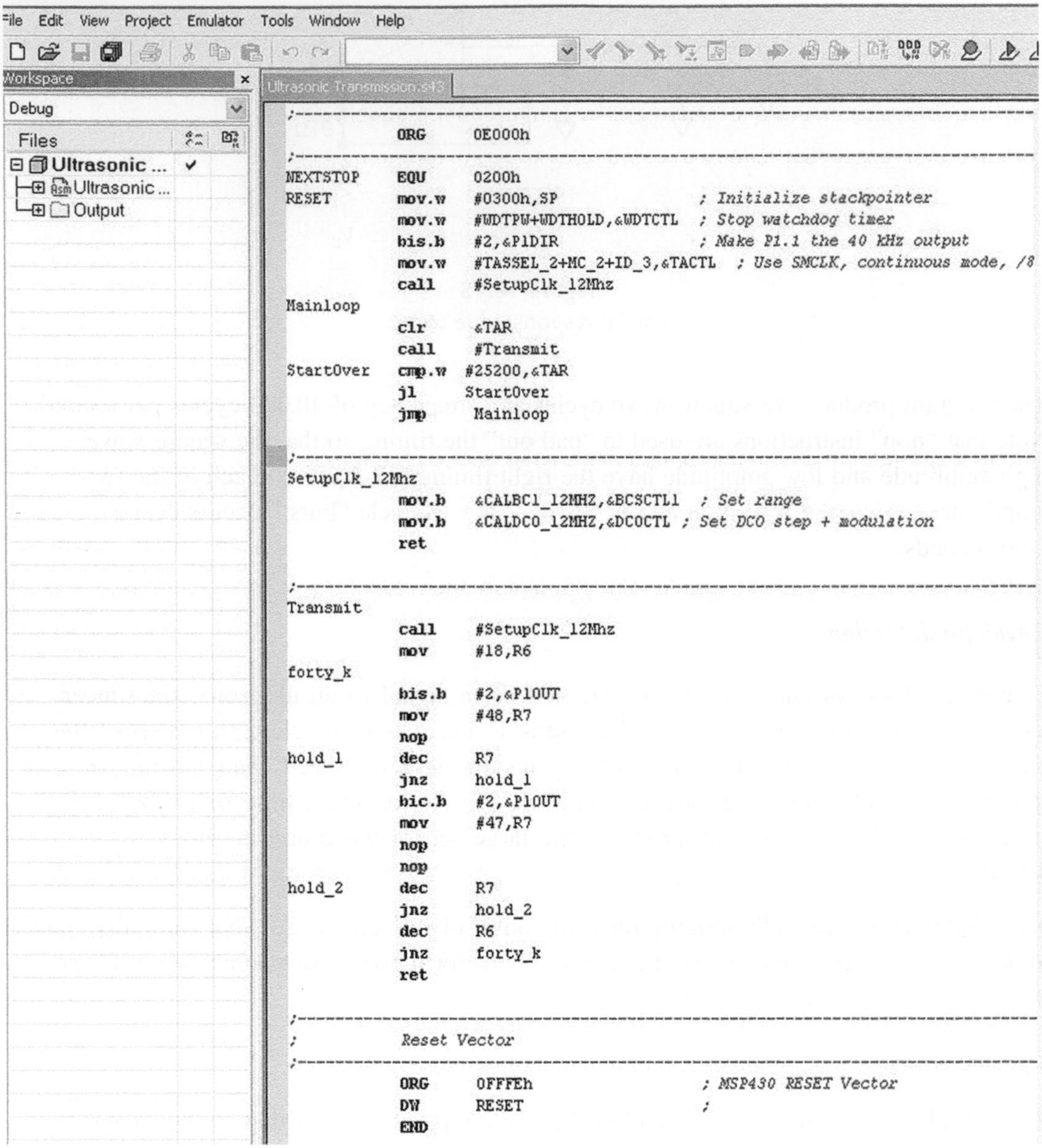

Figure 18.16
The ultrasonic transmission generation program.

m(*t*), is just a 5 kHz sine wave and the carrier is a 40 kHz sine wave. Note that the product, the AM signal, has an *envelope* which is just the original message. Early radio engineers realized that if they could detect this envelope, they could recover the original message!

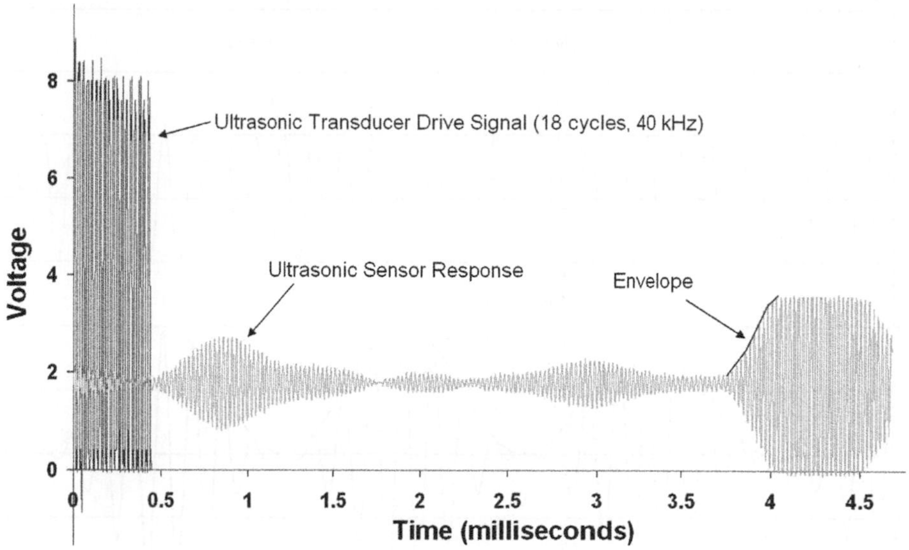

Figure 18.17
Response to ultrasonic transmission.

The circuit for performing such recovery is the diode-filter circuit of Figure 18.19.

The way this circuit works is that, any time v_i is greater than v_o plus the diode drop, v_d, the diode is on (that is, current flows through the diode), charging the capacitor, C, to $v_i - v_d$. Any time v_i becomes less than one diode drop above the capacitor voltage, the diode is off. When the diode is off, the capacitor discharges through the resistor, R. When R and C are chosen appropriately, the circuit output approximates the envelope of the input waveform.

Determining the number of pulses to use

Let's take a look at some measurements of actual ultrasonic transmission and detection. Also, let's see how many pulses are needed and how much gain is needed. For these measurements, the simulated chair leg (2" diameter wooden rod approximately 18" long and 2 feet from the sensor) was used.

Figure 18.20 shows the result of using 15 V for the supply voltage using a bridge-tied driver. There were 18 cycles of square wave driving the transmitter. The composite gain of the two cascaded detection amplifiers was 200. This gain is really too large, as it picks up too much of the clutter and saturates the detection amplifiers for the chair leg reflection (occurring about 4 msec after the start of transmission). Nevertheless, the envelope detector (middle waveform) performs fairly well except for the "detection" just after the transmitter pulses.

Next was tried a single pulse, again with 15 V as the supply voltage (Figure 18.21). The composite gain of the two detection amplifiers was approximately 1000 and was just about

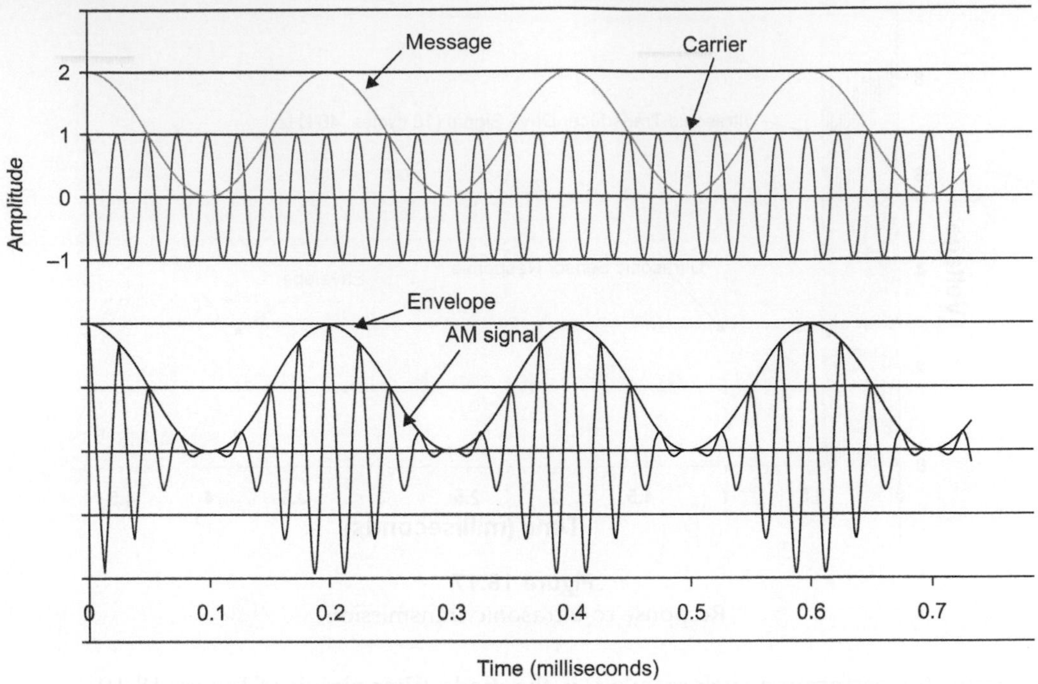

Figure 18.18
AM signal created by multiplying the message with the carrier.

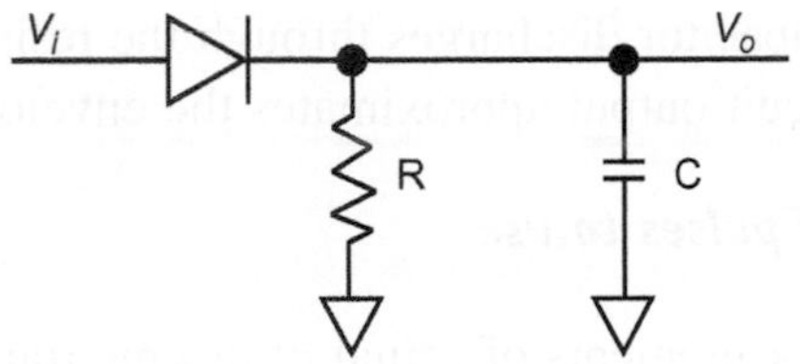

Figure 18.19
AM detector (rectifier-filter).

right for achieving good sensitivity to the chair leg but without showing all the little clutter ahead of the chair leg. Again, there is the false "detection" occurring just after the transmission pulse.

Figure 18.22 used 18 pulses but dropped the supply voltage for the ultrasonic transmitter to 7.5 V. Detector gain is again set to 200 and, as with the 15 V version in Figure 18.20, could stand to be reduced.

Finally, a single pulse was used with 7.5 V on the transmitter driver (Figure 18.23).

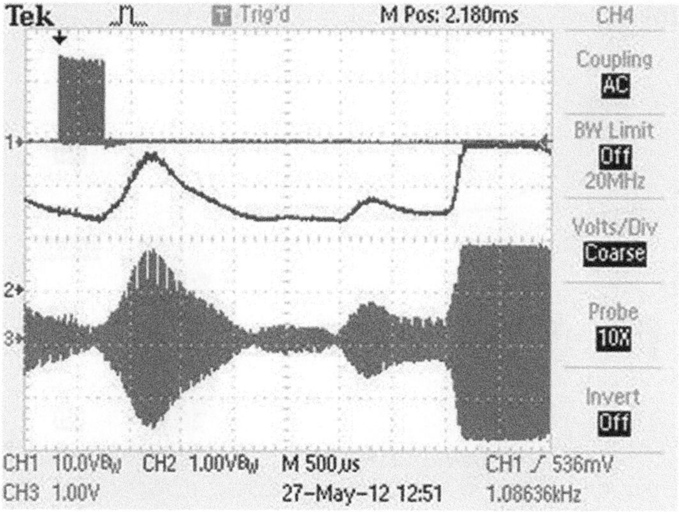

Figure 18.20
Eighteen ultrasonic pulses at 15 V, received envelope, and received waveform.

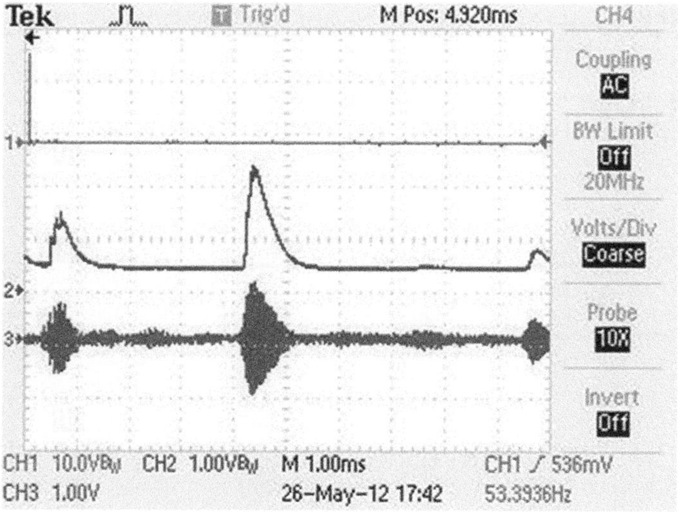

Figure 18.21
Single ultrasonic pulse at 15 V, received envelope, received waveform.

In each of the four examples of reflections just depicted, there is a reflected "burst" that occurs about 1 msec after transmission onset. This is probably due to crosstalk between transmitter and receiver, perhaps due to circuit board coupling between the two devices

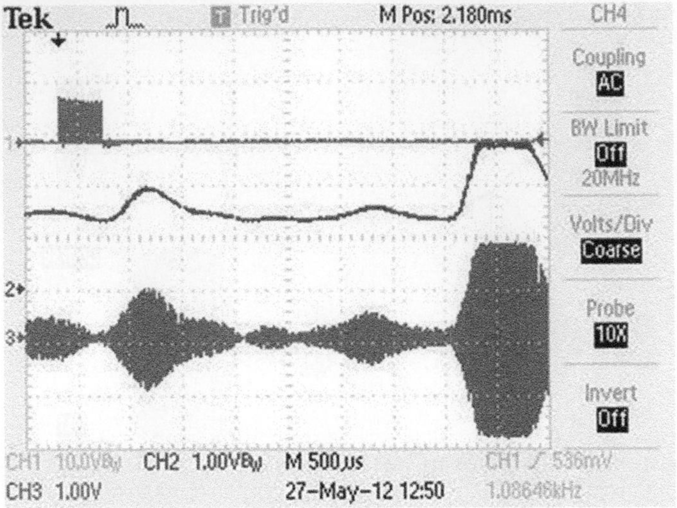

Figure 18.22
Eighteen ultrasonic pulses at 7.5 V, received envelope, received waveform.

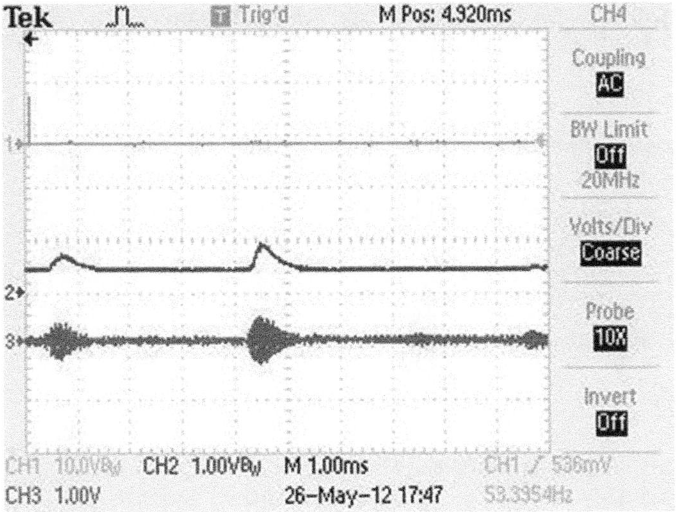

Figure 18.23
Single ultrasonic pulse at 7.5 V, received envelope, received waveform.

(due to the circuit board vibrating). A simple way to avoid falsely responding to this reflection as a perceived obstruction is to set the threshold higher during this first 1 msec, then lower the threshold.

Making decisions about potential obstructions

As mentioned, with TOF measurements, distance is determined entirely by the delay of the reflection with respect to the transmission burst. Amplitude is used only to decide whether a reflection is large enough to qualify as a legitimate obstruction. A threshold is set by the programmer and when the envelope exceeds this threshold, the reflection is judged to be an obstruction, and when below the threshold, not.

The threshold setting is important and leads to a trade-off that is familiar to those who design radars. Setting the threshold high reduces "false alarms" but also may cause the robot to miss legitimate obstructions. Reducing the threshold, of course, means reducing the probability that a legitimate target is missed, but it also increases the probability that the robot reacts to avoid an obstruction that isn't really there (a false alarm).

Optical generation and detection

Another way to detect objects is to use optical energy. Of course, we could use the same idea as with ultrasonic, maybe generate a short pulse or sequence of pulses and then just look for the time-of-flight for the reflections. This is, in fact, done. For example, fiber optic cable is often analyzed for problems using a device known as an optical time domain reflectometer, which launches a short pulse, usually from a laser, and measures the pulses reflected from any imperfections in the glass cable.

The problem with doing this has to do with the fact that light moves so fast. Whereas sound propagates 1 foot in approximately 1 msec, light propagates 1 foot in approximately 1 nanosecond! The circuits needed to measure such extremely short delays are expensive, difficult to build, and use lots of power. For something like our robot – which is supposed to be relatively simple, inexpensive, and run on small batteries – TOF is just out of the question when talking about optical detection. We are left with a couple of alternatives.

Simple LED sensing

We could just simply turn on one or more LEDs and measure the *amplitude* of the signal coming back. In theory, an object further away will produce a smaller amplitude.

There are lots of problems with this simple approach. One problem is that there are many things besides LED-to-object distance that can affect the reflected amplitude. The color and finish of the object change the reflection – a flat black object will reflect less light than a glossy white object. The shape of the object affects the reflection, too – a chair leg reflects differently than a wall, for example.

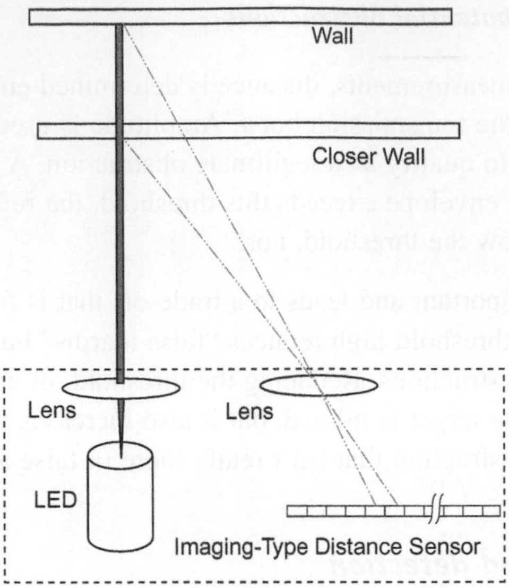

Figure 18.24
Depiction of the Sharp optoelectronic distance-measuring device.

Another problem with the simple LED approach is that it requires that we be able to sense the LED light in the presence of possibly strong background light. Despite these problems, there are some situations where this simple approach can be of use, as we shall see.

Imaging-type distance measurement

Sharp makes an optoelectronic sensor that uses an LED and a position-sensitive detector. This position-sensitive detector is able to determine where, on the sensor, light is hitting. I'm not an expert on this device but I believe Figure 18.24 and the following description are an accurate depiction of how the device works.

The LED output is *collimated* – that is, maintained as a small cylinder of light – by using a lens in its optical path. In this way, a small circle of light illuminates any obstructions in front of the device. The receiver also has a lens, which focuses the circle onto one or a few of the individual sensor elements. Based on which light-sensing element or elements are illuminated, the device is able to make a calculation as to the distance to the obstruction. Versions of this device are able to make accurate distance measurements to about 6 feet away.

This is an interesting way to solve the problem of measuring distance, using light as the energy source, but without using time-of-flight and without relying on the amplitude of the reflected light to determine the distance. However, it did not make the cut for inclusion in this robot design for a couple of reasons.

One was that it consumes 33 mA (typical) to as high as 50 mA (maximum). That's pretty high for our robot and may significantly affect battery life. A second, although lesser, issue is that it operates on 5 V, once a standard logic voltage but, more and more, an oddball voltage. Having to provide 5 V to the distance sensor means adding another power regulator. At any rate, the Sharp device, although an interesting solution to the problem of measuring distance, was supplanted by the ultrasonic solution.

Combining sensors for improved avoidance decisions

Both the Sharp optoelectronic and the ultrasonic distance-measuring system do a good job of determining the distance of impending obstructions. But to make good collision-avoidance decisions, the robot needs to know more than just that there is an obstruction ahead. Figure 18.25 illustrates three possible scenarios. In (a), the obstruction is straight ahead. Depending on how it is programmed, the robot may decide to stop or it may turn to the left or to the right. Neither direction of turning presents an advantage over the other direction.

In the remaining two scenarios, the robot is closing on the obstruction at an oblique angle to the robot. Clearly, in scenario (b), should the robot decide to turn, it would be advantageous to turn left, not right. In scenario (c) it will want to turn right if it makes a turn.

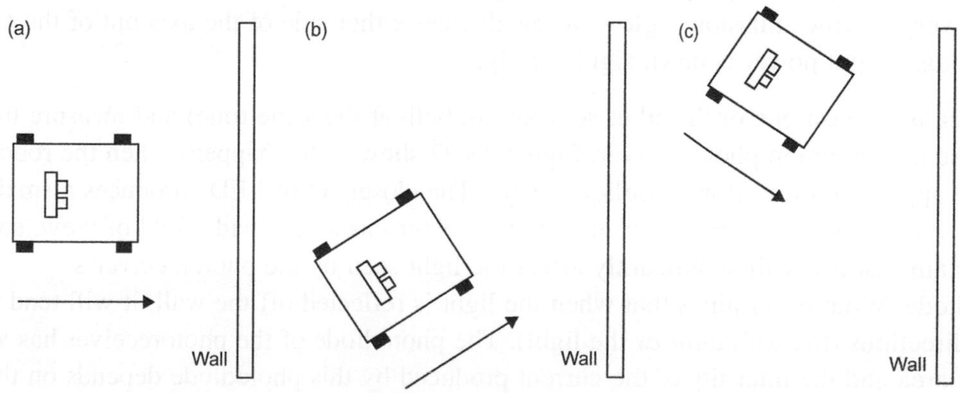

Figure 18.25
(a) Robot moving head-on into obstruction; (b) oblique angle, turn left; (c) oblique angle, turn right.

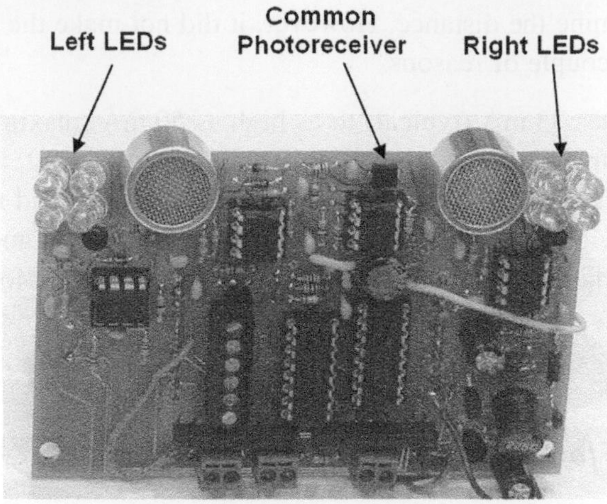

Figure 18.26
A robot controller board showing the two sets of LEDs and the photoreceiver.

Yet, for these second and third scenarios, there is no sensor information provided to help make this decision. If the robot has a single distance-measuring device on board, whether ultrasonic or optoelectronic, that robot has only a single piece of information to rely on. The sensor data will look very nearly the same for all three of the above scenarios (the obstruction is about the same distance away in all three).

What can be done to provide more information? One possibility is to add the simple LEDs mentioned earlier. Figure 18.26 shows the addition of two sets of LEDs, one set on the left and one set on the right. Note that the LEDs are angled about 10° from the normal – the left set of LEDs point slightly left and the right set of LEDs point slightly right. The LEDs have a very narrow emission angle – a few degrees either side of the axis out of the LED, the optical output power is down significantly.

The idea is to flash one or the other set (but not both at the same time) and measure the response at a common photoreceiver. Figure 18.27 shows what happens when the robotic vehicle approaches a wall at an oblique angle. The closer set of LEDs produces a smaller oval of light on the wall. However, the photoreceiver has a very wide field of view, so it is not certain that this will significantly affect the light seen by the photoreceiver's photodiode. What is certain is that when the light is reflected off the wall, it will tend to go in all directions (the wall *diffuses* the light). The photodiode of the photoreceiver has a certain area and the intensity of the current produced by this photodiode depends on the number of photons returned from the wall that are incident on this photodiode area. Just as the light spreads out in going from LED to the wall, it will spread out when going from the wall back to the photodiode. Therefore, fewer photons will be returned from the

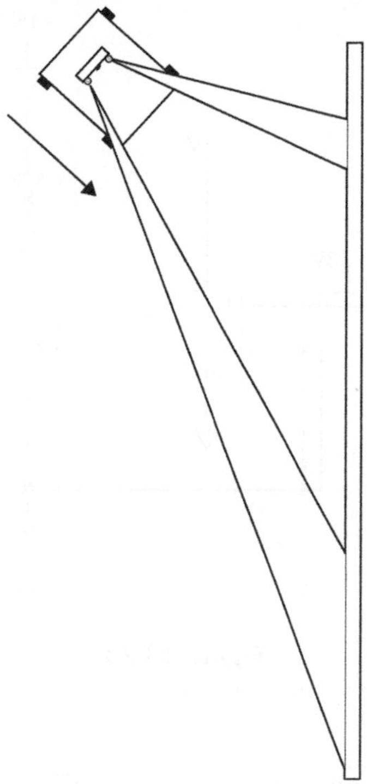

Figure 18.27
Difference in the two light beams from the robot LED/photodiode sensor.

farther-away reflection. As a result, the response of the photoreceiver will be considerably higher for the closer spot than for the farther spot.

What about the problems identified earlier with these simple LED/photodiode sensors? Keep in mind that all we're trying to do here is to compare the photoreceiver signal from each of the two sets of LEDs and roughly categorize as: one set of LEDs produces a significantly larger signal than the other set; or, both sets of LEDs produce about the same return. Since both sets of LEDs are received by the same photoreceiver, we don't have to worry about gain errors from the photoreceiver.

Also, we can produce virtually identical drive currents for the two sets of LEDs by using a precision drive circuit like the one shown in Figure 18.28. Note how this circuit works – the op amp raises its output until the resistor voltage equals the voltage at the noninverting input. So, to produce a current pulse of 100 mA using a 3 V input pulse as shown in the figure, we would make the resistor, R, equal to 30 Ω.

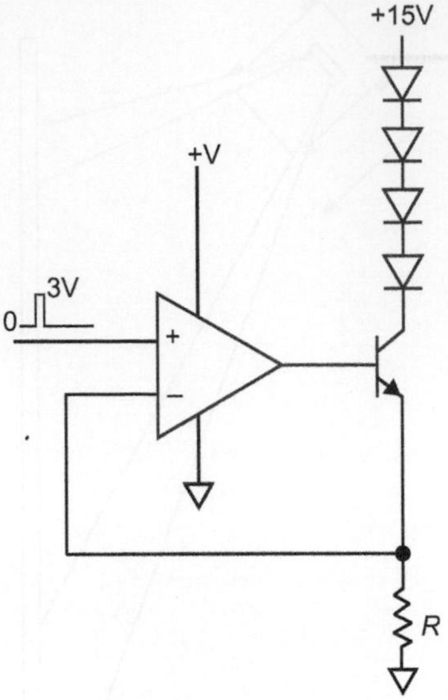

Figure 18.28
Precision driver circuit for set of four LEDs.

Note also that the entire circuit can be turned on and off within microseconds, which provides a huge advantage in conserving battery power. For example, a 100 mA drive can be created that lasts just 20 μsec – long enough for an A/D sample-and-hold and conversion. If such a set of LEDs is turned on at a 60 Hz repetition rate, the current consumed by the set of LEDs (not counting op amp supply current) is just (100 mA) × (20 μsec) × 60 = 120 μA.

Bibliography

[1] J. Proakis, M. Salehi, Communication Systems Engineering, Prentice Hall, 1994.
[2] M. Schwartz, W. Bennett, S. Stein, Communication Systems and Techniques, McGraw-Hill, 1966.
[3] Sharp GP2Y0A02YK0F datasheet. <http://www.sharpsma.com/webfm_send/1487>.

CHAPTER 19

Measuring Speed

Chapter Outline

Many mobile robots have a need to measure speed. This can be for a number of reasons. For one thing, you may want to know where you are – for example, where you are in the world, if the robotic vehicle is some large plane or terrestrial vehicle capable of covering large distances. If you know your starting position and you know what direction you're

MSP430-based Robot Applications.
DOI: http://dx.doi.org/10.1016/B978-0-12-397012-1.00019-9

heading (north, east, south, west), then with the additional knowledge of speed and elapsed time you can determine your current position. This type of navigation is referred to as *dead reckoning*. It has become less used now that people have the global positioning system (GPS) at their disposal, but many vehicles still use it to augment their GPS.

In the case of the robot we're contemplating in this book, a more likely need for measuring speed is that we might want to maintain constant vehicle velocity. Remember that we can use pulse-width modulation to adjust the robot's speed, but the speed is not only a function of what percentage of time the battery voltage is applied to the motor but also a function of the battery voltage itself. As the battery wears down, a 50% duty cycle will run the vehicle slower than a 50% duty cycle with a fresh battery.

If we'd like to maintain a particular vehicle speed we can set up a control loop. Control loops can actually be quite involved, but the one we'll use is simple – just measure the car's speed and, if it's too low, we increase the on-time/off-time ratio of our PWM drive. If the speed's too high, we do the opposite.

Okay, that's simple enough. But, as with all control loops, we'll need some sort of sensor to allow us to make the measurements. We'll cover that topic in this chapter.

Ways to measure vehicle speed

There are at least two methods for measuring vehicle speed. The first is to measure the back EMF voltage from the motor. Do you remember that from the chapter on DC motors? The back EMF is the voltage that the motor produces in its role as a generator. When the PWM drive changes from its "drive" state to its "off" state, the motor drive electronics actually look like an open circuit to the motor and the voltage that appears at the motor pins is solely due to the back EMF. This, in turn, is a function of how fast the motor is turning. So it's possible to perform an A/D conversion on the motor voltage, ideally just after changing the PWM drive for the motor from the "drive" state to the "off" state.

Improving the accuracy of the back EMF voltage measurement

The back-EMF-reading approach to speed measurement has the advantage of electronics simplicity. Its principal disadvantage is that, under the no-load condition of the motor during coasting, the back EMF is (in theory) linearly related to speed:

$$V_{EMF} = K_{EMF} \cdot RPM \tag{19.1}$$

where V_{EMF} is the back EMF voltage, K_{EMF} is the speed constant, and RPM is revolutions per minute. The problem with using this equation to determine vehicle speed is that we

Figure 19.1
Slotted wheel of the R/C car.

don't know the constant of proportionality, K_{EMF}. It is dependent on motor characteristics and will vary with different motors.

It is possible to perform a calibration of some kind on the vehicle, where the vehicle is operated at some known or measured speed and the corresponding EMF voltage is also measured, thus allowing K_{EMF} to be calculated. However, without special equipment (such as a strobe light) such a calibration does not appear to be practical.

An optical method of measuring speed

Most radio-controlled cars have wheels with slots in them, like the one shown in Figure 19.1. Since these wheels are normally chrome-colored, they reflect very nicely.

Using the slots in the wheel to count revolutions

It's not hard to see how this can be turned into a speed sensor. If we simply shine an LED from inside the wheel out (Figure 19.2) and observe the reflected waveform with a photodiode, we'll see something like the waveform in Figure 19.3. When the light strikes the chrome-colored wheel, a strong reflection is returned. When the light travels through the slot in the wheel, very little light is returned.

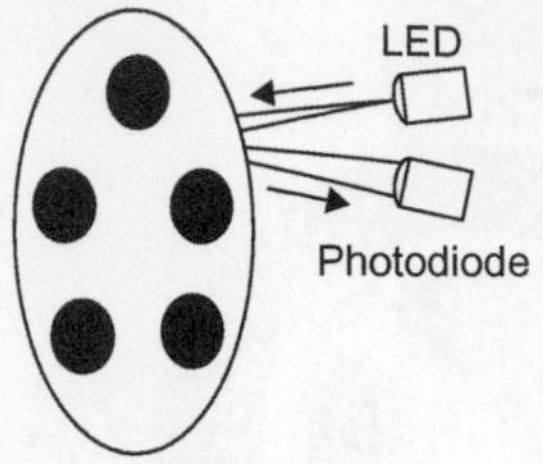

Figure 19.2
LED/photodiode for slot counting.

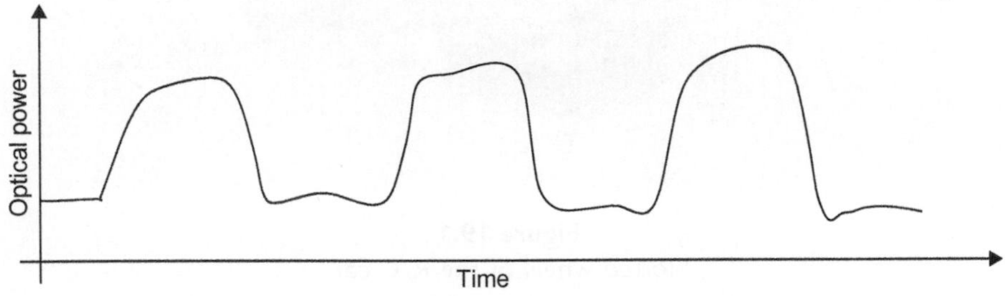

Figure 19.3
Optical power reflected from spinning slotted, chrome wheels.

Note that the light sensed by the photodiode in Figure 19.3 does not go to zero during the intervals when the light is traveling through one of the slots. This is because there is ambient light in the room that makes its way to the photodiode. We'll discuss ways in which to mitigate this problem a little later in the chapter.

Other optical methods for measuring speed

Your car's wheels may not have slots or you may simply wish to use some alternative optical method. One way to do this is to create an insert on the inside of one of the wheels, like the one shown in Figure 19.4. This is a simple example of a *shaft encoder*.

Constructing such an encoder is easy. Start with a word processing or slide presentation or drawing program that allows the creation of circles and triangles. Start out with a black circle in the center of the page. Add to this white triangles. Then put a circle border around the circle to help identify the boundary of the circle for cutting. You can print this out and then paste the circle on cardstock or thin white plastic.

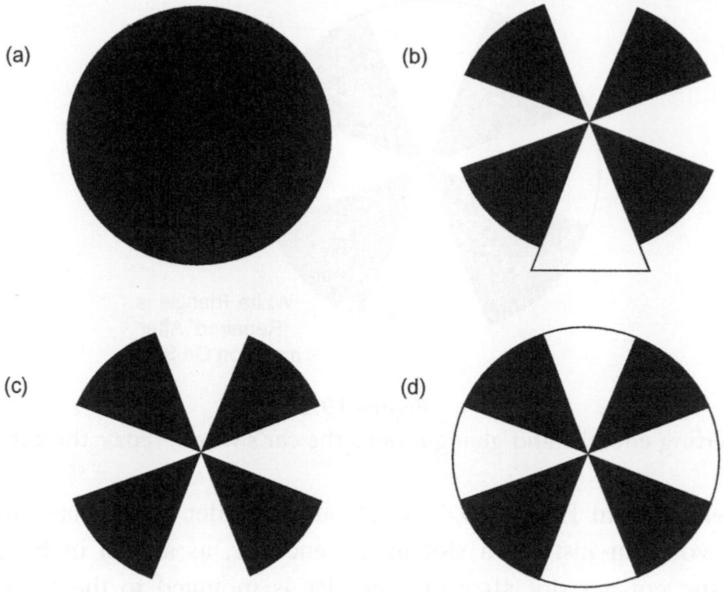

Figure 19.4
(a) Start with a black circle. (b) Add white triangles (black border on bottom triangle is not necessary). (c) Completed encoder image. (d) Adding a circle border will help in cutting the encoder out.

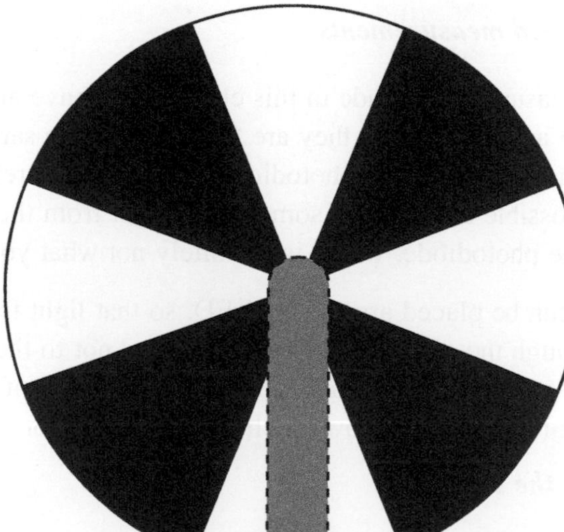

Figure 19.5
Cut out a slot so that the encoder can be slid over the car's wheel shaft.

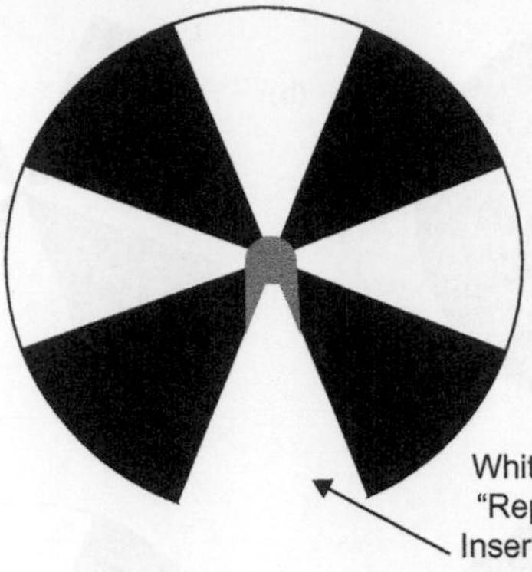
White Triangle Is
"Repaired" After
Insertion On Shaft

Figure 19.6
After inserting encoder and gluing it onto the car's wheel, repair the cut triangle.

The completed encoder in Figure 19.4 is cut out. In order not to have to take the car's wheel off, you can just cut a slot in the encoder, as shown in Figure 19.5, and then repair the cut-out slot after the encoder is mounted to the car's wheel (Figure 19.6). Super Glue will do a nice job of holding the encoder in place on the car's wheel.

Note that in the example above, the encoder is made up of four white and four black triangles, but there is no reason why the encoder can't have more or fewer black/white segments.

Avoiding crosstalk in speed measurements

Typically the types of measurements made in this chapter will have an LED and photodiode located very close to one another, so that they are "looking at" the same surface (in this case the car's wheel). The idea is for the photodiode to observe the reflection off the wheel. However, it is entirely possible that at least some of the light from the LED will be transmitted directly to the photodiode, which is definitely not what you want.

To avoid this, a *shroud* can be placed around the LED, so that light from the LED is only transmitted forward (through the opening of the shroud) and not to the side. The shroud can be made from a short piece of a plastic straw or from a short piece of polystyrene tubing. The example at the end of the chapter shows such a shroud used for the LED.

Calculating speed from the slot count

Calculation of speed from the slot count or black/white segments is straightforward. Let's assign the variable, M, to the number of slots in the wheel. In the case of the wheel shown in Figure 19.1, $M = 5$. Now, the wheel has some diameter, D, and therefore has a circumference, $C = \pi D$. When we count M slots, the wheel has traveled one complete revolution and the vehicle has therefore traveled C feet. By counting the total number of slots, N, over some relatively long period of time, T, we can compute the speed (velocity, v), as:

$$v = \frac{\pi DN}{MT}. \tag{19.2}$$

The wheel in Figure 19.1 has a diameter of 1.85 inches (0.154 ft). Let's say that we count 500 slots (that is, the waveform of Figure 19.3 goes high-then-low 500 times) over a period of 10 seconds. The velocity is then given by:

$$v = \frac{\pi(0.154)(500)}{(5)(10)} = 4.8 \text{ ft/sec}.$$

Sampling rate

The microcontroller will need to sample (perform A/D conversions on) the photoreceiver output at a rate high enough to guarantee that, even when the robotic vehicle is traveling at maximum speed, no transitions from the high state to the low state, or vice versa, in the Figure 19.3 waveform are missed. Let's calculate what this sample rate needs to be.

A maximum speed for a typical R/C car is of the order of 10 ft/sec, which is 120 inches/sec. The wheel of the car is about 2 inches in diameter, which means that the circumference is about 6 inches. So, at top speed the wheel makes $120/6 = 20$ revolutions per second. There are five slots in the wheel of the Figure 19.1 vehicle, so that means that there will be approximately $20 \times 5 = 100$ cycles of the Figure 19.3 waveform in one second when the vehicle is traveling at maximum speed. If we sample at 300 samples per second, we should be able to observe all high-to-low and low-to-high transitions. If we can afford the luxury of an even higher rate, say, 1000 samples per second, that might make things even easier to work right.

Converting optical power to voltage

Photoreceivers convert the current from a photodiode into a voltage. Before discussing photoreceivers further, let's talk first about photodiodes.

Photodiodes

The photodiode is a semiconductor device that has a nearly linear relationship of current to received optical power. The easiest way to think of the photodiode is just as a current source, where the current amplitude is a linear function of optical power incident on the photodiode.

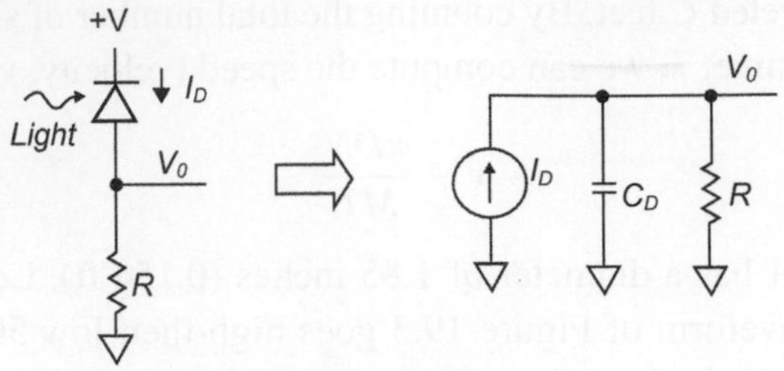

Figure 19.7
A simple photoreceiver and its equivalent circuit.

Photodiodes are generally reverse-biased – the cathode will be operated at a higher voltage than the anode. Photodiodes can also be zero-biased, with both cathode and anode at the same potential. The preference for reverse bias in many circuits is because of the somewhat lower photodiode capacitance that occurs with a non-zero reverse bias.

Photoreceivers

A photoreceiver includes the photodiode plus the circuit that converts the photodiode's current into a voltage. But doesn't a simple resistor convert current to voltage?

Simple photoreceivers

The answer is, yes, a simple resistor can perform the current-to-voltage conversion. Figure 19.7 shows such a photoreceiver.

Using just a resistor as the current-to-voltage converter works but the equivalent circuit in the right-hand circuit of Figure 19.7 shows a potential problem. The photodiode has a natural (that is, parasitic) capacitance which will interact with the resistor to create a slowed RC filter response. Since the current produced by the photodiode is small (generally from nanoamperes to microamperes, depending on the photodiode and the amount of light), the resistor will need to be quite large in order to produce a practical voltage. The photodiode parasitic capacitance can be anywhere from a few pF to more than 100 pF, depending on the particular photodiode used. As a result, the RC time constant can be relatively large – for example, a 1 MΩ resistor and a 50 pF capacitance result in a 50 μsec time constant.

For situations in which we want to use an LED/photoreceiver to make rapid measurements this time constant will be too long. If it turns out that the RC time constant for a simple resistor photocurrent-to-voltage converter is too large, then we will have to look for another solution. Fortunately, there is one and that solution is the transimpedance amplifier.

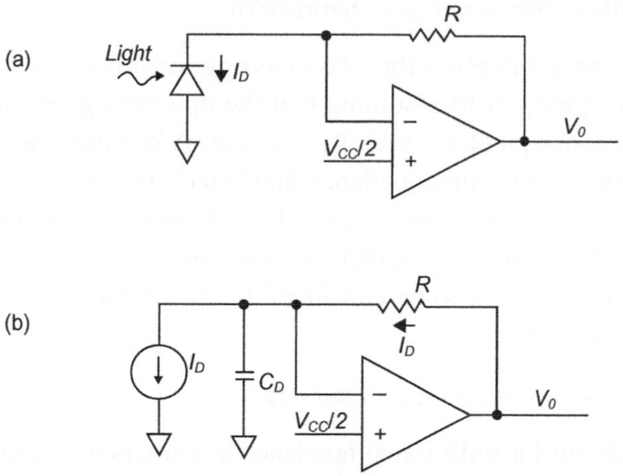

Figure 19.8
(a) Transimpedance amplifier for converting photodiode current to voltage. (b) Transimpedance amplifier with equivalent photodiode electrical model.

Transimpedance amplifier

A faster way to convert the photodiode current to voltage is to use a transimpedance amplifier. This circuit is shown in Figure 19.8. Assume that the op amp is operating linearly. If that is the case, then the op amp's inverting input is at approximately the same voltage as the non-inverting input, $V_{CC}/2$. Thus, the photodiode is reverse-biased.

The current flowing through the photodiode must flow through the resistor, R (remember that the op amp's inputs are extremely high impedance, so virtually no current flows into or out of these inputs). Ignoring, for the moment, the current flowing into the parasitic capacitance the response of the amplifier can be computed using Kirchoff's voltage law around the feedback loop of the amplifier. Since the op amp's inverting input is at $V_{CC}/2$, then V_O is:

$$V_O = \frac{V_{CC}}{2} + I_D R \tag{19.3}$$

Okay, so what's so great about this response versus the response of the simple resistor-only circuit? Remember that the amplifier is assumed to be operating in its linear region and that the inverting input is therefore nearly the same voltage as the non-inverting input ($V_{CC}/2$). Therefore, the voltage across the capacitor is held at very nearly a constant $V_{CC}/2$. That means no charging or discharging of the capacitor. The photodiode capacitance is effectively eliminated from the circuit, thereby greatly speeding up the circuit.

Transimpedance amplifier shortcomings – bandwidth

In the above analysis, the assumption that the inverting input is at the same potential as the non-inverting input is equivalent to assuming that the op amp's gain-bandwidth (GBW) product is infinite. Of course, that's not really the case. A complete analysis of how the op amp's GBW affects the transimpedance amplifier's performance is beyond the scope of this book. However, for those interested in additional reading on this topic, references 1 and 2 are excellent articles that explain more about this. Suffice it to say that the op amp's GBW will need to be considerably greater than the desired bandwidth of the transimpedance amplifier.

Transimpedance amplifier shortcomings – stability

Another thing to watch out for with transimpedance amplifiers is instability. Reference 1 gives considerable detail about transimpedance amplifier stability. As that article points out, two ways to improve stability are: (1) add a capacitor across the feedback resistor (the article explains what the value of the capacitor must be to guarantee stability; and (2) use an op amp with greater gain-bandwidth. Note that adding a capacitor across the feedback resistor reduces the overall transimpedance amplifier bandwidth, so keeping this capacitor value as small as possible is usually a good idea.

When is the transimpedance amplifier not linear?

In this chapter we've assumed that the transimpedance amplifier is operating in its linear region. What could cause the amplifier to behave nonlinearly? The answer is that the amplifier could *saturate*. This happens when the amplifier output reaches its maximum or minimum extreme. Since the photodiode/transimpedance amplifier responds to all light reaching the photodiode, this could happen as the result of bright lights in the room. This serves as a good tie-in to the next section – what effect does ambient light have on the photoreceiver?

Ambient light characteristics

There are three main types of ambient light that our robot might have to operate in. These are sunlight, incandescent light, and fluorescent light. Before we talk about these different types of light, it's important to talk about the wavelength of light and how it affects our photoreceiver.

Light wavelength

Light is actually electromagnetic energy that occurs at very high frequencies. Like lower-frequency electromagnetic energy, it can be categorized by the different sine-wave frequencies that make up a particular light source. However, by tradition, light is generally

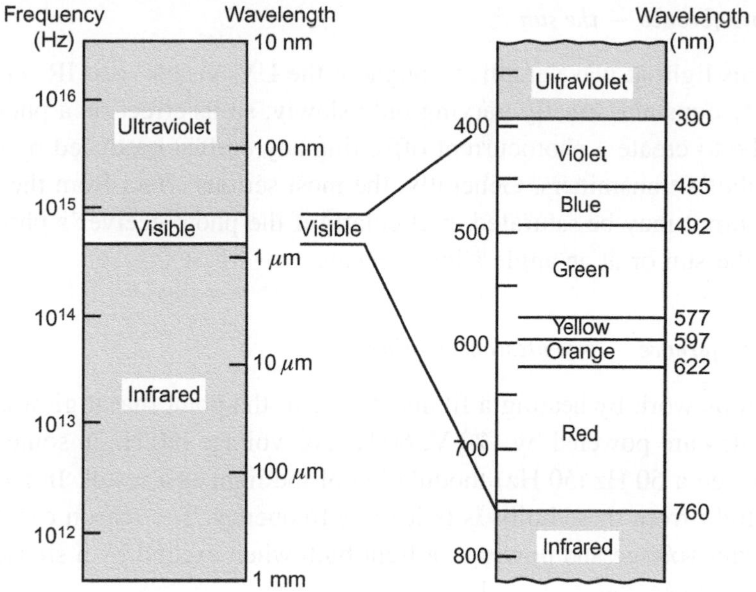

Figure 19.9
Optical frequencies and wavelengths (from Saleh and Teich).

categorized by its wavelength rather than its frequency. Remember from our discussion of piezoelectric ultrasonic sensors that wavelength has an inverse relationship to frequency, given by:

$$\lambda = vT = \frac{v}{f} = \frac{c}{f} \tag{19.4}$$

where λ is the wavelength, f is the frequency, and v is the velocity of propagation which, for light, is usually written as c.

The wavelength is what causes our eyes and our brain to perceive different colors. Figure 19.9 is a diagram showing the correspondence of color to wavelength. From this we can see that light that we perceive as yellow has a wavelength of approximately 580 nanometers.

Note also that the electromagnetic energy that we classify as light includes spectral components that our human eyes can't see. This invisible light includes ultraviolet (UV) (light with wavelength below 390 nm) and infrared (IR) (light with wavelength above 760 nm). This infrared portion of the spectrum, in particular, is important for robot applications, because many of the light sources and photodiodes that we might consider for sensors operate in this region, particularly in the near-infrared, from about 850 nm to 1 μm.

Ambient light interference – the sun

The sun produces light at wavelengths throughout the UV, visible, and IR regions. Its intensity is fairly constant, usually varying only slowly, so its effect on a photoreceiver in our robot will be to create a photocurrent offsetting any current produced by the optical pulses of the robot's transmitters. Generally, the most serious effect from the sun is that the photoreceiver output may be saturated, particularly if the photoreceiver's photodiode is "looking" into the sun or at an angle slightly away.

Ambient light interference – incandescent lighting

Incandescent bulbs work by heating a filament wire to the point that it glows. Because incandescent bulbs are powered by 120 V, 60 Hz AC voltage (50 Hz in some countries) we might expect to see a 60 Hz (50 Hz) modulation of the light as a result. In fact, the modulation of light from these bulbs is twice this frequency. The reason can be understood by considering the voltage and power to a light bulb when excited by a sinusoidal source.

The power at any instant in time (the instantaneous power) is given by:

$$P = \frac{V^2}{R} \tag{19.5}$$

where V is the voltage and R is the light bulb's resistance. The "120 V" for a 120 V AC source refers to the RMS (root mean square) voltage. Such a voltage actually has an instantaneous value, at time, t, given by

$$V(t) = \sqrt{2}\cdot 120\cdot \sin[2\pi(60t) + \varphi] \tag{19.6}$$

where φ is an arbitrary phase constant offset.

Combining Eqs. 19.5 and 19.6, the instantaneous power at any time, t, is found as:

$$P(t) = \frac{2\cdot 120^2\cdot \sin^2[2\pi(60t) + \varphi]}{R} = \frac{14400\cdot[1 - \cos(2\pi(120t) + \varphi)]}{R} \tag{19.7}$$

where we make use of the trigonometric identity:

$$\sin^2\alpha = \tfrac{1}{2}[1 - \cos(2\alpha)]. \tag{19.8}$$

Think about that last equation. The fact that, to calculate power, the voltage is squared, and therefore the sine function is squared, is what takes us from the line frequency (60 Hz) to twice the line frequency (120 Hz).

Figure 19.10 is a plot of the 120 V (RMS) waveform, along with the calculated power at each time instant. The resistance, R, is assumed to be 240 Ω. With this value for resistance, the light

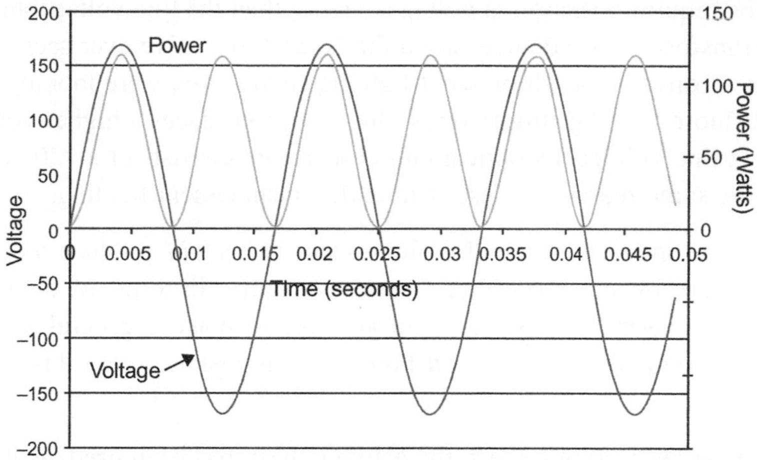

Figure 19.10
Voltage and power associated with a 60 W incandescent light bulb.

bulb consumes 60 W of *average* power. The incandescent light bulb produces its light output as a function of the electrical power in, so its light exhibits 120 Hz modulation, not 60 Hz.

The incandescent light bulb is a very powerful source of infrared energy. In fact, by far the largest part of the light output from an incandescent bulb is in the infrared region, not the visible region. The light bulb is a great source of heat and infrared energy and only a weak source of visible light, which is why it's so inefficient for the task of lighting.

Fortunately, the 120 Hz component of the incandescent's output is greatly attenuated. This is because, even though the electrical power into the bulb is changing at a 120 Hz rate, the bulb's filament temperature can't vary very much at such a high rate due to the thermal inertia of the filament. This inability of the filament to heat up and cool down at this relatively high frequency serves to filter most of the light modulation of the bulb. Consequently, most of the light from incandescent bulbs is constant optical power, due to the average electrical power input, and only a small component of the light from the bulb will be varying at the 120 Hz rate. The problem that incandescent bulbs' output will cause our robot, if they cause a problem at all, will be one of creating a current offset in our photoreceiver and, in extreme cases, saturating the photoreceiver's output.

Ambient light interference – fluorescent lighting

It is fluorescent lighting that is most likely to exhibit significant AC modulation of the light. Unlike incandescent bulbs, fluorescent bulbs typically exhibit significant modulation of the light. Our eyes and brains don't respond to modulation above a few tens of cycles per second, so we aren't aware of it, but it's definitely there.

Fluorescent bulbs require operation at voltages higher than the line voltage and this is produced by a transformer-like device called the *ballast*. If you've ever seen the rectangular electrical box in a conventional fluorescent lighting fixture, you were looking at the ballast. In conventional fluorescent lighting fixtures, this device produces a higher voltage version of the 60 (50) Hz line voltage. Its optical output therefore consists of a 120 (100) Hz component for the same reasons as described under incandescent lighting.

However, unlike the incandescent bulbs, fluorescent bulbs *will* produce a strong 120 Hz optical output. There are many types of fluorescent bulbs, in terms of their optical spectra (cool white, warm, etc.), so it is not possible to make a general statement about how much of their output is infrared and how much is visible, since this varies with the type of bulb.

In recent years, a newer type of ballast, the active ballast, has been used, as it allows the use of smaller ballast components. However, it switches at frequencies in the tens of kHz, which can be particularly difficult to counter.

Reducing the effect of ambient lighting

In order to allow our robot's optoelectronic sensors and controls to work in lighted areas, we'll need to come up with ways in which to counter the effect of ambient lights. There are at least five ways that we can do this. They are:

- Optical filtering
- Analog electrical filtering
- Digital filtering
- Synchronization to the interfering signal
- Directionality of the photoreceiver

Optical filtering to reduce ambient light effects

One straightforward way to reduce the effects of ambient light interference in optoelectronic systems is to limit the photoreceiver's optical spectrum to just the narrow range of wavelengths that includes the LED's wavelengths. For example, an OSRAM SFH4550 LED emits light over a narrow range of wavelengths (Figure 19.11).

Matched filters

The light from this LED peaks at around 860 nm and has virtually no output below 775 nm and above 900 nm. If we could optically filter the light incident on the photoreceiver to this narrow range of wavelengths, then, assuming that the ambient light source has its light spread over a very large range of wavelengths and does not have a

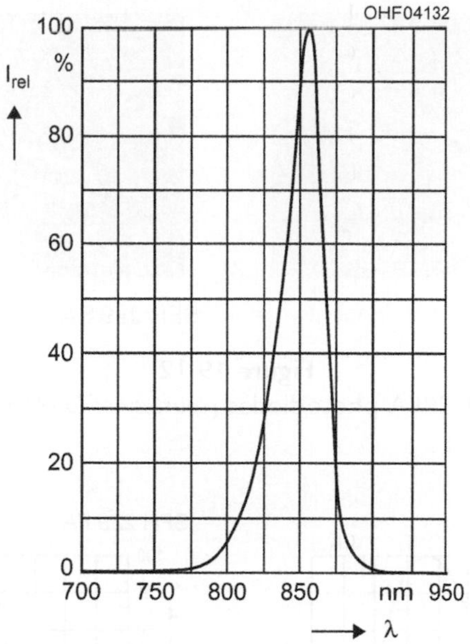

Figure 19.11
Relative spectral emissions of the SFH4550 LED *(courtesy of OSRAM Opto Semiconductors).*

lot of energy over this narrow range, the photoreceiver would see mostly the intended signal and very little of the interfering ambient signal. In effect, this greatly increases the signal-to-noise ratio. Another way of looking at this is that we have *matched* the receiver to the signal. The optical filter is a very simple form of what is referred to as a *matched filter.*

But how do we optically filter light? Actually, you've probably seen optical filters, although you may not have been aware that that's what they were. For example, stage lights at a play or concert will sometimes use a colored "gel" in front of the light to transmit a particular color. A red gel will transmit wavelengths, say, in the 620 nm and above, while absorbing the yellow, green, blue, etc. of the stage light's wavelengths below 620 nm. Red Jello will similarly absorb most colors while transmitting red.

Okay, but where do we get optical filters for this narrow range of near-infrared wavelengths? Fortunately, some photodiodes come with optical filters built in. For example, the OSRAM company makes photodiodes, the SFH229 and the SFH229FA with and without optical filtering, as shown in Figure 19.12.

These two photodiodes are identical except for the dark plastic lens on the SFH229FA versus the clear plastic lens on the SFH229. The effect of these different materials is shown in Figure 19.13.

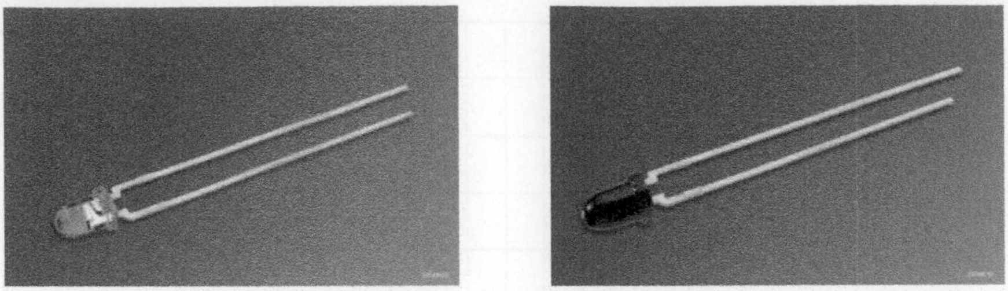

SFH 229 SFH 229 FA

Figure 19.12
The SFH229 and SFH229FA photodiodes *(courtesy of OSRAM Opto Semiconductors)*.

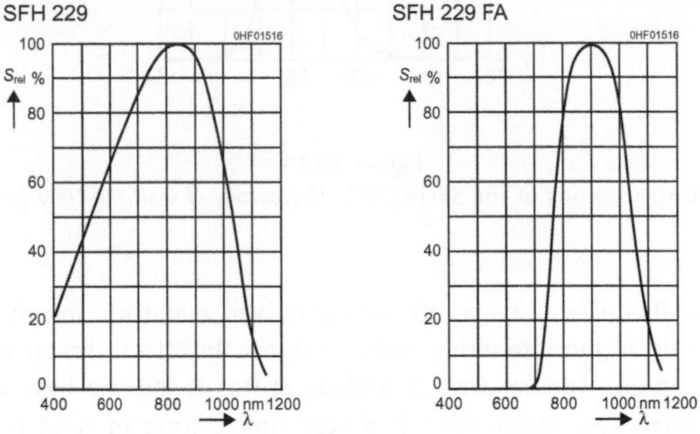

Figure 19.13
SFH229 and SFH229FA photodiode relative spectral sensitivity *(courtesy of OSRAM Opto Semiconductors)*.

Whereas the SFH229 has response over the entire visible range (in addition to the near-infrared region) the SFH229FA eliminates virtually all wavelengths below 700 nm. A simple part number change in our design ensures that optical filtering occurs.

Optical filtering is an excellent way to reduce interference from ambient lighting, but the ambient light sources will contain at least some energy in the near-infrared region, so additional measures are often needed to remove interference from these sources.

Analog electrical filtering to reduce ambient light

Is there a way to create a spoke counter signal that can be *electrically* matched to the photoreceiver? That is, can we create in the electronics a modulation of the optical waveform that will help differentiate it further from the background interference? If we can do that, we'll increase signal-to-noise still more and thereby improve chances that our circuit behaves the way we intended.

Remember that, at the beginning of this chapter, we talked about shining an LED at the rotating chrome wheel of the R/C car. The implication was that the LED would be left on all the time, like a flashlight. But there's no reason to leave the LED on continuously. Leaving it on continuously wastes power and, as we'll see, turning it on for just brief periods helps to create a signal that can be differentiated from the background.

We've already determined that our spoke counter needs a sample rate of at least 300 samples/second in order to reliably count the slots going by on the rotating wheel and that a rate as high as 1000 samples per second would possibly be even better. Each such sample consists of a sample-and-hold followed by the actual A/D conversion. Remember, from the chapter on A/D conversion, that this entire process is handled by the MSP430 A/D converter.

We simply set up the converter at the beginning of the robot's program and, after that, just set the appropriate bits (ENC and ADC10SC in the register ADC10CTL0) to initiate each A/D sample. If we select the MSP430's built-in A/D clock, ADC10OSC (remember, this is approximately a 5 MHz clock), set the sample-and-hold interval for 64 clock cycles and operate the other instructions at about a 1 MHz clock rate (MCLK set to its default values), then the entire conversion process will take roughly 30 μsec (based on measurements I've made). We need only have the LED on from the beginning of this conversion time until the end of this conversion time.

Besides saving lots of LED power (the LED is only on about 3% of the time with a 30 μsec conversion time performed every 1 msec), this means that we can AC-couple the output of the resistor or transimpedance amplifier, and transfer mostly signal and very little of the low-frequency ambient light.

Digital filtering to reduce ambient light

There are digital filtering techniques that can be used to reduce interference effects. Most of these are either advanced techniques beyond the scope of this book or require a computer with more "horsepower" than the MSP430.

There is one digital filtering technique that is both simple and for which the MSP430 is well-suited. Figure 19.14 shows an LED drive pulse (top waveform) and a photoreceiver's response. The photoreceiver is receiving this signal in the presence of strong 120 Hz fluorescent light interference.

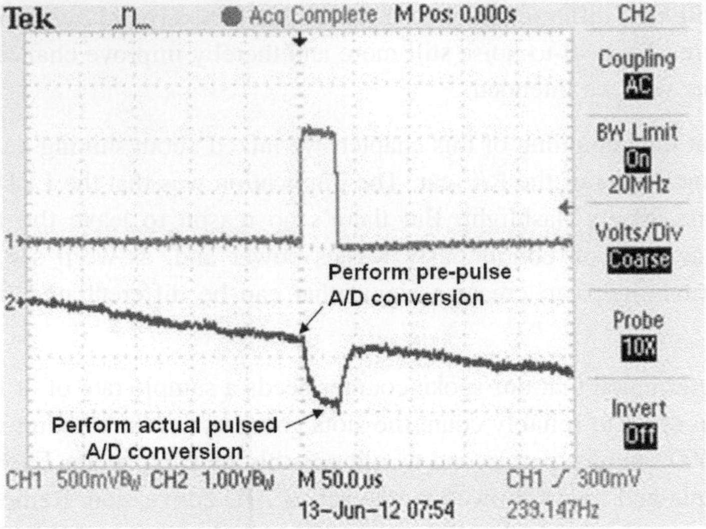

Figure 19.14
LED drive pulse (top waveform) and received signal in the presence of 120 Hz interference.

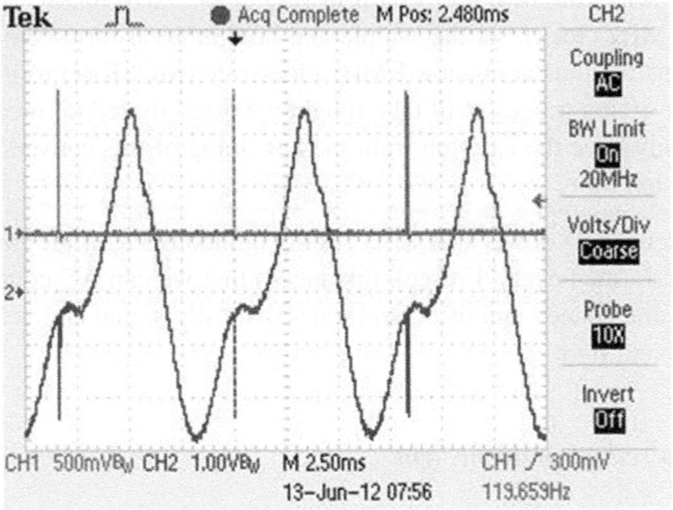

Figure 19.15
Signal synchronized to strong ambient interference.

By making an A/D conversion prior to pulsing the LED, a measurement of just the response to the ambient light is made. Since the interfering ambient light varies at a rate that is low relative to the pulse width, the change in ambient light level is very small during the pulse

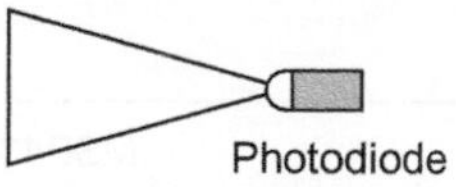

Figure 19.16
For a signal whose general direction is known, narrow-field-of-view photodiodes reduce interference.

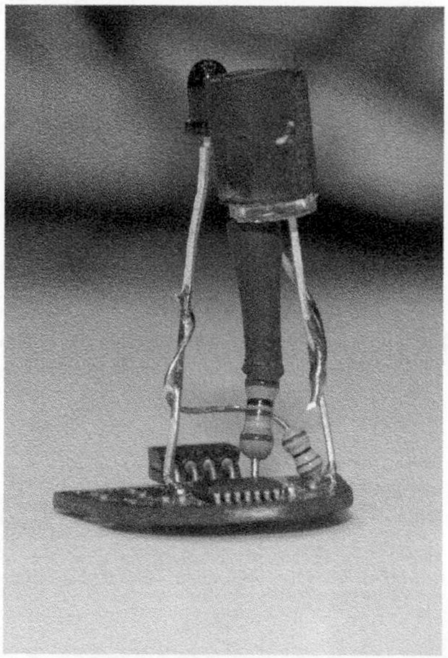

Figure 19.17
The TI EZ430-T2012 circuit card with LED and photodiodes added.

interval. Subtracting the pulsed measurement from the unpulsed measurement therefore eliminates almost all of the ambient light effect.

Synchronizing the samples with the interference

Another technique that works well with interfering ambient lighting is to synchronize measurements with the interfering source. For example, if we know that the main source of optical interference is going to be 120 Hz from fluorescent lights, we could make measurements at rates of 120 Hz and submultiples of 120 Hz (60 Hz, 30 Hz, etc.). This won't work for the spoke counter, where the sampling rate needs to be higher than 120 Hz, but for some of the applications that we'll talk about in subsequent chapters, this technique works quite well. Figure 19.15 shows another situation in which a pulsed LED signal

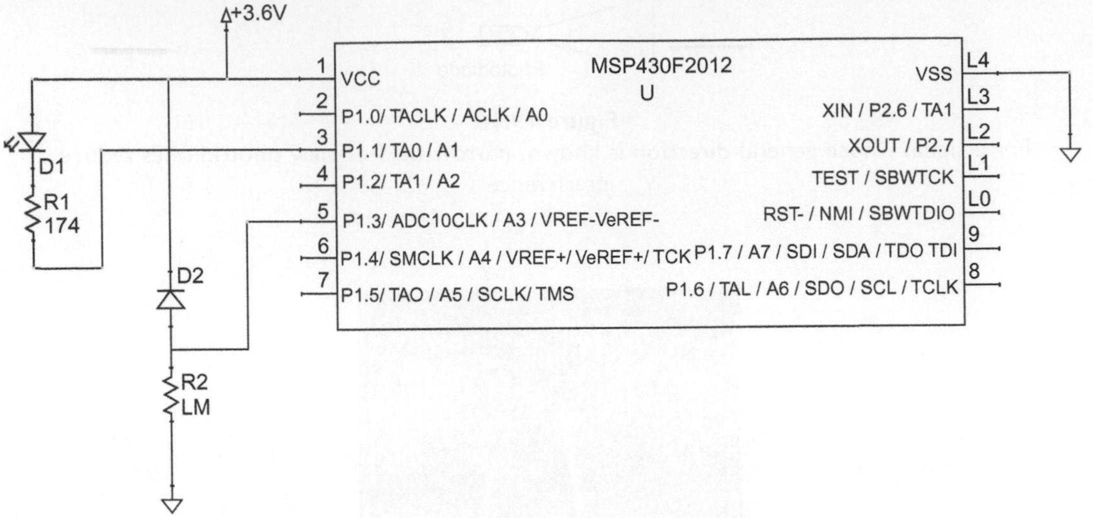

Figure 19.18
The EZ430-T2012 board with added components.

Figure 19.19
EZ430 circuit showing the black/white encoder wheel.

(top waveform) was received in the presence of huge 120 Hz interference. In spite of this extremely large interference, the measurements, at each instance, can be compared relative to one another, since the interference level is always virtually unchanged.

Reducing the photodiode's field of view to reduce interference

If the direction that the signal will be coming from is known, then there is yet another way to reduce optical interference from ambient lighting. This is to choose a photodiode with a narrow field of view (see Figure 19.16). In this way, we use the additional information about the signal light's direction to help reduce interference.

(a)

```
;
#include  "msp430f2012.h"
;------------------------------------------------------------------------------
LogicState  EQU     0200h
NmbrCycls   EQU     0202h
NmbrTrns    EQU     0204h
Speed       EQU     0206h
Threshold   EQU     0208h
ThrshL      EQU     020Ah
ThrshH      EQU     020Ch
Previous    EQU     020Eh
;------------------------------------------------------------------------------
            ORG     0F800h                  ; Program Reset
;------------------------------------------------------------------------------
RESET       mov.w   #0280h,SP               ; Initialize stackpointer
            call    #Initialize
Loop
            clr     &TAR
            bic.w   #ADC10IFG,&ADC10CTL0
            bic.b   #2,&P1OUT               ; Turn the LED on
            bis.w   #ENC+ADC10SC,&ADC10CTL0 ; Start the conversion
Flag
            bit     #ADC10IFG,&ADC10CTL0
            jz      Flag
            mov     &ADC10MEM,R6
            bic     #ENC+ADC10SC,&ADC10CTL0
            bis.b   #2,&P1OUT
            call    #count
delay
            cmp.w   #1000,&TAR
            jl      delay
            clr     &TAR
            jmp     Loop                ; Again
```

Figure 19.20
(a) Speed-measurement program (first part).

Speed measurement example

As a wrap-up to this chapter, here's a demonstration of the speed measurement techniques. This demo uses a tiny circuit board, the Texas Instruments EZ430-T2012, which can be used with the EZ430-F2013 development system and which is shown in Figure 19.17. The EZ430-F2013 is similar to the LaunchPad board that TI sells, but doesn't accept DIP versions of the microcontroller.

Note the shroud for the LED, which was made from a piece of ¼" polystyrene tubing painted flat black.

The simple circuit for this version of the microcontroller is shown in Figure 19.18. D1, the LED, is an OSRAM SFH4550 and D2, the photodiode, is an OSRAM SFH229FA. The SFH229FA photodiode has a 13 pF capacitance and a $\pm 17°$ field of view.

(b)

```
count
            ; count is the subroutine to determine the number of black/white
            ;    in approximately one second
            ; LogicState looks at the A/D reading and assigns one of two
            ;    logic levels based on that reading - 1, 0
            clr     &LogicState
            cmp     &Threshold,&ADC10MEM
            jl      Zero
            mov     &ThrshL,&Threshold
            inc     &LogicState
            tst     &Previous
            jnz     ComputeSpd
            inc     &NmbrTrns
            jmp     ComputeSpd
Zero
            mov     &ThrshH,Threshold
ComputeSpd
            inc     &NmbrCycls
            cmp     #1024,&NmbrCycls
            jl      Finish
            mov     &NmbrTrns,&Speed
            clr     &NmbrTrns
            clr     &NmbrCycls
Finish
            mov     &LogicState,&Previous
            ret

;-------------------------------------------------------------------------------
SetupADC
            mov.w   #SREF_0+ADC10SHT_3+ADC10SR+ADC10ON,&ADC10CTL0
            mov.b   #8,&ADC10AE0              ; Input on P1.3
            mov.w   #INCH_3,&ADC10CTL1
            ret

;-------------------------------------------------------------------------------
Initialize
            mov.w   #WDTPW+WDTHOLD,&WDTCTL    ; Stop watchdog timer
            bis.b   #0FFh,&P1DIR              ; Make all port 1 bits outputs
            bic.b   #1,&P1OUT
            mov.w   #TASSEL_2+MC_2,&TACTL
            call    #SetupADC
            clr     &NmbrCycls
            clr     &NmbrTrns
            clr     &Previous
            mov     #0D0h,ThrshL
            mov     #120h,ThrshH
            ret

;-------------------------------------------------------------------------------
;           Interrupt Vectors Used MSP430x2013
;-------------------------------------------------------------------------------
            ORG     0FFFEh                    ; MSP430 RESET Vector
            DW      RESET                     ;
            END
```

Figure 19.20

(b) Speed-measurement program (second part).

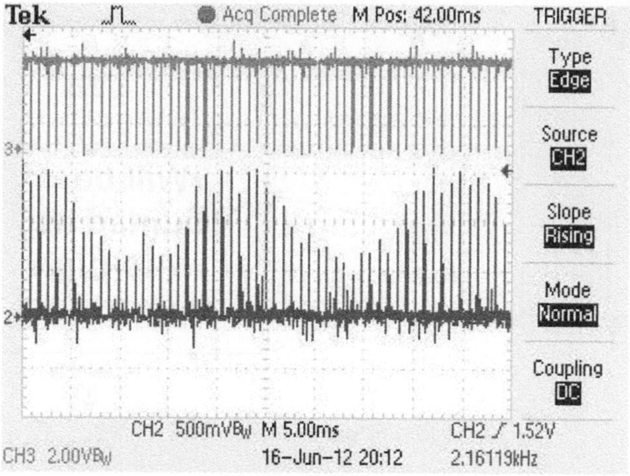

Figure 19.21
P1.1 (LED driver) waveform and P1.3 (photoreceiver) waveform.

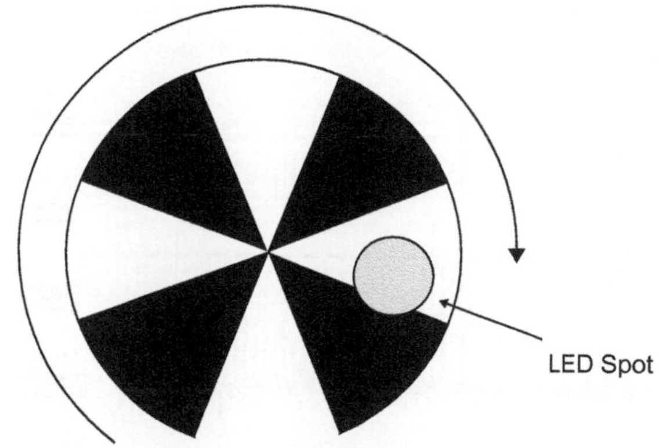

Figure 19.22
LED spot causes slow transitions.

The EZ430-T2012 board, along with the EZ430-F2013 development system, is shown in Figure 19.19. Note that this particular speed-measurement system uses an encoder wheel, like the one in Figures 19.4 through 19.6, although this one has five black/white segments, rather than the four black/white segments of those figures.

The program written to count the number of segment changes going by is in Figure 19.20. When this program is run, it produces the LED drive waveform (top) and the photodiode response (bottom) (Figure 19.21).

One thing to note about this is that the response waveform does not change immediately from its high state to low state or vice versa (that is, the envelope is not a square wave) but changes gradually and looks sinusoidal. Why is that?

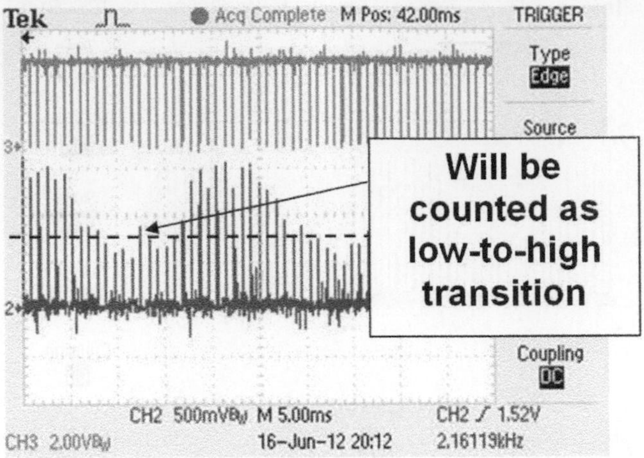

Figure 19.23
Noise in the speed response causes extra counts.

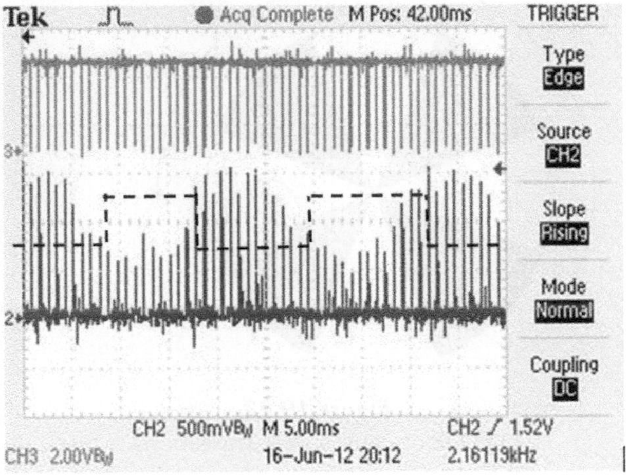

Figure 19.24
Hysteresis avoids false transition detections.

The reason is that the LED spot is large enough that it overlaps segments. Therefore transitions from white to black, or vice versa, cause the reflected signal to change slowly (Figure 19.22).

One other thing to note about the Figure 19.21 waveforms – the photodiode response envelope is not as smooth as might be expected. This is most likely due to the motor's occasional hesitation or perhaps due to some vibration that causes the LED spot to move. Whatever the cause, it presents some problems in trying to produce a simple algorithm for

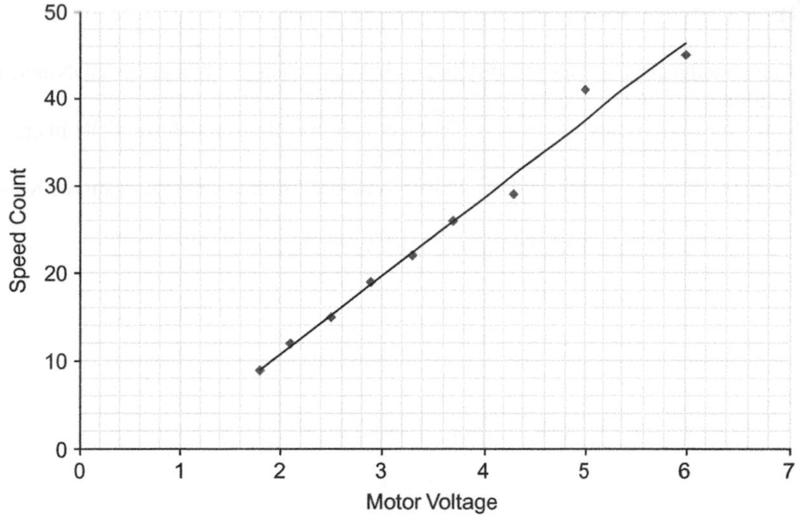

Figure 19.25
DC motor voltage vs. speed count.

counting wheel revolutions. Let's say that we are counting low-to-high transitions and that we set the threshold for what constitutes a low state and what constitutes a high state as in Figure 19.23.

Hysteresis

A straightforward solution to this problem is what is known as *hysteresis*. With hysteresis, the threshold is varied depending on the state that the output is estimated to be in. This can be seen in Figure 19.24, where the Figure 19.21 data has been taken and a hysteretic threshold superimposed on it. When the first low-state region is found on the oscilloscope screenshot, the threshold is increased to a fairly high level. Noisy "glitches" thus do not change the output state.

To test the speed counting, the circuit of Figure 19.18 along with the Figure 19.20 program were used while running the motor at various DC voltages. The Figure 19.25 data shows a linear relationship between motor DC voltage and the speed count.

Note that the speed count is the number of black-to-white transitions counted. Since there are five such transitions per wheel revolution and since there are approximately two wheel revolutions per foot traveled, the highest speed count of 45 corresponds to approximately 4.5 ft/sec.

Bibliography

[1] B. Baker, Transimpedance-amplifier stability is key in light-sensing applications. EDN magazine, September 4, 2008.

[2] B. Baker, Photo-sensing circuits: the eyes of the electronic world are watching. EDN magazine, August 7, 2008.

[3] R. Otte, L. de Jong, A. van Roermund, Low-Power Wireless Infrared Communications, Kluwer Academic, 1999.

[4] B. Saleh, M. Teich, Fundamentals of Photonics, John Wiley & Sons, 1991.

CHAPTER 20

Creating High Voltage

Chapter Outline

In previous chapters we've discussed voltages of, for example, 15 V and higher. Yet, the radio-controlled cars that we will modify typically use 6 V or 7.5 V battery packs. Where did the extra volts come from?

Recall that inductors and capacitors have the unique properties of maintaining current (in the case of inductors) or voltage (in the case of capacitors). That is, an inductor, operating at some given current, will resist changing that current and a capacitor, operating at some given voltage, will resist changing that voltage. We can use these components with those particular properties to design circuits capable of creating output voltages higher than the input voltage.

The circuits described in this chapter are referred to as DC-DC converters. They take a DC input and convert that voltage to a supply voltage that is higher or lower than the input voltage (in the case of the robots in this book, we are interested in those DC-DC converters that increase the voltage).

The charge pump

The charge-pump circuit uses capacitors to achieve higher voltages. The simplest such circuit is a *voltage doubler*. The circuit has two states, which it continually switches between. The first state (the one depicted in Figure 20.1) is the charging state. In this state capacitor *C1* (sometimes referred to as the flying capacitor) charges to V_{IN}.

MSP430-based Robot Applications.
DOI: http://dx.doi.org/10.1016/B978-0-12-397012-1.00020-5

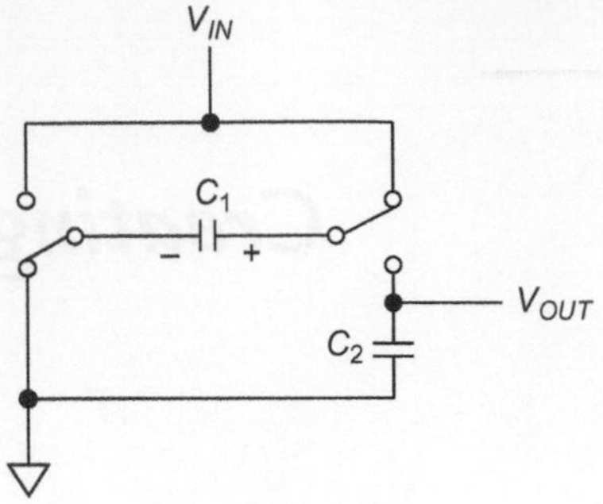

Figure 20.1
The voltage doubler.

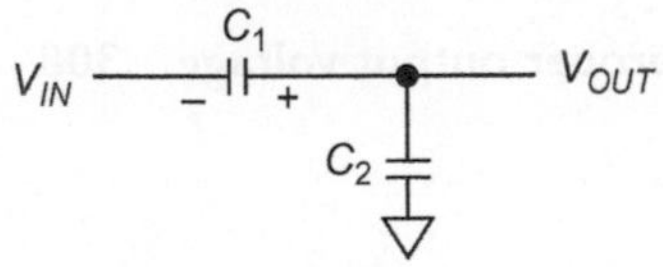

Figure 20.2
Voltage doubler, charge transfer state.

The second state, the charge transfer state, has the two switches in their opposite configuration (left switch up, right switch down). What happens in that case? Figure 20.2 depicts the circuit. Assume that C_2 is initially at 0 V. C_1, which had V_{IN} volts across it prior to the switch, transfers some of its charge to capacitor C_2. As a result, the voltage across C_2 rises while the voltage across C_1 falls. The output voltage (the C_2 voltage) rises to a value between V_{IN} and $2 \cdot V_{IN}$.

The circuit is then commanded to revert to its charging state (Figure 20.1) and the voltage across C_1 is replenished to V_{IN} (the voltage across C_2 remains at its charged value, assuming no load). When it is then commanded to the charge transfer state, C_1 again transfers some of its charge to C_2. After this charge transfer, the voltage across C_2 is higher than during the previous charge transfer state. As this process continues, the output voltage gradually approaches its final value of $2V_{IN}$. Once the circuit gets past the initial transient voltage buildup, this circuit maintains the $2V_{IN}$ output value.

Of course, we assumed no load on V_{OUT} – not a very practical assumption for a circuit intended to provide a supply voltage to some load. Nevertheless, as long as the load is relatively light, the circuit will maintain very nearly double the input voltage.

Charge-pumped circuits similar to the voltage doubler can be built to provide higher voltages, usually integer-multiples of the input voltage (that is, 2×, 3×, etc.). There are also many integrated circuit solutions available for implementing this circuit.

Inductor-based boost circuits

Inductor-based switching converters are today extremely common. There are many different types of inductor-based DC-DC converters but the one that is the focus of this chapter (and used in the robot project) is the boost regulator. This circuit is shown in Figure 20.3. This circuit, like the charge pump discussed in the last section, takes the input voltage and boosts it to a higher voltage. How does it do that?

Remember that the inductor increases its current according to the equation:

$$\Delta I = \frac{V_L}{L}\Delta T = \frac{V_{IN}}{L}\Delta T. \tag{20.1}$$

So, if the switch in Figure 20.3 is closed at $t = 0$ and if the initial current in the inductor is 0 then, after ΔT seconds, the current has increased to:

$$I_P = \frac{V_{IN}}{L}\Delta T. \tag{20.2}$$

Now the switch is opened. What happens? I_L initially has no path to ground, so it starts to decrease. A decreasing I_L causes V_L to become negative. But, since V_L is referenced with its positive side to V_{IN}, this means that the voltage on the " – " side of V_L – the junction of the inductor, diode, and switch – becomes positive with respect to V_{IN}. If the sequence just described seems confusing, take some time to go through the Figure 20.3 diagram, keeping in mind the way the voltage and current are referenced.

Once the switch opens, the diode's anode voltage rapidly rises until it is one diode drop above the capacitor voltage, at which point the diode turns on and the inductor "dumps" its current into the capacitor. This switch-on/switch-off cycle is repeated continuously.

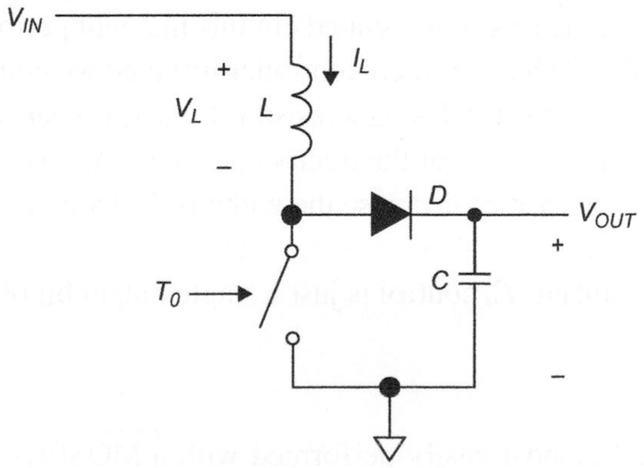

Figure 20.3
An inductor-based boost circuit.

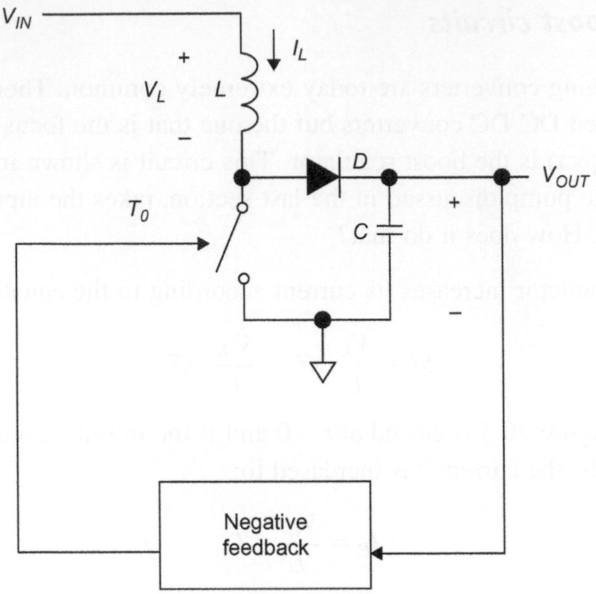

Figure 20.4
Boost circuit with negative feedback.

Once the energy delivered from the inductor equals the energy consumed by the load plus any additional losses, the output is in equilibrium.

If the input voltage changes somewhat or if the load current demand changes, we'd still need the boost circuit to maintain the proper voltage. To do that, the output voltage must be sensed and the time-on, T_O, adjusted to produce a lower or higher output voltage. That calls for a negative feedback loop, like the one shown in Figure 20.4.

Basically, negative feedback, in this case, means reducing T_O if V_{OUT} is too high and increasing T_O if V_{OUT} is too low. There are lots of integrated circuits that will perform this function. However, since we will already have a microcontroller installed we could really just perform this function with the microcontroller. Using a resistor divider, we can simply divide the output voltage down to a voltage in the range of the microcontroller's A/D converter, then make the decision about whether to increase or decrease the width of T_O based on whether the converted analog voltage is less than or greater than some stored target number corresponding to the voltage that we wish to maintain. T_O control is just a single output bit of the microcontroller.

The switch

The switch of Figure 20.4 is most easily performed with a MOSFET. Keep in mind that, even for an application that delivers only a few milliamperes to the load, the peak switch current may reach of the order of 1 ampere. Of course, this will only last a few

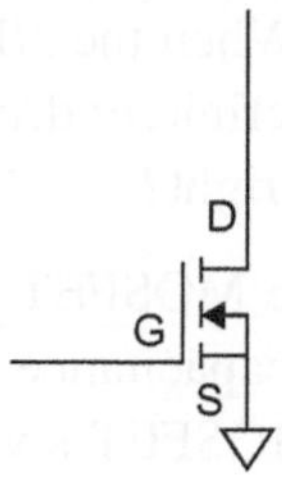

Figure 20.5
The N-channel MOSFET, a candidate for the switch.

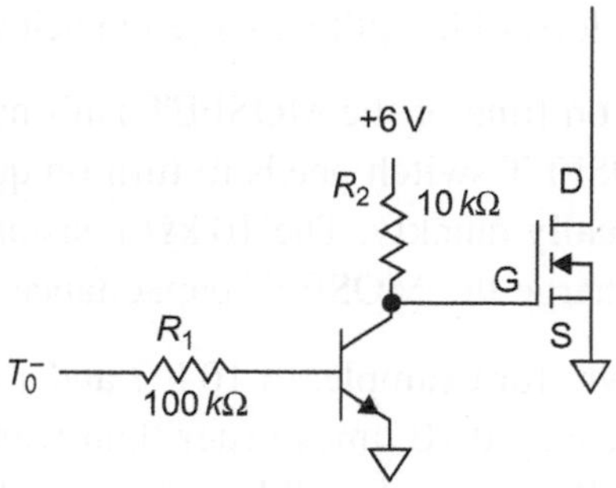

Figure 20.6
Adding an NPN transistor driver stage.

microseconds typically, so the device does not get hot, but it's important to choose a MOSFET with relatively low on-resistance, so that the full (or nearly full) V_{IN} is applied across the inductor. Figure 20.5 shows an N-channel enhancement-mode MOSFET that has the makings of such a switch.

When the gate (G) of this transistor is driven high with respect to the source (S), the drain-to-source path closes. When the gate is driven to ground, so that it's at the same voltage as the source, the drain-to-source path opens. Wow! That's exactly what we were looking for. Unfortunately, as with lots of things in electronics design, it's not quite that simple.

For one thing, most MOSFETs with low on-resistance and capable of handling high levels of current will need to have their gate voltage taken to a voltage well above the 3 or so volts of the MSP430 high logic state. Okay, we could fix that by adding an NPN transistor stage (Figure 20.6).

In this case, the logic signal from the MSP430 is the inverse of T_0, that is, T_0-, since a low on this signal drives the MOSFET switch to its on state. When this signal is high, about 20 μA will flow into the NPN transistor's base. If the NPN transistor's β is 30 or higher, it will be driven into saturation and the NPN transistor's saturation voltage will be low enough to cause the MOSFET to open. When the NPN is off (by setting the T_0- logic signal to zero) the gate will go to 6 V, definitely driving the MOSFET output low, with very low on-resistance. Problem solved, right?

Well, there is an additional problem. The MOSFET gate has considerable capacitance. It isn't just the MOSFET's gate-to-source capacitance. There is a gate-to-drain capacitance as well and that, in combination with the MOSFET's voltage gain, forms something called a Miller's capacitance. The topic of Miller's capacitance is beyond the scope of this book but suffice it to say that it produces an effective capacitance much larger than just the gate-to-drain capacitance. The upshot is that a MOSFET gate, while presenting an extremely high DC resistance, looks like quite a large capacitance.

In some applications, the turn-on time of the MOSFET isn't much of an issue, but in the case of a boost circuit the MOSFET switch needs to turn on quickly (for example, a few microseconds) and turn off equally quickly. The 10 kΩ resistor in the Figure 20.6 circuit just isn't going to be able to charge the MOSFET capacitance in those kinds of time.

R_1 and R_2 could be scaled down, for example, to 10 kΩ and 1 kΩ respectively, so that the MOSFET gate capacitance is charged 10 times faster than with the Figure 20.6 circuit. But keep in mind that the MOSFET switch will be off a very large percentage of the time (>99%). And to keep the MOSFET switch off, the NPN transistor will need to be on (saturated). So this drive circuit needlessly wastes battery energy to hold the MOSFET off.

A fairly simple solution to this problem is to use a PNP transistor to drive the MOSFET gate (Figure 20.7). The PNP, when driven with sufficient base current, will rise quickly to its saturation emitter-to-collector voltage (approximately 0.5 V), so that the MOSFET gate is pulled up to about 5.5 V.

Like the NPN version of the gate driver, this one dissipates no energy in one state (in this case, when T_0- is high) and dissipates a relatively large amount of energy in the other state

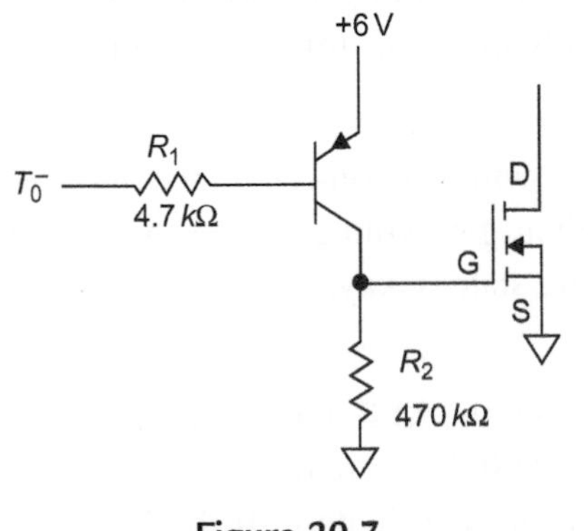

Figure 20.7
The PNP transistor as MOSFET gate driver.

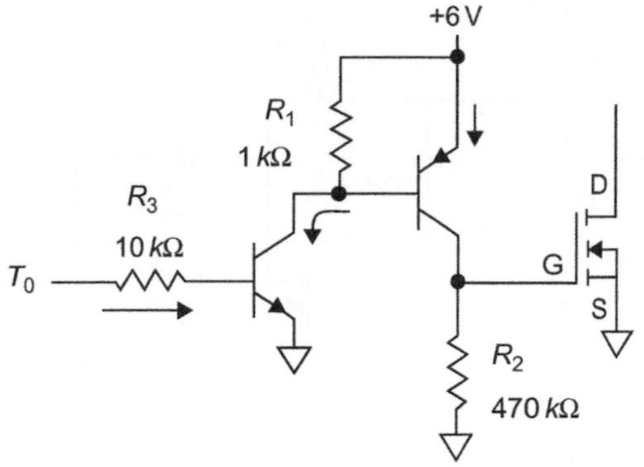

Figure 20.8
Combining an NPN and a PNP transistor as MOSFET gate driver.

(the current will be approximately 12 mA when T_0- is low and, therefore, the gate is driven high). However, in this case, the high-power-dissipation state is the high-gate-voltage state, which is the one that's only on for a very low percentage of the time (typically $< 1\%$).

Well, there's a problem with this arrangement as well – when we turn the PNP transistor off, the signal T_0 must be at or near $+6$ V. But T_0 is generated by the MSP430 microcontroller and its high state won't be much above 3 V. What to do? Let's combine an NPN and a PNP (Figure 20.8).

This circuit meets all of our requirements – it can be driven on and off from the low-voltage MSP430, it drives the MOSFET gate to greater than 5 V, it charges and discharges the gate capacitance quickly, and it dissipates very little power when the duty cycle is low. Note that, for this circuit, the MOSFET gate is high when T_0 is high.

Also, note that the pull-up resistor, R_1, doesn't serve much of a role in this configuration – its only function is to turn the PNP transistor off faster than would occur if we left R_1 out (Figure 20.9).

Note that capacitors C_1 and C_2 are each shown as a single capacitor. In actual practice, each of these capacitors is made up of a 100 μF aluminum electrolytic capacitor in parallel with a 10 μF ceramic capacitor. The reason for this has to do with how different capacitors behave over frequency.

Figure 20.10 shows the impedance of an aluminum electrolytic capacitor as a function of frequency. The impedance of an ideal capacitor obeys the equation:

$$Z_C = \frac{1}{2\pi f C} \tag{20.3}$$

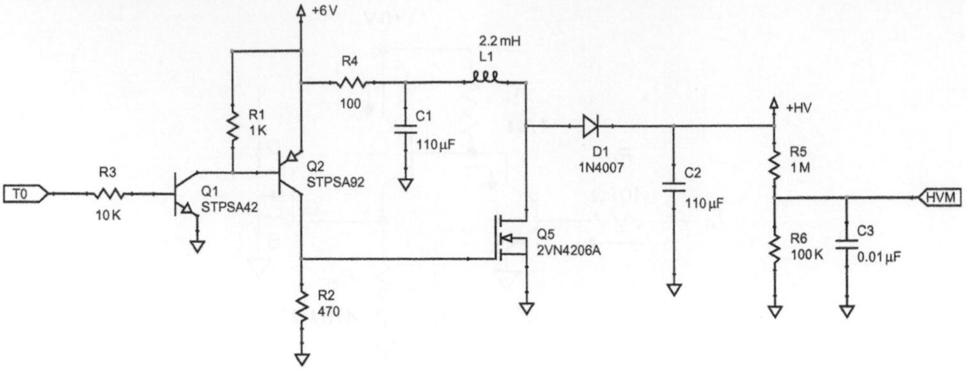

Figure 20.9
Entire boost circuit.

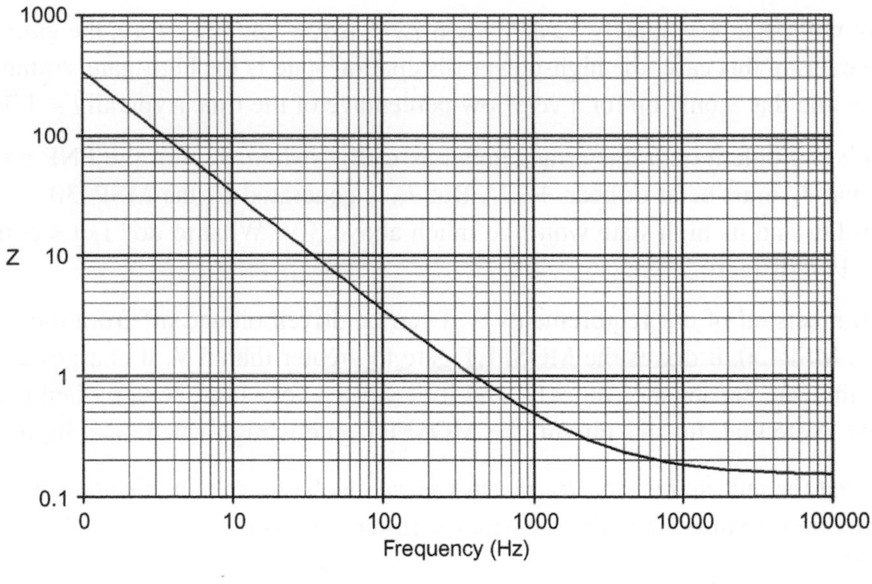

Figure 20.10
Impedance of a 470 µF capacitor as a function of frequency *(courtesy EPCOS).*

for frequencies from 0 to infinity. If a capacitor follows this equation then the impedance will be inversely proportional to frequency. But, as frequency is increased, real capacitors eventually stop obeying Eq. 20.3. The Figure 20.10 graph shows the impedance of a 470 µF capacitor. The capacitor follows Eq. 20.3 fairly well up to about 10 kHz. Past that frequency, the impedance starts to deviate from the predicted behavior and, by 100 kHz, is completely flattened out (a flat region means that the capacitor basically looks like a small resistance at those frequencies).

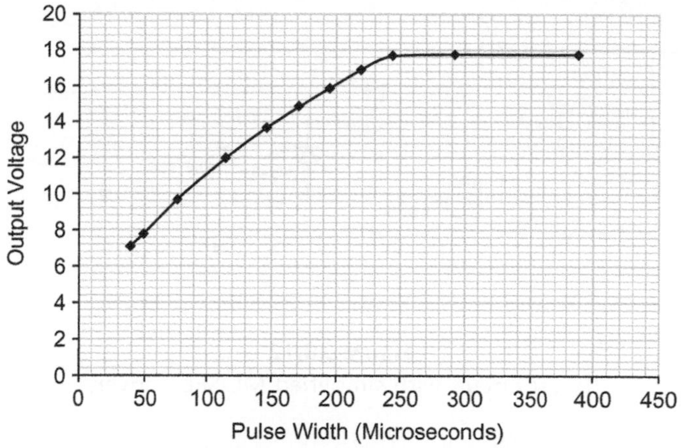

Figure 20.11
Output voltage as function of input pulse width.

The aluminum electrolytic capacitor is a very inexpensive critter but, unfortunately, where high-frequency response is important (and our boost circuit is such an application) it won't hack it. Ceramic capacitors can generally behave like, well, a real capacitor, to much higher frequencies. But a ceramic 100 μF capacitor capable of handling voltages of the order of 20 V will set you back about $15 to $20! To get something close to the ceramic 100 μF performance at a small fraction of the price, an inexpensive aluminum electrolytic 100 μF capacitor is operated in parallel with an inexpensive ceramic 10 μF capacitor to form a combination that has good high-frequency response without breaking the bank.

Boost circuit performance

The Figure 20.9 circuit was operated with various pulse widths, to determine how its output voltage varied with pulse width. The only additions to the circuit were an 18 V Zener diode and a 20 kΩ resistor, both in parallel with C2. The results of the test are shown in Figure 20.11.

The limit of 18 V at pulse widths greater than 240 μsec is the result of the Zener diode, which was added to hard limit the output in the event that the pulse width is inadvertently allowed to be on too long.

A simpler MOSFET driver

The two-transistor driver circuit explained in the previous section has a number of things going for it, including:

- Low cost
- Low power
- Fast operation

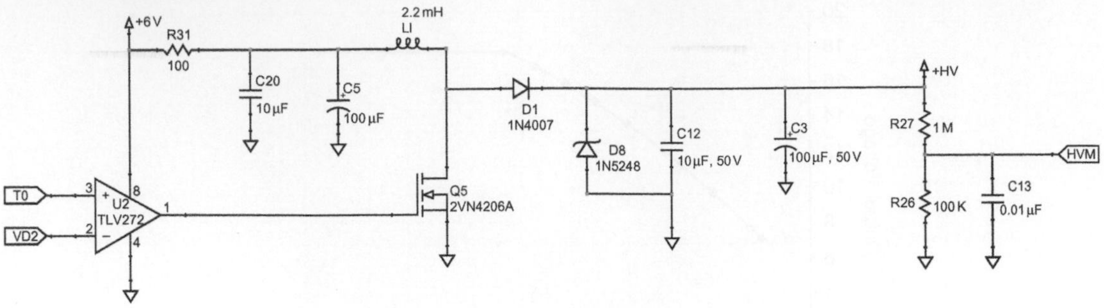

Figure 20.12
Boost circuit with simplified MOSFET driver.

However, it is not as simple as it might be. If we think about the problem of driving the MOSFET, this is really no different from driving the ultrasonic transmitter. In both cases, what we really need is to *level-shift* the logic voltages, from 0-to-3 V at the input side, to 0-to-something-higher at the output side. A solution that costs a little more, uses a little more power and is a little slower to turn on and off, but which is much simpler, is shown in Figure 20.12. Instead of the combination NPN/PNP transistor driver, this circuit uses half of a TLV272 dual op amp to drive the MOSFET gate. *VD2* is a voltage that is half the microcontroller voltage (approximately 1.5 V). The op amp acts as a comparator, driving the gate to 6 V whenever T_O is greater than *VD2*.

The TLV272 output can change its output at a maximum rate of about 2 V/μsec. This maximum rate of change is known as the op amp's *slew rate*. Such a slew rate means that the op amp will require about 3 μsec to turn the MOSFET on and an equal amount of time to turn the MOSFET off.

Note that this version of the schematic shows the large capacitors as the actual combination of ceramic and aluminum electrolytic capacitors. It also shows the Zener diode that is added to hard limit the output in the event that T_O is on too long.

Negative feedback maintains the proper output voltage

Figure 20.11 shows how the boost circuit's output voltage changes with the input pulse width. Of course, this was for a constant output load of 20 kΩ and with a constant 6 V input voltage. If we always had exactly this load resistance and this input voltage, we could just look things up on the Figure 20.11 plot, and choose the appropriate pulse width to achieve the desired output voltage (for example, 175 μsec if the output voltage is supposed to be 15 V).

Unfortunately, things aren't quite that easy (if you haven't noticed, this is a common theme when talking about electronics design). The load will vary and the input voltage will be some range of voltages (the circuits shown in this book were developed for operation at 6 V, but the design is intended to accommodate higher voltages, such as 7.5 V). We will need to vary the pulse width as the load changes and as the input voltage changes. There are a lot of integrated circuits available to handle this control issue but, fortunately, we have a microcontroller in the design that is eager to do the job.

As mentioned in a previous chapter, control systems can be very complex. There are a number of very important issues, including stability, speed of response, overshoot, etc. that have to be considered in many applications. Our application is simple enough that we can get away with just monitoring the high voltage, comparing that to a target voltage, then increasing or decreasing the pulse width to increase or decrease the high voltage. The program is shown in Figure 20.13.

Before looking at the control loop part of the program, let's look at how the microcontroller's peripherals are set up:

- The digital output, T_O, is generated at the microcontroller P1.5 pin
- The clock is set up to generate a calibrated 8 MHz
- The timer divides the clock by 8, so that the input is 1 MHz
- The A/D
  - Samples channel 4 (on P1.4)
  - Uses the 2.5 V internal reference
  - Uses the internal 5 MHz ADC10CLK clock
  - Samples the input for 64 clock cycles

The control loop starts at the "MainLoop" label of the program. The first thing that is done is to zero out the timer (the "clr &TAR" instruction). We'll need the timer to determine when our 16.67 msec interval is over.

Next, the A/D conversion is performed. The program doesn't advance until the ADC10IFG flag is set in the ADC10CTL0 register. Note from the circuit schematics that we're looking at a divided-down version of the high voltage. So the voltage that is sampled is:

$$HVM = \frac{100k}{1M + 100k} HV \tag{20.4}$$

If we desire HV to be 15 V, then we should be looking for HVM to be 1.364 V. So, we will want to increase the T_O pulse width if HVM is less than 1.364 V, otherwise we should decrease the pulse width.

But what digital number does 1.364 V correspond to? As we saw, the A/D is set up to use a 2.5 V reference. This is the maximum voltage that we can convert, so this voltage

(a)

```
;********************************************************************************
#include  "msp430G2452.h"
;--------------------------------------------------------------------------------
Current    EQU     0200h
;--------------------------------------------------------------------------------
           ORG     0F800h                      ; Program Reset
;--------------------------------------------------------------------------------
RESET      mov.w   #0280h,SP                   ; Initialize stackpointer
           mov.w   #WDTPW+WDTHOLD,&WDTCTL      ; Stop watchdog timer
           bis.b   #020h,&P1DIR                ; P1.5 output direction
           mov.w   #TASSEL_2+MC_2+ID_3,&TACTL
           mov.w   #SREF_1+ADC10SHT_3+ADC10SR+REF2_5V+REFON+ADC10ON,&ADC10CTL0
           mov.w   #INCH_4,&ADC10CTL1          ; A/D input is on P1.4
           mov.b   &CALBC1_8MHZ,&BCSCTL1
           mov.b   &CALDCO_8MHZ,&DCOCTL
           mov     #1,&Current
MainLoop
           clr     &TAR
Read_V
           bic     #ADC10IFG,&ADC10CTL0
           bis     #ENC+ADC10SC,&ADC10CTL0 ; Start the conversion
TestADC
           bit     #ADC10IFG,&ADC10CTL0        ; Check conversion status
           jz      TestADC
           cmp     #022Eh,&ADC10MEM            ; 22E hexadecimal corresponds to 15V
           jge     GreaterThan
           inc     &Current
           jmp     CloseSwitch
GreaterThan
           dec     &Current
CloseSwitch
           call    #Clipping
           mov     &Current,R8
hold_high
           bis.b   #020h,&P1OUT                ; Set P1.5 output
again
           dec     R8                          ; Decrement R8
           jnz     again
           bic.b   #020h,&P1OUT                ; Clear P1.5 bit
           call    #delay                      ; Wait for next 60 Hz start
           jmp     MainLoop                    ; Jump to the top

;--------------------------------------------------------------------------------
```

Figure 20.13

(a) First part of the high-voltage program. (b) Second part of the high-voltage program.

(b)
```
;------------------------------------------------------------------------------
Clipping
                cmp     #2,&Current
                jge     TestForTooLarge
                mov     #2,&Current
                ret
TestForTooLarge
                cmp     #2000,&Current
                jge     ClipHigh
                ret
ClipHigh
                mov     #2000,&Current
                ret

;------------------------------------------------------------------------------
delay
                cmp.w   #16600,&TAR
                jl      delay
                ret
;------------------------------------------------------------------------------
;               Interrupt Vectors Used MSP430x2013
;------------------------------------------------------------------------------
                ORG     0FFFEh                      ; MSP430 RESET Vector
                DW      RESET                       ;
                END
```

Figure 20.13
(*Continued*)

corresponds to the maximum A/D number of 1023. Therefore, 1.364 V corresponds to the digital number:

$$\text{Threshold} = \frac{1.364}{2.5}1023 = 558_{10} = 22E_{16} \tag{20.5}$$

That just about takes care of things except that we need to place limits on the pulse width. The reason is that the loop can become unstable during power-up or large changes in the load. How can that happen? Let's say that our control system has some overshoot (it does). When power is turned on, the control loop causes the pulse width to keep increasing. But once it gets past the threshold that we set, it starts decreasing the pulse width. The program, without additional logic, isn't smart enough to stop decreasing the pulse width after it gets to zero. So the program can actually compute a negative value for the pulse width! That's pretty crazy and, as you can imagine, the control loop starts behaving in a very weird manner at that point.

To solve this, we simply limit the pulse-width value to a low value. If the loop counter (this variable is called "Current" in the program) is small (in the program it is checked for

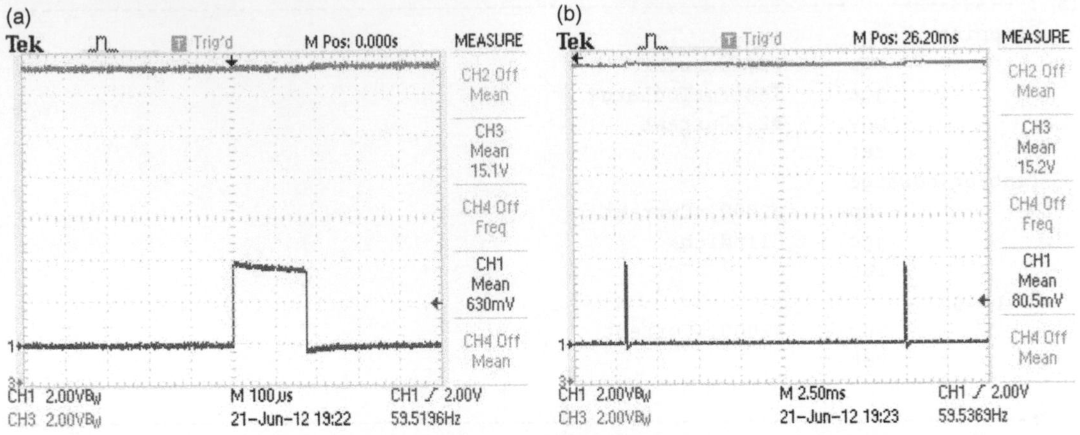

Figure 20.14
Output voltage and T_0 waveform at two different time scales.

the value 2) then it is just kept at that low value. And, while we're at it, we also limit the upside value for the pulse width. This is what the "Clipping" subroutine does.

As you can see from Figure 20.14, the program does a very nice job of keeping the high voltage at 15 V.

Note the slight jump in the high voltage just after each T_0 pulse. This is typical for a switching type of power supply.

Bibliography

[1] A. Pressman, Switching Power Supply Design, McGraw-Hill, 1991.
[2] J.A. Starzyk, Ying-Wei Jan, A. Fengjing Qiu., DC-DC charge pump design based on voltage doublers, IEEE Trans. Circuits. Syst. 48(3) (2001) 350–359.

CHAPTER 21

Remote Controls

Chapter Outline

Way back, at the beginning of this book, the definition for the kind of robots that we'd be talking about included the attributes "mobile and *autonomous*". So why are we discussing remote controls?

Well, the robot being developed in this book will be autonomous. But we need to have some way to turn the thing on and, especially, to turn it off. If the robot runs slowly enough, turning it off could be done by picking it up and turning off the power switch. But the hope here is that the robot will be able to travel at fairly high speeds, all the while avoiding obstacles in the room. Catching the little critter may be difficult, not to mention making you feel silly.

It turns out that developing this remote will make this whole project just a tad too much to include at this time. But it's worth talking about, as you will likely want to add this function at some point in the future.

RF remote controls

Using the existing radio-controlled car controller

Your radio-controlled car came with an RF radio remote control that probably had two joysticks – one for forward/backward and one for left/right. On the hand-held transmitter is

MSP430-based Robot Applications.
DOI: http://dx.doi.org/10.1016/B978-0-12-397012-1.00021-7

probably an indication of the frequency at which the transmitter operates. For example, a 27 MHz transmitter generates a 27 MHz sinusoidal carrier waveform that is sent out in all directions.

When a particular joystick on the controller is operated, a series of on/off intervals corresponding to that particular command are generated, with the carrier being transmitted during the on periods and nothing being transmitted during the off periods. The original electronics board that came with your radio-controlled car includes logic to decode these commands. The logic drives the power electronics on the board to turn the drive motor forward or backward, or to turn the front wheels right or left.

Since we are interested only in making the car stop or making it start, we could use one or the other joystick to perform this function. The decoded logic signals could be connected to an open I/O port on the microcontroller and sensed.

"Hacking" an existing electronic circuit like this is a legitimate way of creating a new electronic function. However, it is not the way we'll do things in this chapter. Part of the reason is that there are lots of different RC decoders out there, with different voltage levels, etc., which makes trying to discuss the exact details of what to do nearly impossible. The other reason is that we can use this opportunity to learn more about wireless optical communications.

RF network solutions

Quite a few companies have developed evaluation modules and other pre-assembled circuits that are intended to demonstrate how wireless nodes can be used. Many of these are quite powerful in terms of their ability to create an RF wireless network and perform important wireless networked tasks, such as wireless sensors. For example, Texas Instruments makes a line of such evaluation boards under the name Chronos and you may wish to take a look at these for future projects.

We won't be using these either, mainly because of the cost but also because using these sophisticated tools for something as simple as issuing a Start/Stop command is a bit like using a supercomputer to add "1 + 1".

Optical remote controls

Optical wireless remotes are really nothing new. You've probably been controlling the TV in your home for years with such a device. In its most basic form, it's simply an LED that turns on in the transmitter and a photodetector in the receiver that detects the presence of this light. We've already discussed such circuits with regard to the optical speed sensor and with regard to the collision avoidance augmentation circuit.

Okay, sounds simple. But, as we've seen in previous chapters, such a simple-minded approach won't work except perhaps in the instance where the transmitter and receiver are just inches from one another. There are just too many sources of optical interference, mainly in the form of the sun (for outdoor applications) and indoor lighting.

So how is the TV remote control able to work in such environments? Knowing a little more about this device will help us design our own remote. Of course, the light is modulated on and off to produce a logic 1 or a logic 0. But that, by itself, isn't going to cut it. The light from the remote's LED is relatively weak compared to room lights. A 120 Hz interference pattern from overhead fluorescent lights could look like a valid sequence of ones and zeros to the receiver.

Remote control designers fixed this problem long ago by modulating the on time at some constant frequency, such as 30 kHz (common frequencies are given in Ref. 1). So a sequence of on/off/on might look like Figure 21.1.

How does this help with respect to the interference problem created by ambient lighting? At the receiver (inside your TV), after the transimpedance amplifier converts the light into voltage, is a *bandpass filter*. This bandpass filter passes only that part of the signal with information modulated at or near the filter's *center frequency*. So, for the signal of Figure 21.1, the bandpass filter has a center frequency set at 30 kHz.

The effect is shown in Figure 21.2. Whereas the remote's signal, added to the 120 Hz ambient light, looks to the receiver like the signal in Figure 21.2, the signal, after passing

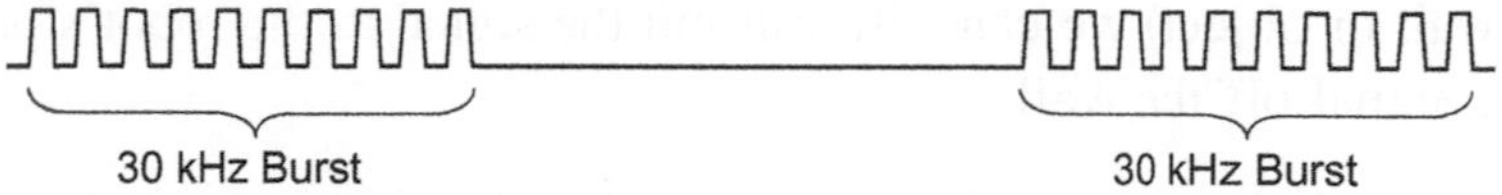

Figure 21.1
A one/zero/one sequence for a TV remote.

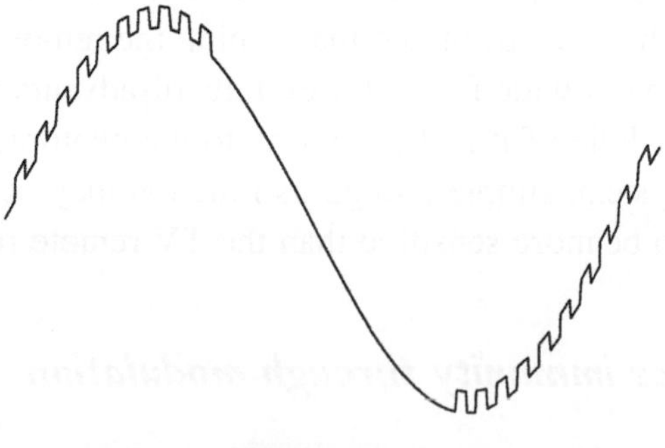

Figure 21.2
The one/zero/one sequence in the presence of strong 120 Hz interference.

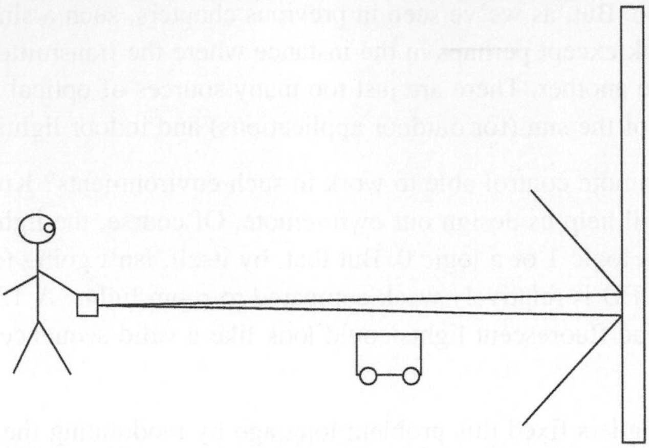

Figure 21.3
Sending the remote signal by bouncing the signal off a wall.

through the bandpass filter, looks like the Figure 21.1 signal again. The bandpass filter virtually eliminates all of the 120 Hz interference.

Actually, the problem that we are faced with is more difficult than that of the TV remote. The TV remote is intended for use as a line-of-sight (LOS) optical communications device. By that is meant that the TV remote is pointed directly at the TV's receiver.

In the case of our Start/Stop remote, we would like to be able to point the remote at a wall or object in front of the robot (Figure 21.3). That way, if the robot is traveling away from us (toward the wall or object) we can still transmit the signal to the robot's photoreceiver by bouncing the signal off the wall.

This type of bounced-light transmission is discussed in detail in Refs. 2, 3, and 4. Note that the light, when reflecting from the surface of the wall is *diffused* – the light spreads out in all directions. This is different from the *specular* reflection that we see from a mirror or other optical surface, where the light's angle of reflection is determined by its angle of incidence. Such diffusion has the advantage that aiming the remote is not critical (assuming the robot's photoreceiver has a wide field-of-view). Its disadvantage is that the light is going everywhere, so very little of it makes its way to the photoreceiver. As a result, our optical communications system, simple though its function may be, will be quite a challenge and will need to be more sensitive than the TV remote receiver.

Improving interference immunity through modulation

We've already seen how the TV remote receiver is able to eliminate most of the low-frequency interference by modulating its presence-of-light intervals with a relatively high-frequency

(30 kHz to 60 kHz) envelope at the transmitter, then sending the received signal through a bandpass filter at the photoreceiver. There are ICs that can perform the entire TV remote receiver function in a single package. Unfortunately, these ICs are designed to accept the LOS signals from a transmitter pointed directly at the photoreceiver and probably don't have the sensitivity to respond to the very low signal levels that will occur with our bounced-off-the-wall (diffused) signals.

Nevertheless, we will want to use the idea of modulated light as a way of separating signal from background interference. In this case, we will use 30 Hz as the modulation frequency. When we press the button on our remote, the remote will produce 30 cycles of 30 Hz light modulation, so that the total length of time on will be 1 second.

But why 30 Hz? It's not a coincidence that this happens to be a submultiple of our 120 Hz background nemesis. We will devise a somewhat unusual filter called the "integrate-and-dump" for the 30 Hz signal. In theory, this filter can completely eliminate any even-multiple-frequency of 30 Hz (60 Hz, 120 Hz, etc.). But before we do that, let's talk about the transmitter.

Robot remote transmitter

The remote transmitter is pretty straightforward. When the transmitter needs to send out a Start/Stop signal, we want it to modulate the light on and off at a 50% duty cycle and a 30 Hz rate for a period of 1 second (30 cycles). When it's not doing that, we want it to do nothing, including (ideally) not consuming any battery current.

An easy way to do this is to just use a TI LaunchPad board as the basis of the transmitter. For this simple application, any DIP version of the MSP430 will do. I used an MSP430F2001 simply because they're inexpensive and I happened to have a bunch sitting around.

Basically, we're going to use the P1.3 switch that the LaunchPad already has on the board (don't worry if you've removed R34 and C24 from the board previously – they aren't needed). The program sets up P1.3 as an input pulled up through its internal pull-up resistor. The program continually monitors this input and, when it sees that it goes to logic 0 (as the result of the switch being pushed) it then proceeds to generate the 30 Hz signal for 1 second.

The schematic for the transmitter is shown in Figure 21.4. It's very simple. There is a connection to an external battery pack (two AAA batteries), and a circuit to send 100 mA through the SFH4550 LED. By itself, an MSP430 output cannot produce anywhere near 100 mA. In fact, an MSP430 output isn't intended for outputs much past 1 mA. So to drive the LED with that much current, an external transistor is used. But the

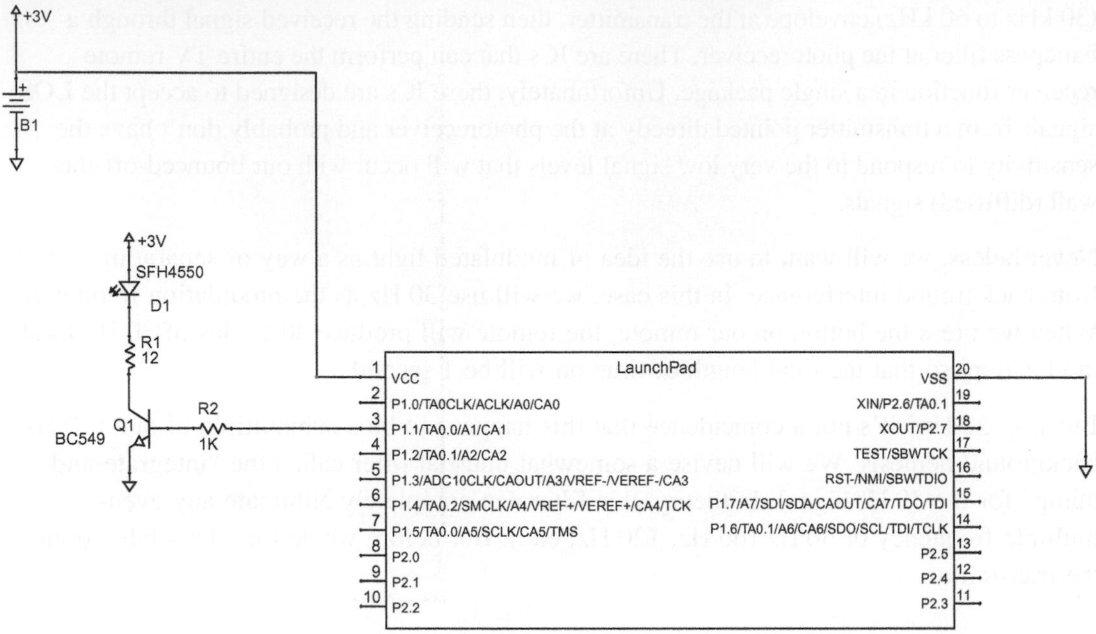

Figure 21.4
The robot remote transmitter circuit.

external transistor, in order to turn 1 mA into 100 mA, will need a β in excess of 100. Fortunately there are transistors that have such high current gain.

The BC549 NPN transistor has the following characteristics:

- $\beta > 100$
- $v_{CE,sat} \cong 0.3$ V at $I_C = 100$ mA
- $v_{BE} \cong 1.1$ V at $I_C = 100$ mA

Using the 3 V as the battery voltage from the two AAA batteries, the base current to the transistor will be:

$$I_B = \frac{3 - 1.1}{1K} = 1.9 \text{ mA}. \tag{21.1}$$

In fact, the base current will be somewhat lower than this, due to the resistance of the MSP430 P1.3 output. Nevertheless, the base current should be above 1 mA. With a base current greater than 1 mA and with the intent to drive the LED with no more than 100 mA, the transistor will be in saturation (since the β is greater than 100), so we can calculate the LED current based on this assumption.

Figure 21.5
The robot remote transmitter.

The typical forward voltage of the SFH4550 LED is 1.5 V, so the current through the LED will be:

$$I_{LED} = \frac{3 - 0.3 - 1.5}{12} = 100 \text{ mA}. \qquad (21.2)$$

Since the additional components required to build the Figure 21.4 circuit are so few, the extra components (the LED, transistor, and two resistors along with the battery pack) can be built on a "piggyback" board like the one shown in Figure 21.5.

The extra components are simply soldered together on a vector board, which has the connectors soldered to it to mate to the LaunchPad board.

The assembly language program needed to run this transmitter is shown in Figure 21.6. After initializing the digital I/O (Port 1), the timer, and setting the MCLK and SMCLK at a calibrated 1 MHz (divided down to 128 kHz), the main loop is started. This part of the

```
;******************************************************************************
#include  "msp430F2001.h"
;------------------------------------------------------------------------------
            ORG     0F800h                  ; Program Reset
;------------------------------------------------------------------------------
RESET       mov.w   #0280h,SP               ; Initialize stackpointer
            mov.w   #WDTPW+WDTHOLD,&WDTCTL  ; Stop watchdog timer
            ; Make all the Port I/O outputs except P1.3
            mov.b   #0F7h,&P1DIR
            mov.b   #8h,&P1REN             ; Enable the pull-up resistor on P1.3
            mov.b   #8h,&P1OUT             ; Pull up on P1.3
            ; Run MCLK and SMCLK at calibrated 128 kHz
            mov.b   &CALDCO_1MHZ,&DCOCTL
            mov.b   &CALBC1_1MHZ,&BCSCTL1
            mov.b   #SELM_0+DIVM_3+DIVS_3,&BCSCTL2
            ; run the timer using SMCLK, continuous mode
            ;  Timer input clock is divided by 8, so timer is updated at 15625 Hz
            mov     #TASSEL_2+MC_2+ID_3,&TACTL
CheckSwitch
            bit.b   #8h,&P1IN
            jnz     CheckSwitch
            ; At this point, generate a 30 Hz square wave for 1 second
            bic.b   #2,&P1OUT
            clr     &TAR
            clr     R5
            mov     #60,R6
Generate30Hz
            xor.b   #2,&P1OUT
            add     #260,R5
Dwell
            cmp     R5,&TAR
            jl      Dwell
            dec     R6
            jnz     Generate30Hz
            jmp     CheckSwitch

;------------------------------------------------------------------------------
;           Interrupt Vectors Used MSP430x2013
;------------------------------------------------------------------------------
            ORG     0FFFEh                  ; MSP430 RESET Vector
            DW      RESET                   ;
            END
```

Figure 21.6
Assembly language program for the robot remote transmitter.

program simply monitors the P1.3 input pin. Note that P1.3 is pulled up by an internal resistor to the +3 V supply voltage, so it will be a high logic input except when the P1.3 pushbutton switch (in the lower left-hand corner of the board in Figure 21.5) is pressed. When that happens, the P1.3 input reads low.

A logic low on P1.3 causes the main loop program to proceed to generating a 30 Hz square wave at P1.1 which drives the Q1 transistor (and therefore turns the SFH4550 LED on). The 30 Hz square wave repeats for 30 cycles (one second).

Robot remote receiver

To get started, let's think about the situation where we want to detect a 30 Hz signal. Figure 21.7 shows the general approach. Of course, it starts with a photoreceiver, which is just the photodiode and transimpedance amplifier that is already used for the optical collision-avoidance augmentation circuit. A conventional filter then takes that signal from the photoreceiver, passes the 30 Hz signal and filters out some of the interference at higher frequencies.

The conventional filter is taken from Ref. 5 and is referred to by the author of that book as the KRC design. It is a second-order lowpass filter, so it attenuates frequencies well beyond the cut-off frequency according to the equation:

$$\text{Attenuation} = \left(\frac{f_c}{f}\right)^2, \; f>>f_c \qquad (21.3)$$

where f is the frequency of the signal and f_c is the cut-off frequency. So, for a second-order lowpass filter with cut-off frequency of 30 Hz, it attenuates 120 Hz background by a factor of $(1/4)^2 = 1/16$.

The design from Ref. 5 is shown in Figure 21.8 and the complete circuit is shown in Figure 21.9. The details of how such filters are designed are beyond the scope of the book (there are, in fact, entire college courses on filter design). So, we are taking this design in a "cookbook" approach – taking someone else's word that this design will work without completely understanding the operation. It's always better to understand how things work

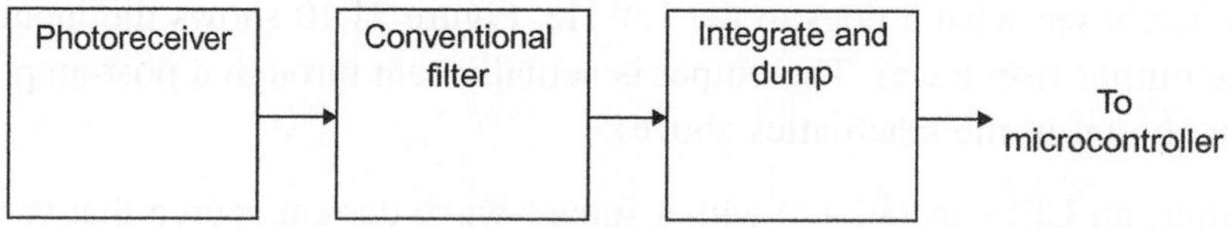

Figure 21.7
The remote signal receiver block diagram.

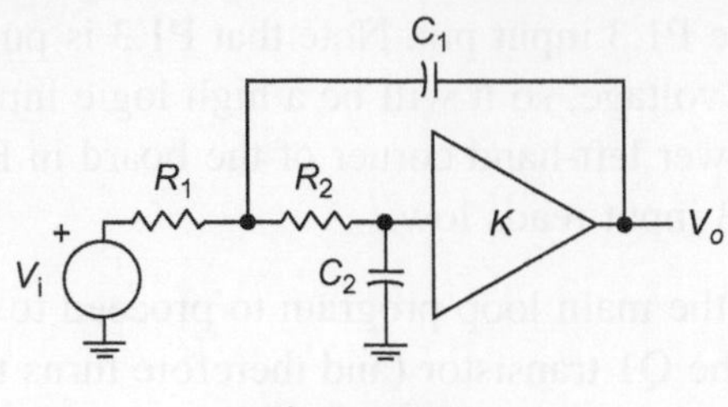

Figure 21.8
A second-order filter *(from Passive and Active Network Analysis and Synthesis* by Aram Budak*)*.

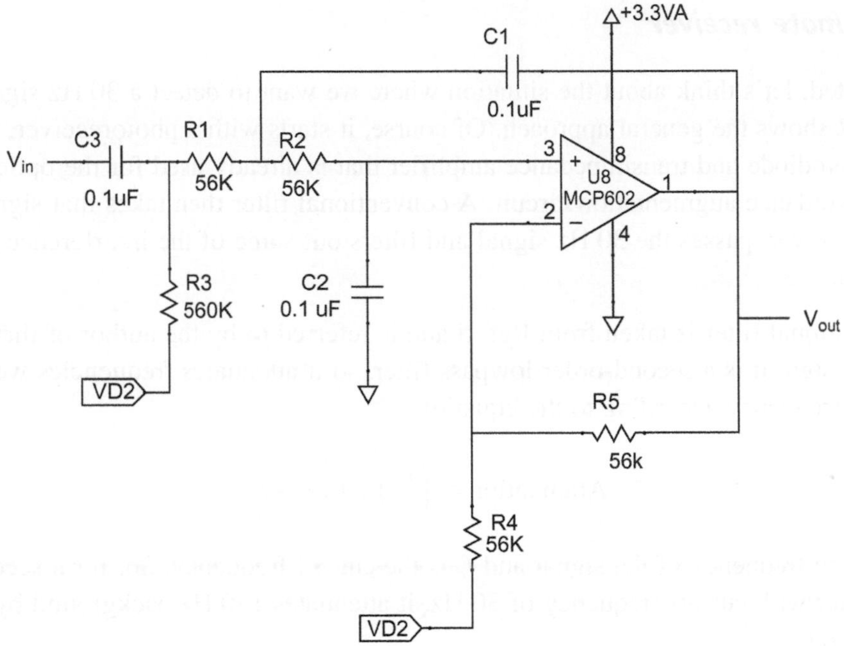

Figure 21.9
Complete conventional 30 Hz second-order filter with gain of 2.

but it's okay, particularly when starting out, to use circuits in this cookbook way. Of course, we will need to verify that the circuit actually does what it's supposed to do.

The gain used for this filter is $K = 2$ ($R_5/R_4 + 1 = 2$). The filter passes 30 Hz virtually unattenuated. Let's see what it does to the 120 Hz. Figure 21.10 shows the input (bottom trace) and the output (top trace). The output is actually sent through a post-amplifier with gain of 5 (not shown in the schematics above).

To test the filter, an LED, modulated with a square-wave current source that varies between 100 mA and 0 mA at a 30 Hz rate, was pointed at a wall about 3 feet away from the LED. The transimpedance amplifier / 30 Hz filter circuit was also about 3 feet away from the

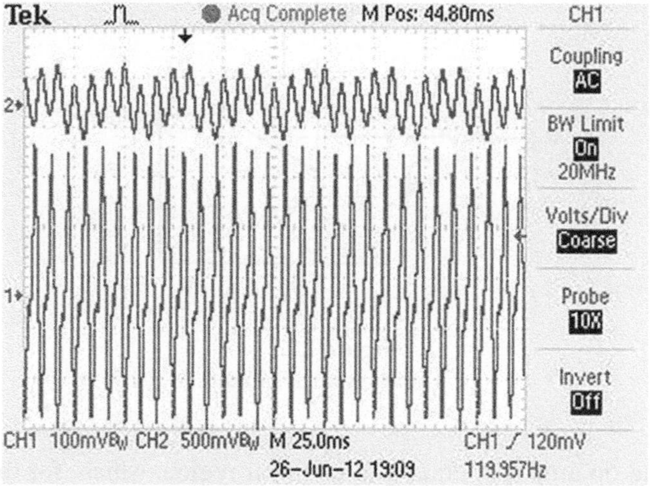

Figure 21.10
Input and output waveforms for 30 Hz, second-order, conventional filter.

wall. The experiment was performed in a room with fluorescent lights known to produce strong 120 Hz content.

As Figure 21.10 shows, the 120 Hz background component is huge compared to the 30 Hz signal. In fact, the 30 Hz envelope is barely perceptible in the bottom trace. After the filter, there is still lots of 120 Hz, but it's been attenuated by about a factor of 16 (remember that the filter itself has a gain of 2 and there is a gain-of-5 post-amplifier that is not shown in the schematics). The 30 Hz component has approximately 20 mV peak-to-peak in the bottom trace and 200 mV peak-to-peak in the top trace, which, after accounting for the gain-of-10 amplification, means that the 30 Hz experiences little attenuation.

Okay, so the good news is that the second-order filter significantly attenuated the 120 Hz interference from the fluorescent lights. The bad news is that, even with this attenuation, trying to determine that the output waveform has a 30 Hz component is still tough. Fortunately we have the "integrate-and-dump" filter as the ultimate 120 Hz eliminator.

What is an integrator?

To understand how the integrate-and-dump works, we'll need to understand what an integrator does. Normally such an explanation would involve calculus, but the reader is not assumed to have encountered calculus yet, so we'll go about this from a little different angle. An electronic integrator can be implemented using the Figure 21.11 circuit. Let's analyze this circuit for DC values of v_{IN}.

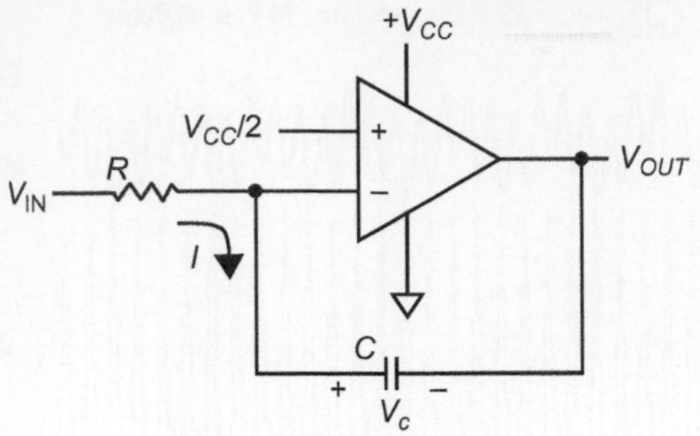

Figure 21.11
The integrator circuit.

Let's assume that the op amp is operating in its linear region, which, for this application, simply means that the op amp output is not saturated either at its high value or its low value. Since it's operating in its linear region, the inverting input is at the same voltage as the non-inverting input (typically, there is a slight difference between the two, called the *offset voltage*, but we can ignore that for this analysis). The non-inverting input is set at half the DC supply voltage ($V_{CC}/2$) so that's the operating voltage of the inverting input as well.

If v_{IN} is greater than $V_{CC}/2$, then the current, I, will be:

$$I = \frac{v_{IN} - \frac{V_{CC}}{2}}{R} \tag{21.4}$$

Since the inputs of an op amp are extremely high impedance, the current's only path is through the capacitor. From the beginning electronics chapter, we know that the voltage across the capacitor changes value as:

$$\Delta v_C = \frac{I}{C}\Delta T = \frac{v_{IN} - V_{CC}/2}{RC}\Delta T. \tag{21.5}$$

Let's try an example. Let's say that:

- $R = 10\ \text{k}\Omega$
- $C = 0.1\ \mu\text{F}$
- $V_{CC} = 3.6\ \text{V}$
- $v_{IN} = 2.2$ V during Δt interval, 1.8 volts thereafter
- $\Delta T = 1$ msec

Then, according to Eq. 21.5, the voltage across the capacitor at the end of the Δt interval is:

$$\Delta v_C = \frac{2.2 - 1.8}{10^{-3}}10^{-3} = 0.4\ \text{V}. \tag{21.6}$$

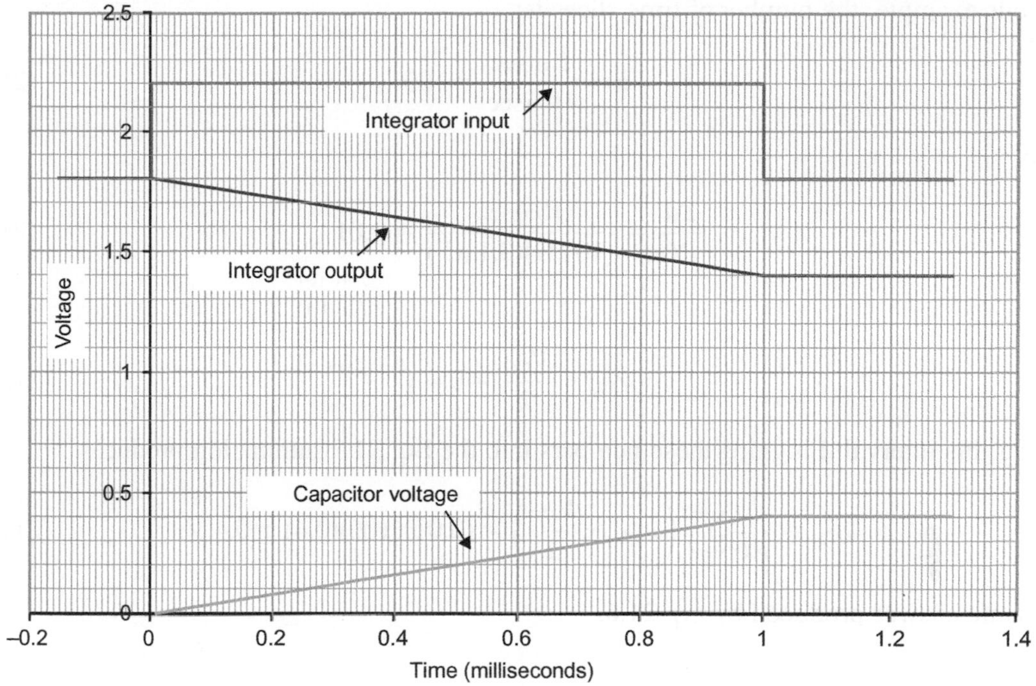

Figure 21.12
Integrator output for a pulse input.

Now, observe how v_C is referenced in the Figure 21.11 diagram. The capacitor voltage, as referenced, subtracts from the voltage at the inverting input, which is $V_{CC}/2 = 1.8$ V. So the output voltage, v_{OUT}, is therefore $1.8 - 0.4 = 1.4$ V after 1 msec. Note that the integrator output remains at a constant level once the input voltage equals the integrator's reference voltage of 1.8 V. The input, output, and capacitor voltages are shown in Figure 21.12.

Numerical integration (summation of samples)

One way that we can look at the integrator is to divide the input into small time slots. For example, in Figure 21.12, we could look at 10 μsec (0.01 msec) slices of the input (the grid of Figure 21.12 is marked in 10 μsec increments). We could write the capacitor voltage at any slice number, n, as:

$$v_{C,n} = \frac{\Delta t}{RC}\sum_{i=1}^{n}\left(v_{IN,i} - \frac{V_{CC}}{2}\right) \tag{21.7}$$

Here, Δt is the time slice interval (for this example, 10 μsec), $v_{IN,i}$ is the input voltage at time slice, i, and $v_{C,n}$ after n intervals.

For our example, the number of time slices is:

$$n = \frac{1 \text{ msec}}{10 \text{ µsec}} = 100 \tag{21.8}$$

Also,

$$\frac{\Delta t}{RC} = \frac{10 \text{ µsec}}{(10\ k\Omega)(0.1\ \mu F)} = 0.01 \tag{21.9}$$

and

$$v_{IN,i} - \frac{V_{CC}}{2} = 0.4 \text{ V} \tag{21.10}$$

during this interval. Plugging these numbers into the Eq. 21.7 summation produces:

$$v_{C,n} = \frac{\Delta t}{RC}\sum_{i=1}^{n}\left(v_{IN,i} - \frac{V_{CC}}{2}\right) = (0.01)(100)(0.4) = 0.4\ V, \tag{21.11}$$

the same answer produced by Eq. 21.6.

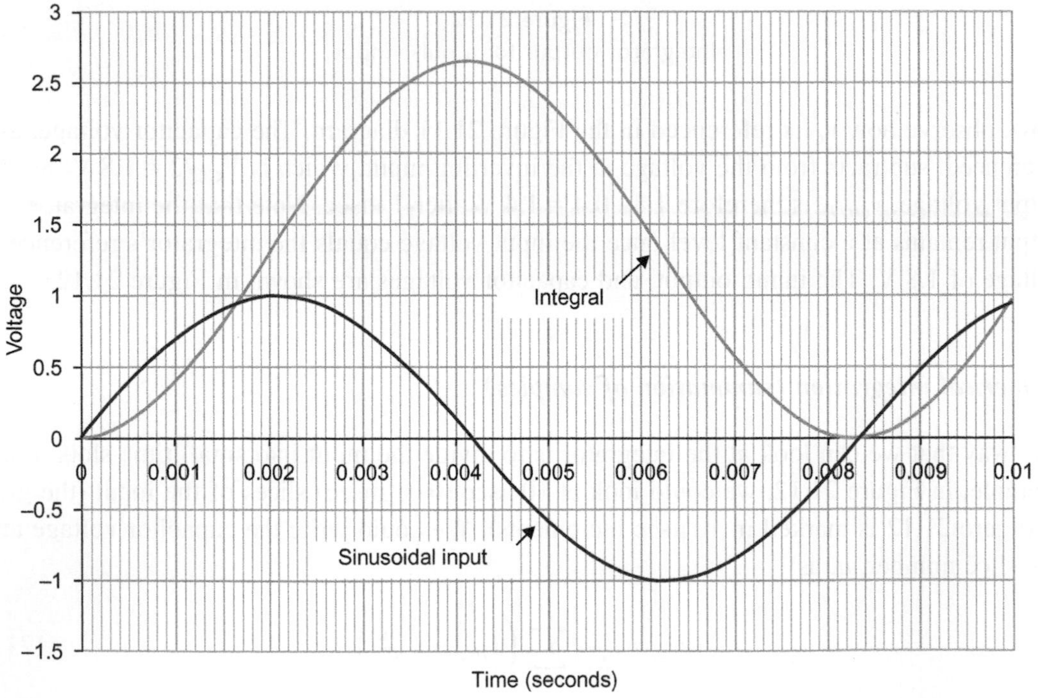

Figure 21.13
A 120 Hz sinusoidal input and integrator output.

Using the summation of samples approach might seem like a lot more work than the electrical circuit approach that we used to come up with Eq. 21.5, but it will prove useful when the input waveform is something other than a simple pulse.

Think about a sinusoidal input waveform, like the one in Figure 21.13. The 1.8 V offset is left out of the input and output waveforms for clarity. During the positive part of the sinusoidal input (up to just past 4 msec), the integral rises, which we should expect from Eq. 21.7. Once the sinusoidal input goes negative, the integral begins to decrease in value. *When exactly one cycle of the sinusoidal input has gone by (at 8.33 msec) the integral is again 0.* This is key to how the integrate-and-dump works – if we integrate for exactly one period of a 120 Hz waveform, then any 120 Hz interference that exists vanishes.

But Figure 21.13 assumed that our integrate-and-dump somehow knew where the start of the sine wave was. In practice, that won't happen. So what happens if we integrate over the 120 Hz period starting at some point other than 0? Figure 21.14 shows a 120 Hz waveform that is 30° advanced with respect to the receiver timing. In spite of starting at a voltage of 0.5 V, the integral still goes to 0 at exactly one period after starting the integration.

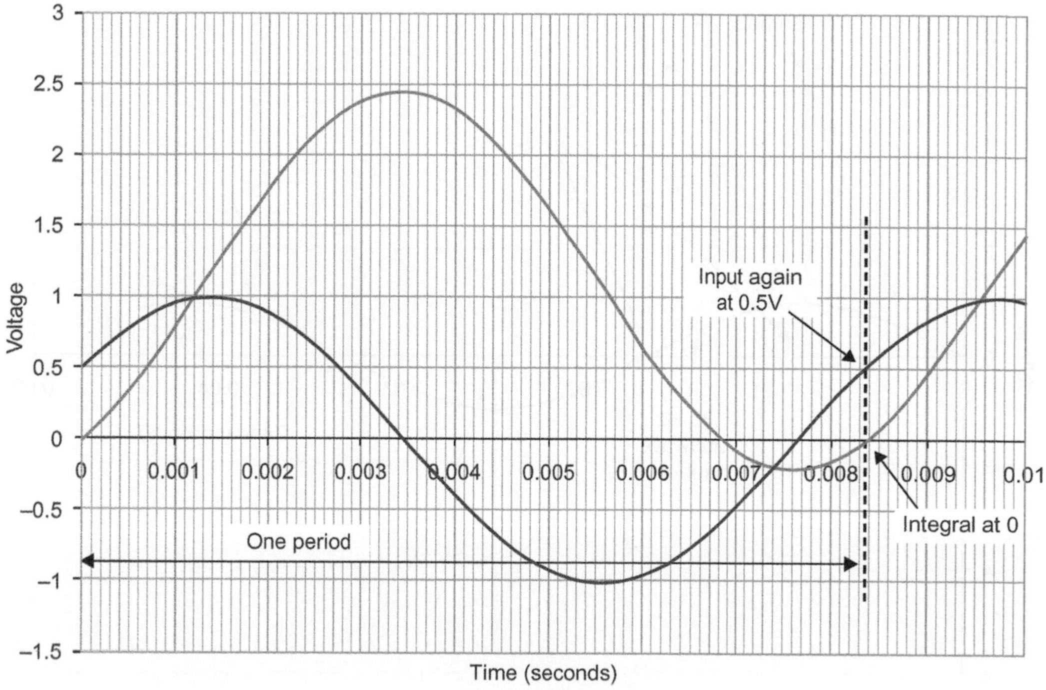

Figure 21.14
A phase-shifted sinusoidal input and integrator output.

The integrate-and-dump

What the above graphs tell us is that, if we integrate for one period of the 120 Hz interference signal, then zero out the integrator (dump), and repeat this process over and over, we'll eliminate any trace of the interfering signal (at least in theory). But what does this circuit do to the actual signal that we're trying to recover? Figure 21.15 shows what the integrate-and-dump, with 8.33 msec repetition rate, does to the 30 Hz signal.

Note how this circuit works – it integrates for 8.33 msec, then an A/D conversion samples the integrated signal, followed by a zeroing of the integrator. We get two very healthy positive analog samples, followed by two very healthy negative analog samples. If we sum the first two samples, then subtract the sum of the third and fourth samples, we get a result that is an extremely large number and one that has eliminated the 120 Hz interference (actually, we need to scale the integrator's RC time constant down in this example, since the ± 5 V peaks created in Figure 21.15 would saturate the electronics).

There is, however, a little bit of a problem. If the 30 Hz signal is shifted in time, the output changes. Figure 21.16 shows what happens when the 30 Hz signal is advanced by 8.33 msec relative to the 30 Hz signal of Figure 21.15.

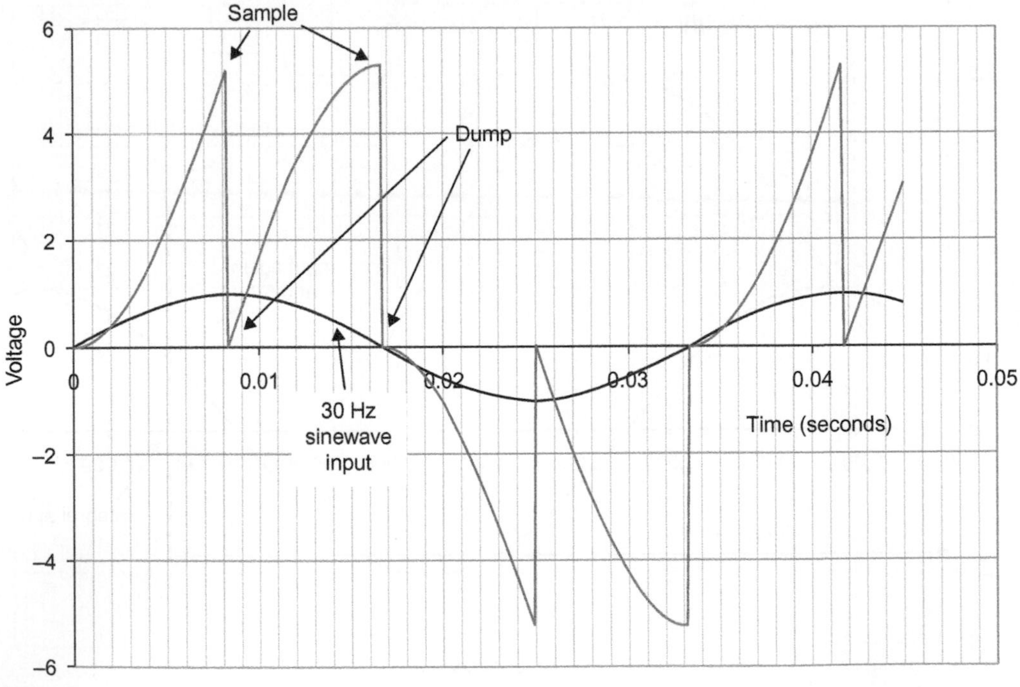

Figure 21.15
Response to 30 Hz input of integrate-and-dump operating at 120 Hz.

To recover this signal fully, the sum of the second and third samples should be subtracted from the sum of the first and fourth samples. The Figure 21.15 waveform had, as its optimal sequence of operations: add/add/subtract/subtract. This sequence produces the *in-phase* version of the demodulation. The waveform of Figure 21.16 has, as its optimal sequence of operations: add/subtract/subtract/add. This sequence of operations produces the quadrature-phase version of the demodulation.

As it turns out, by computing both the in-phase and quadrature-phase version of the demodulation and combining these in a prescribed way, the amplitude of the 30 Hz signal can be recovered, regardless of its phase with respect to the receiver's demodulating circuits.

The integrate-and-dump circuit

But what does the integrate-and-dump circuit look like? It's actually quite simple. It's just the integrator from Figure 21.11 with a switch across the capacitor to dump (zero) the capacitor's charge at the end of each integration period. Figure 21.17 shows the circuit. A simple MOSFET works just fine for the switch.

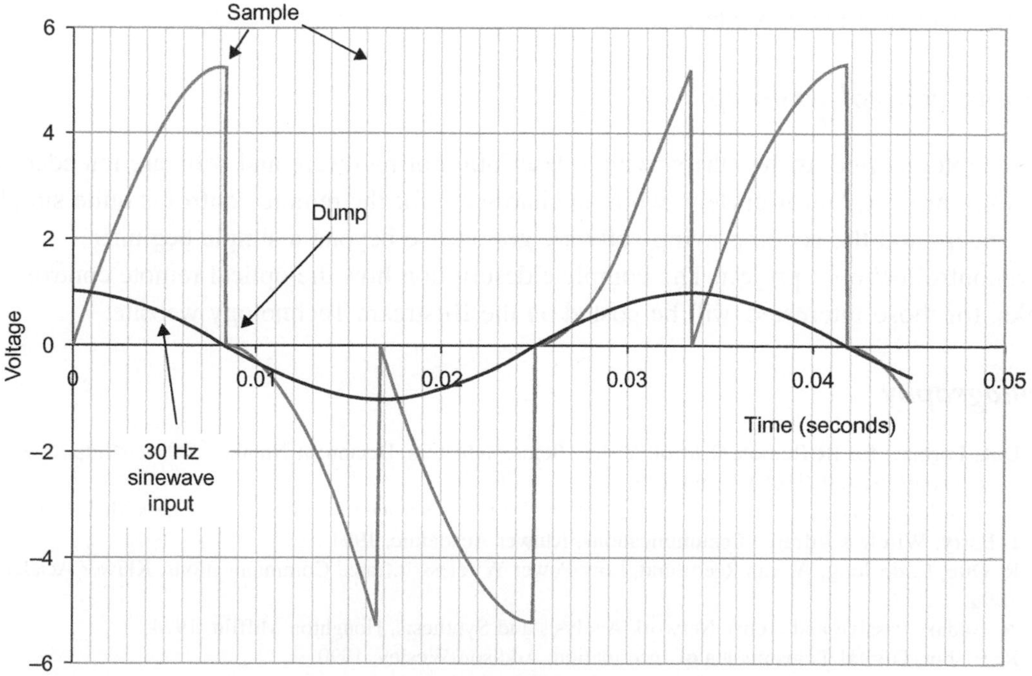

Figure 21.16
Integrate-and-dump output for 30 Hz sinewave advanced in time 8.33 msec.

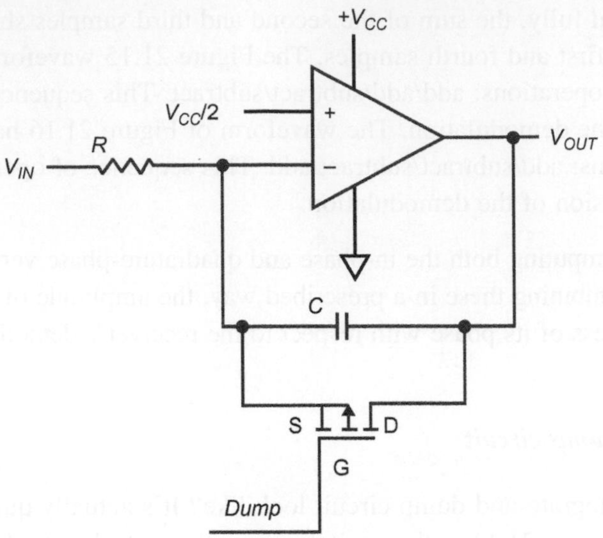

Figure 21.17
The integrate-and-dump circuit.

The dump signal is a very short pulse, only long enough to short-circuit the capacitor and thereby discharge the capacitor.

Remote control wrap-up

This chapter started as the others, with a description of hardware and software intended for our robot project. Unfortunately, while the hardware for the remote control is quite simple, the concept and the accompanying software grew to be beyond a simple beginner microcontroller/robot project. The complete description how the optical remote control works, for those interested, will be posted on the Bitstream Technology website.

Bibliography

[1] Data Formats for IR Remote Control. Vishay Semiconductors, Document Number 80071. <http://www.vishay.com/docs/80071/dataform.pdf>.

[2] S. Hranilovic, Wireless Optical Communications Systems, Springer Science and Business Media, 2005.

[3] J. Barry, Wireless Infrared Communications, Kluwer Academic, 1994.

[4] R. Otte, L. de Jong, A. van Roermund, Low-Power Wireless Infrared Communications, Kluwer Academic, 1999.

[5] A. Budak, Passive and Active Network Analysis and Synthesis, Houghton Mifflin, 1974.

[6] R. Blahut, Digital Transmission of Information, Addison-Wesley, 1990.

CHAPTER 22

Troubleshooting

Chapter Outline

Let's say that you've just built a fairly complex circuit, either the one that we've been talking about in this book, or some other project that you've designed or copied. You're ready to test it out. You turn on the switch and – nothing. Now what?

You'll need to think about this because nearly every circuit consisting of more than three integrated circuits and nearly every program of more than 35 lines of code does not work initially. Okay, I made those statistics up. But my experience has been that even moderately complex circuits and programs usually have some problem that keeps them from working after being built.

Learning to troubleshoot is invaluable

Troubleshooting is one of those skills that sets a great engineer apart from a good engineer. I've seen lots of bright engineers who seem to have the right educational background for the job but who avoid the unpleasant, sometimes frustrating job of troubleshooting. They make excuses for not sitting down at the bench and getting a circuit or a program working.

MSP430-based Robot Applications.
DOI: http://dx.doi.org/10.1016/B978-0-12-397012-1.00022-9

But coming up with good design ideas usually doesn't count for much if you can't put those ideas into practice.

This chapter provides some advice on how to get started troubleshooting. It is divided into two sections – the first is on general troubleshooting strategies and the second is on specific things that can happen for this robot project.

Strategies for troubleshooting

Make good use of the integrated development environment

The integrated development environment (IDE), in combination with the LaunchPad board, is more than just a compiler and a way to load your program from personal computer to MSP430. It is also an *in-circuit emulator* – that is, it can run the program but can do additional things, like allowing breakpoints to be set and allowing contents of memory and registers to be examined.

Let's write a program, with a deliberate error included, and see how this IDE can provide the troubleshooting path for discovering the error. The program is very simple – it looks at the analog signal on P1.1 every 1 msec and outputs, on P1.2, a 1 if the signal is above 1.8 V, otherwise it leaves P1.2 at 0. The program is run on an MSP430G2231 microcontroller that is inserted in a LaunchPad board.

The program is given in Figure 22.1. The main loop consists of clearing the timer counter, performing the A/D conversion, making the decision about whether the A/D input is greater than or less than the $V_{CC}/2$ (512 for a 10-bit A/D), and then setting or resetting the output bit based on that decision.

The input analog signal at P1.1 could be any voltage between 0 and V_{CC} (3.6 V) but for purposes of illustration I chose to use a triangle wave at slightly more than 100 Hz (slow enough for the 1 kHz sample rate to keep up). Ideally, the input and output waveforms will look like the Figure 22.2 waveforms.

Of course, the actual digital output will not be exactly lined up with the $V_{CC}/2$ voltages of the triangle wave due to the fact that it updates its digital output only every 1 msec.

Okay, so, after building the Sampling Example program of Figure 22.1, you download it to the LaunchPad board containing the MSP430G2231 microcontroller and you hit Go on the Debug pulldown tab and ... nothing. Well, something, but it's just a constant voltage at the P1.2 output. What's wrong?

IAR Information Center for MSP430 | Sampling Example.s43

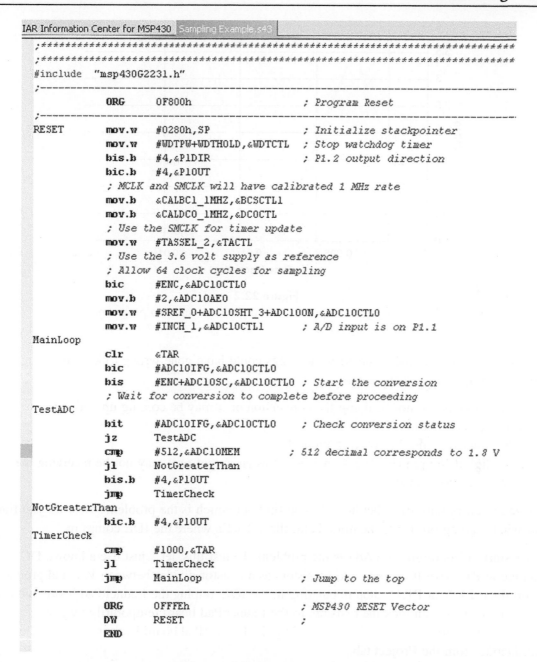

```
;*****************************************************************************
;*****************************************************************************
#include  "msp430G2231.h"
;-----------------------------------------------------------------------------
            ORG     0F800h                  ; Program Reset
;-----------------------------------------------------------------------------
RESET       mov.w   #0280h,SP               ; Initialize stackpointer
            mov.w   #WDTPW+WDTHOLD,&WDTCTL  ; Stop watchdog timer
            bis.b   #4,&P1DIR               ; P1.2 output direction
            bic.b   #4,&P1OUT
            ; MCLK and SMCLK will have calibrated 1 MHz rate
            mov.b   &CALBC1_1MHZ,&BCSCTL1
            mov.b   &CALDCO_1MHZ,&DCOCTL
            ; Use the SMCLK for timer update
            mov.w   #TASSEL_2,&TACTL
            ; Use the 3.6 volt supply as reference
            ; Allow 64 clock cycles for sampling
            bic     #ENC,&ADC10CTL0
            mov.b   #2,&ADC10AE0
            mov.w   #SREF_0+ADC10SHT_3+ADC10ON,&ADC10CTL0
            mov.w   #INCH_1,&ADC10CTL1      ; A/D input is on P1.1
MainLoop
            clr     &TAR
            bic     #ADC10IFG,&ADC10CTL0
            bis     #ENC+ADC10SC,&ADC10CTL0 ; Start the conversion
            ; Wait for conversion to complete before proceeding
TestADC
            bit     #ADC10IFG,&ADC10CTL0    ; Check conversion status
            jz      TestADC
            cmp     #512,&ADC10MEM         ; 512 decimal corresponds to 1.8 V
            jl      NotGreaterThan
            bis.b   #4,&P1OUT
            jmp     TimerCheck
NotGreaterThan
            bic.b   #4,&P1OUT
TimerCheck
            cmp     #1000,&TAR
            jl      TimerCheck
            jmp     MainLoop                ; Jump to the top
;-----------------------------------------------------------------------------
            ORG     0FFFEh                  ; MSP430 RESET Vector
            DW      RESET                   ;
            END
```

Figure 22.1

Program to sample P1.1 and set or reset P1.2 based on that sample.

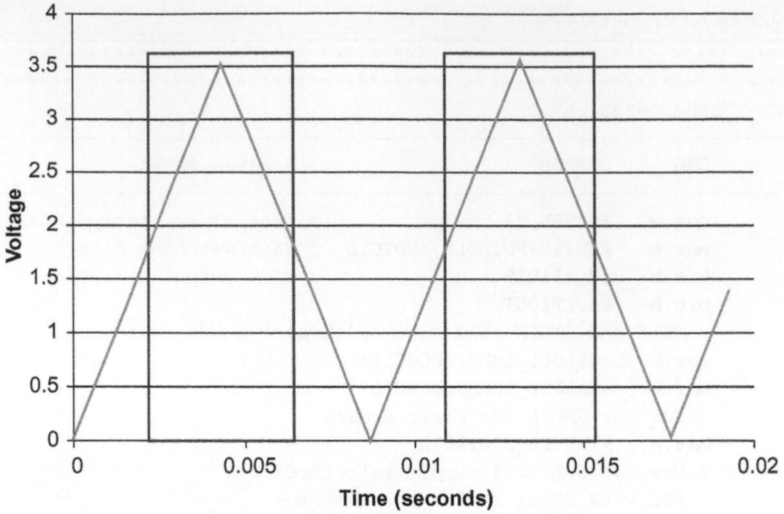

Figure 22.2
Expected digital output for a triangle wave input to program.

First thing to do is to ask yourself what things could have gone wrong to keep the program from running. Let's list them here:

- The A/D may be not be doing the conversion or it may be coming up with an unexpected result.
- The timer may not be working properly.
- The digital I/O pin (P1.2) may not be set as an output or it may not be receiving the correct value to display.

Those are all possibilities, but how do you find out which is the problem? How do you find out what's going on inside the microcontroller? That's where the IDE comes in.

Let's start off assuming the A/D is the problem. To test this, let's just use a known DC voltage at P1.1 (we'll use two 4 kΩ resistors as a resistor divider between V_{CC} and ground to produce 1.8 V at P1.1). Stop debugging in the IDE, unplug the LaunchPad board and add the resistors (you can tack the resistors to the LaunchPad board temporarily by just soldering them to the appropriate pins). Plug the LaunchPad board back in and Download and Debug from the Project tab.

Now, let's run to just before execution of the CMP #512,&ADC10MEM instruction. Do this by clicking your computer's mouse so that the cursor is just ahead of the instruction, in the same line as the instruction. Now select Run to Cursor from the Debug menu in the IDE. The IDE breaks when it gets to the CMP instruction – that is, it stops execution of the program. Now's the opportunity to see what the A/D thinks the 1.8 voltage is.

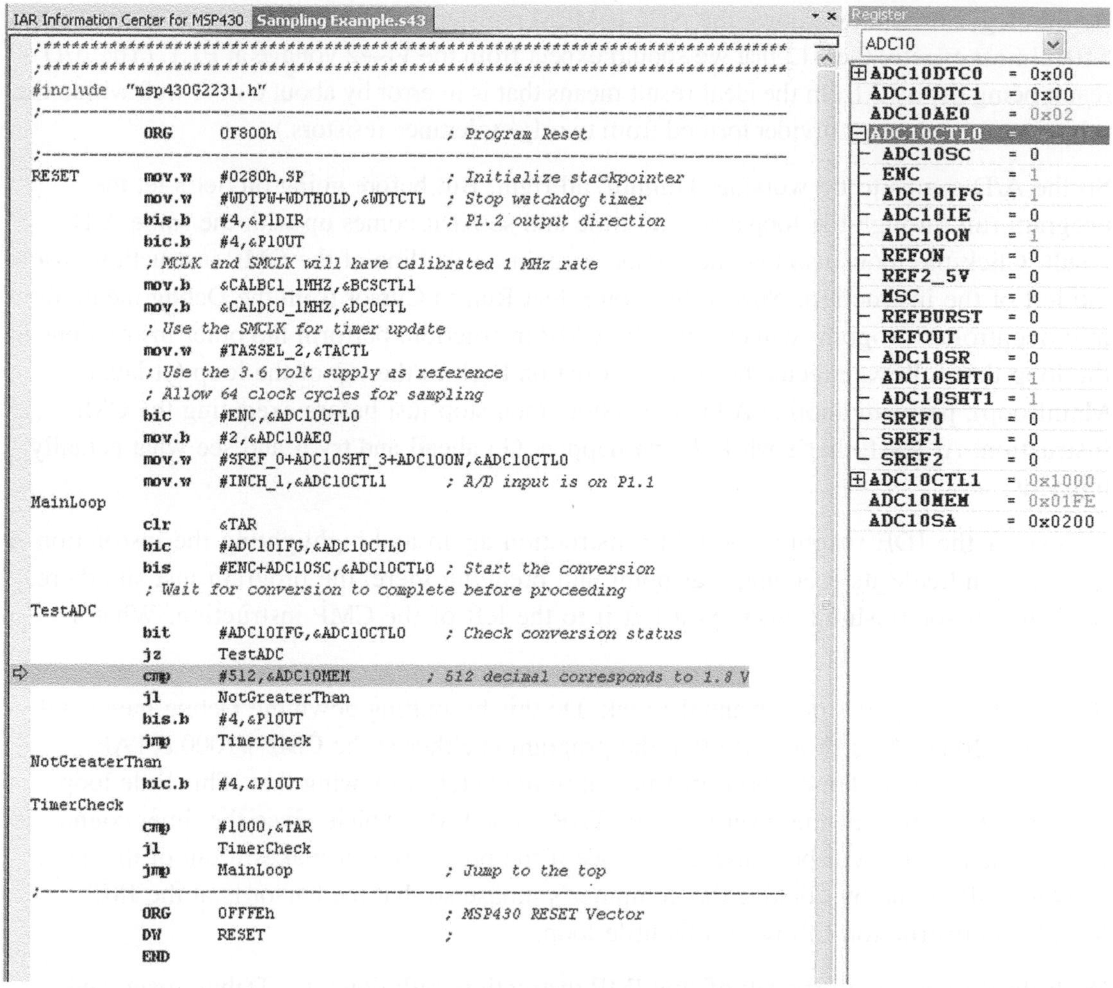

Figure 22.3
Sampling Example program with ADC10 register contents shown.

To see this, pull down the View menu and select Register. With your mouse, you can slide the Register window to the left, by the code window, so that it's more easily read. Now, in the Register window you should see a pulldown menu that initially says CPU Registers. Pull this down and select ADC10. There will be a number of ADC10 control registers' contents shown, as in Figure 22.3.

We are interested at this point in the A/D result, stored in the control register ADC10MEM, and in the control register bits, most of which are in ADC10CTL0 control register. Note that we can expand the contents of the ADC10CTL0 register into the individual bits by clicking the " + " to the left of the ADC10CTL0 label.

As the Register window shows, the ADC10MEM register contains $1FE_{16}$, which, in decimal, is 510, almost exactly the 512 that we should expect from the $V_{CC}/2$ voltage at P1.1. (The A/D result being off by 2 from the ideal result means that is in error by about 6 mV, well within the tolerance for a resistor divider formed from two 1% tolerance resistors.)

So the A/D seems to be working. Hmmm, all right. But before going on, let's let the program run through the loop a second time and see if it comes up with the same A/D result. Click the mouse so that the cursor is again on the line of the CMP instruction, just to the left of the instruction. Now, when you select Run to Cursor from the Debug menu, the microcontroller program will execute the CMP instruction, perform the other instructions in the loop that follow, execute the JMP instruction back to the top of the loop (at label MainLoop), perform another A/D conversion, then stop just before executing the CMP instruction. At least, that's what *should* happen. Go ahead and try it and see what actually happens.

Instead of the IDE reaching the CMP instruction again and highlighting the instruction in green, to indicate its reaching that point and breaking there, the program just sits there, with the cursor flashing where you left it to the left of the CMP instruction. What's going on?

Well, to find out, let's do a manual Break. Do this by pulling down the Debug menu and selecting Break. You should see that the program is either at the CMP #1000,&TAR statement or the JL TimerCheck instruction immediately following that. This little loop is supposed to wait until the timer counter, TAR, is at 1000, which, since the timer counter is clocked at 1 MHz, will be 1 msec. Let's see if the program ever makes it out of this loop. This is easily done by clicking the computer's mouse so that the cursor is at the JMP MainLoop instruction following this little loop.

With the cursor just to the left of this JMP instruction, pull down the Debug menu and select Run to Cursor. You'll see that the program never gets to this point. Why not?

It's time to look at the timer and its control registers. Go to the Register window and in the pulldown menu select Timer A2. At this point you will need to do a manual Break in order to see the contents of these registers (memory and registers are only available in the IDE after a break). The timer control registers' contents will likely look like those of Figure 22.4. As with the ADC10 control registers, we can expand certain control registers, such as TACTL, to show the individual bits.

The very first thing to note is that the timer counter, TAR, has a value of 0000. Of course, that could be a coincidence (it could be counting and just happened to have stopped at 0000) but we can easily verify that that's not the case by selecting Go from the Debug

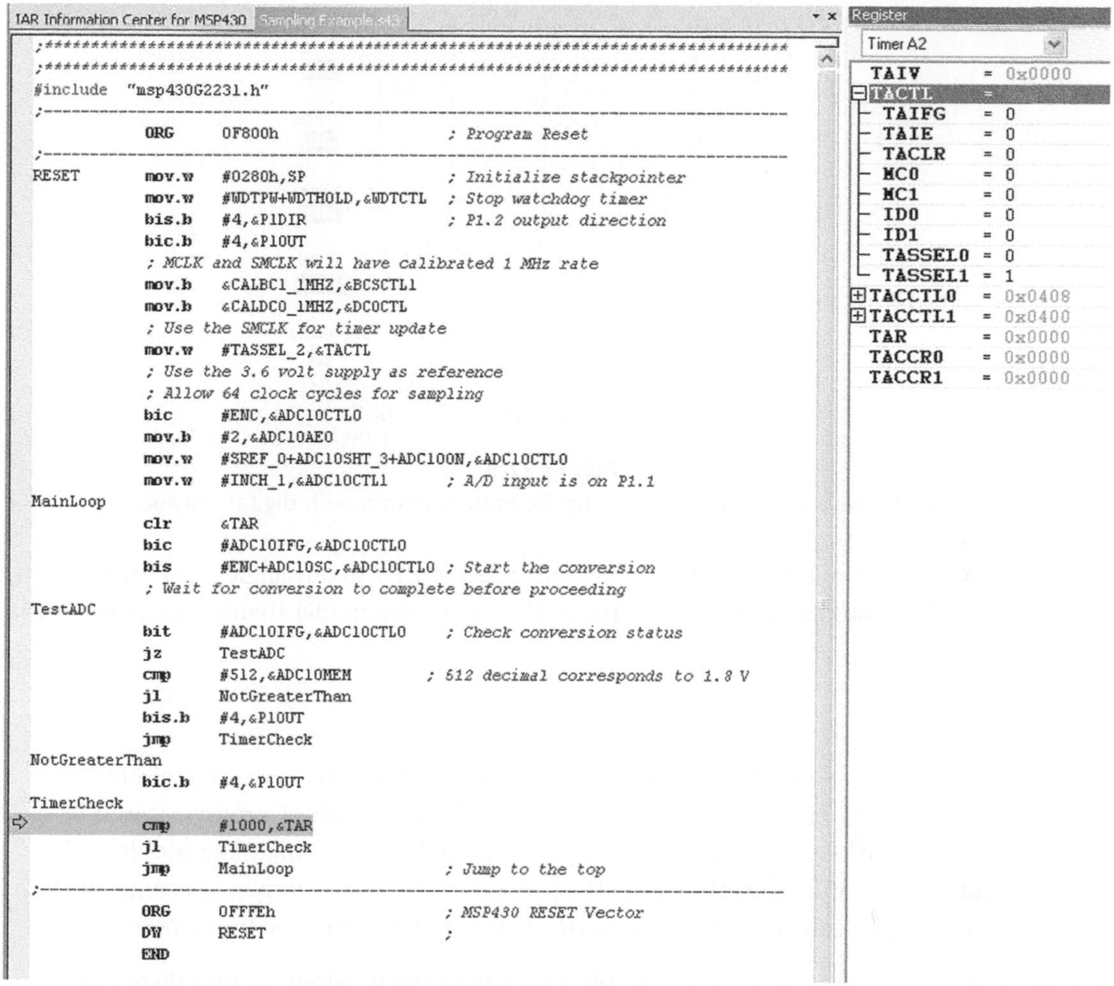

Figure 22.4
Sampling Example showing the timer control register contents.

pulldown menu, then selecting Break and seeing that the TAR contents are again at 0000. It turns out that the timer is not counting!

So why is that? If we examine the contents of TACTL, we see that the mode control bits, MC0 and MC1, have a value of 00. That corresponds to the Stop Mode! In the Stop Mode, the timer counter ignores the input clock and does not count. *The timer should be in the Continuous Mode*. This is easily corrected by changing the MOV.W #TASSEL_2,&TACTL to the instruction, MOV.W #TASSEL_2 + MC_2,&TACTL. Make this change in the program, Rebuild the program, Download and Debug, and then run the program.

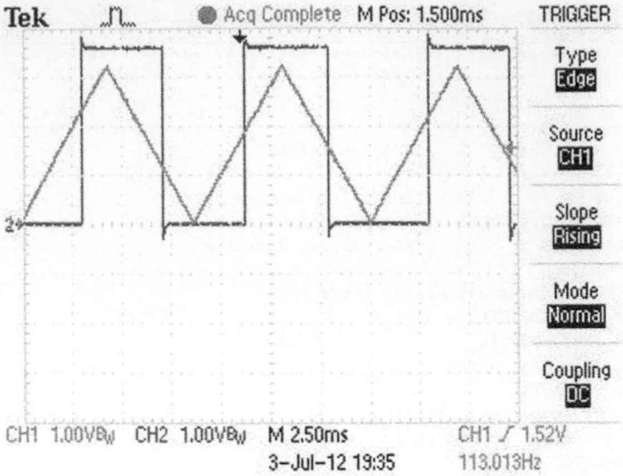

Figure 22.5
Triangle wave input to the Sampling Example program with digital output.

With this change, the microcontroller, when connected at P1.1 to a triangle waveform like the one in Figure 22.5, produces a digital output at P1.2 as shown in that figure. Problem solved.

Simplify!

One of the obstacles to troubleshooting both circuits and programs is that the circuit or program may be quite large. Erroneous behavior within the circuit or program could be due to a large number of potential problems and it can be difficult to isolate the problem. The way to tackle this is often to build a circuit or program that just incorporates some small part of the overall circuit or program, to verify that that particular function works.

For example, even the very simple example program in the previous section displayed interactions between timer and A/D that made isolating the problem tricky. One thing that could have been done to verify that the A/D was set up correctly was to start with the original program and eliminate all instructions not related to the A/D process. Then run this stripped-down program and verify that the A/D was working, using some known DC voltage, like the resistor-divider output that we used in that example.

We could also have simplified the original sample program to just timer-related instructions, eliminating all A/D-related instructions, as in the program in Figure 22.6. All instructions having to do with A/D operations are stripped from the program. In addition, the instructions involving setting or resetting P1.2 are stripped and replaced with a single instruction (XOR.B #4,&P1OUT) that toggles the P1.2 output pin each time through the loop.

In a similar manner, circuits can often be simplified, so that only one or two ICs are involved in the stripped-down function.

IAR Information Center for MSP430 | Sampling Example Minus A2D.s43

```
;*******************************************************************************
;*******************************************************************************
#include  "msp430G2231.h"
;-------------------------------------------------------------------------------
            ORG     0F800h                  ; Program Reset
;-------------------------------------------------------------------------------
RESET       mov.w   #0280h,SP               ; Initialize stackpointer
            mov.w   #WDTPW+WDTHOLD,&WDTCTL  ; Stop watchdog timer
            bis.b   #4,&P1DIR               ; P1.2 output direction
            bic.b   #4,&P1OUT
            ; MCLK and SMCLK will have calibrated 1 MHz rate
            mov.b   &CALBC1_1MHZ,&BCSCTL1
            mov.b   &CALDCO_1MHZ,&DCOCTL
            ; Use the SMCLK for timer update
            mov.w   #TASSEL_2+MC_2,&TACTL
MainLoop
            clr     &TAR
            xor.b   #4,&P1OUT
TimerCheck
            cmp     #1000,&TAR
            jl      TimerCheck
            jmp     MainLoop                ; Jump to the top
;-------------------------------------------------------------------------------
            ORG     0FFFEh                  ; MSP430 RESET Vector
            DW      RESET                   ;
            END
```

Figure 22.6
Example of simplified version of program, eliminating A/D operations.

Use the IC/software vendors' support services

As of the writing of this book, Texas Instruments provides a useful support service for MSP430 microcontroller users. Users can send email questions and code samples to the applications engineers there and receive, usually within a day or two, a reply. This can be very helpful. One note – you'll definitely want to send stripped-down versions of your program, as the applications engineers don't have the time to wade through your 200-line program. To get this help go to the TI website or the URL http://www-k.ext.ti.com/sc/technical-support/email-tech-support.asp?MCU.

Look around

This may sound like strange advice but, when you've got a circuit problem and can't find what's causing it, start looking around at points on the circuit other than those directly related to the function that's not behaving as expected. Probe all the pins of the IC that are associated with the function, not just the IC pins directly related to the problem function. Ask yourself if the signal that you see at each pin makes sense. This is advice that was given to me by an older engineer when I was just starting out and it's been helpful a number of times.

Specific troubleshooting tips

Here are some troubleshooting tips specific to this project and to related electronics/microcontroller projects. This is by no means a complete list of everything you'll run up against, but it includes some of the more common problems.

Options set incorrectly in IDE

When you assemble your program, most error messages that the IAR Workbench IDE returns are straightforward. One error message that's definitely not is the one shown in Figure 22.7.

That first error message at the bottom looks like some sort of code monkey gibberish. What actually caused the error is that, when I created the Project in the IDE, I forgot to set the MSP430 type to the MSP430G2211 that I was using (I did manage to include MSP430G2211 in the "Include" statement at the top of the program). This got the assembler confused, which resulted in this very odd response.

Specifying a decimal number when a hexadecimal number is intended

Let's say that in the program shown in Figure 22.6, we decided to specify the constant 1000_{10} as its hexadecimal equivalent, $3E8_{16}$, instead. The program in Figure 22.8 shows the replacement, but notice that no indication was given to the assembler that this was intended to be a hexadecimal number rather than a decimal number. When no "0x" prefix or "h" suffix is used for a constant, the assembler assumes that the number is decimal, in which case the only valid numerals are 0 through 9. Seeing an "E", the assembler immediately knows something isn't right.

So, the assembler picked up our error. But what if we had intended to make the constant 398_{16}? In that case, the assembler again assumes the number is supposed to be a decimal constant (because it doesn't find a "0x" prefix or an "h" suffix) so it treats the number as 398_{10} and doesn't alert us to an error (it can't, since it has no way of knowing that we really wanted 398_{16} instead of 398_{10}).

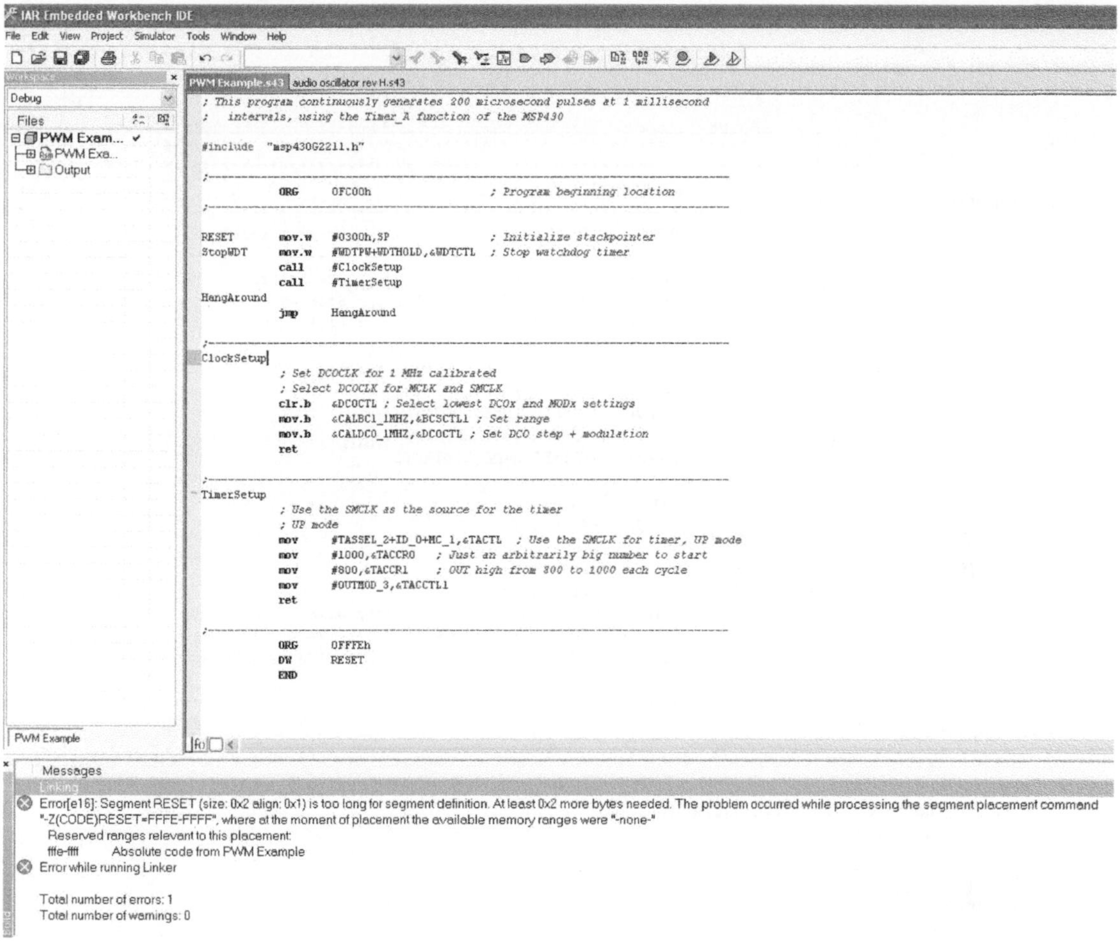

Figure 22.7
Error message resulting from options not being set properly.

Differences between the BIS and MOV instructions

Let's say that the register R5 has contents 0101 1101 1011 0011_2. So what is the difference between the following two instructions:

```
BIS     #4,R5
```

and

```
MOV     #4,R5?
```

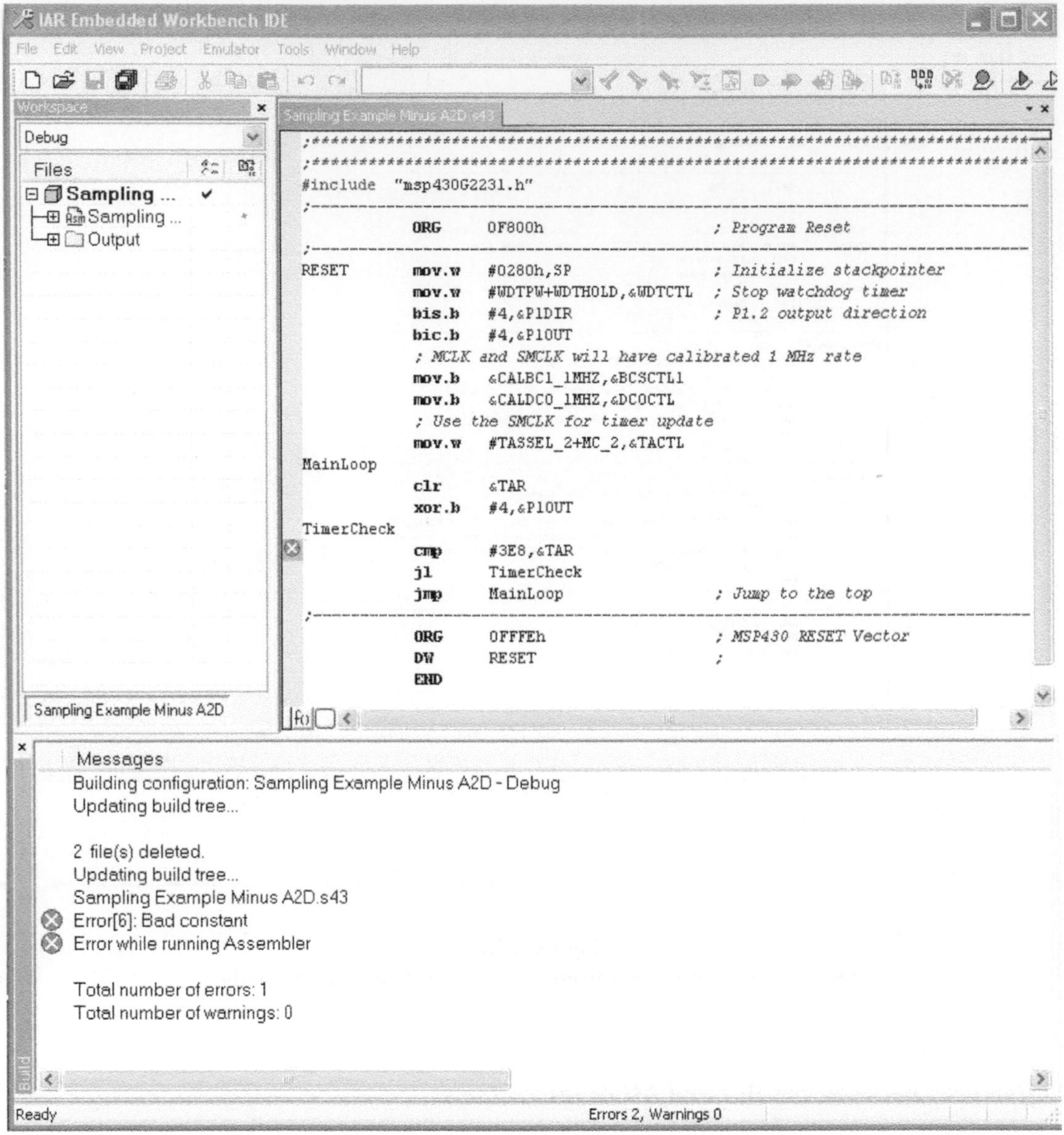

Figure 22.8
Error message due to hexadecimal numeral when decimal is assumed.

The answer, of course, is that register R5 will contain 0101 1101 1011 0111_2 after the BIS instruction but will contain 0000 0000 0000 0100_2 after the MOV instruction. The BIS instruction affects only those bits corresponding to a 1 in the source operand, while the MOV affects all 16 bits (or 8 bits, if the instruction is a byte instruction). This may seem obvious, but it's a common source of error.

Forgetting to use a "#" before the label for a CALL instruction

This is a common error and one that the assembler will not flag. The program will execute the CALL instruction and send the Program Counter off to unexpected places within the microcontroller's memory.

Unexpected resets

Unexpected resets can occur in a number of ways. "Brown-out" resets can occur if a hardware fault causes the microcontroller's supply voltage, V_{CC}, to sag due to excessive current being drawn during the fault.

Unexpected software-induced resets can occur in a number of ways. One way this can happen is that the program can overwrite the RAM locations where the stack is storing Program Counter contents for subroutine calls. Remember that the microcontroller stores the Program Counter contents before executing a CALL instruction, so that it knows where to return to when the subroutine's RET instruction is executed. It stores this on the "stack", which is really just a section of memory that the programmer designates for that purpose. The stack is normally the last few words of RAM. If the subroutine should inadvertently write over the stack locations, the microcontroller will use the corrupted stack contents and load them into the Program Counter upon subroutine return, sending program execution to some bogus part of memory. The microcontroller will "execute" these meaningless instructions in this unintended program memory. Eventually, program execution will make its way to the Reset location of the program (for example, $E000_{16}$ in the MSP430G2452 or $F800_{16}$ in the MSP430G2231).

To find that the microcontroller is doing this, you could single-step through the instructions (using the Single Step command from the Debug pulldown menu of the IDE) and step through until the Reset occurs, indicating that the last instruction before Reset is the cause. Or you can "comment out" groups of instructions, using the Block Comment command in the Edit pulldown menu of the IDE, then rebuild the program and run the program without those instructions, to narrow down to the offending instruction. Once things are fixed, you can do a Block UnComment from the Edit pulldown menu, rebuild, and download the fixed program.

ADC10 encode bit

Note that, once you set the ENC bit in the ADC10CTL0 control register of the ADC10 function of the MSP430, there are a whole bunch of other ADC10 control register bits, both in the ADC10CTL0 and the ADC10CTL1 control registers, which cannot be changed. Your program can go ahead and try to change these bits but the MSP430 microcontroller hardware will ignore such changes.

Troubleshooting wrap-up

The suggestions and examples in this chapter are intended as a start to the process of troubleshooting. As you will find out, troubleshooting is an important skill, one that can be accelerated by advice from experienced individuals and manufacturers, but also one that requires just getting in there and puzzling over things until they're solved.

Bibliography

[1] Email Semiconductor Design Support, Texas Instruments. <http://www-k.ext.ti.com/sc/technical-support/email-tech-support.asp?MCU>.

CHAPTER 23

Creating a Real-Time Operating System

Chapter Outline

We've got pretty much all the functions that we will need to make the robot work. We've got:

- Ultrasonic transmission and reception for obstacle distance measurement
- LED driver and reception to augment this obstacle detection
- High voltage (15 V) to power the LEDs and the ultrasonic transmitters
- Motor drive
- Steering drive

In addition, we'll need to add the tasks of:

- Decision-making (using the sensor data to make decisions about speed and direction)
- Do nothing (there will be periods when we just want the microcontroller to idle)

It's time to decide how to put these together.

MSP430-based Robot Applications.
DOI: http://dx.doi.org/10.1016/B978-0-12-397012-1.00023-0

This might seem like a simple task, maybe nothing more than just combining all the shorter programs into one longer program. But there's more to it than that. Some of these tasks need to occur at precise intervals. For example, the ultrasonic transmitter is energized with a 40 kHz square wave for a short interval and then the reflected signal is sampled at regular intervals. The presence of a signal above threshold at a particular sampling point allows the calculation of the distance to the obstacle creating the reflection. *We can precisely calculate the distance only if we precisely know the time between transmitted signal and reflection occurrence.*

We also need to be aware of timing for the purpose of setting the duty cycle of the motor drive. Remember that the motor drive varies the amount of power delivered to the motor by turning the voltage on or off at a particular duty cycle. Thus, we may wish to set things up so that we vary the duty cycle from, say, 10% on (very slow) to 100% (maximum speed). So awareness of time is important for this function as well.

Juggling multiple tasks using a real-time operating system

I hesitate to use the phrase real-time operating system (usually abbreviated RTOS), simply because it sounds like a big, scary thing. As you're probably aware, your home computer has an operating system (OS), which controls how and when the different computer tasks are performed. It must be very complicated because it requires hundreds of megabytes on the computer and can cost a lot of money. Fortunately, the RTOS that we'll develop is relatively simple. But before we discuss what's needed, it's a good idea to figure out what the difference is between an operating system and a real-time operating system.

Just what is a real-time operating system?

The important word to notice in the phrase real-time operating system is *time*. This is what sets an RTOS apart from just an OS. The RTOS must keep track of time in the external world that it is sensing or driving. So an accurate measure of time is essential (fortunately, we have this with the MSP430 timer/counter and a calibrated clock).

An RTOS needs to be able to do the following three things:

- Schedule tasks
- Dispatch tasks
- Provide intertask communications

In this chapter, we'll develop a structure that allows the MSP430 to perform these functions.

Scheduling

We have a number of tasks that need to be performed and we need to determine first of all whether we can perform all these tasks in the allotted time. Let's list the tasks and then think about how often each needs to be executed.

Ultrasonic transmission

We need to transmit a square-wave set of pulses through the ultrasonic transmitter for roughly 1 to 20 cycles (25 to 500 μsec), then measure the reflections. How often do we need to repeat this transmit process? The robotic vehicle will have a maximum speed of the order of 10 feet per second, so 120 inches per second.

If we begin a new transmit and measure process every 1/60 of a second, then, at top speed, we will be updating the distance measurements every 2 inches. Given the fact that the robot will be operating in a static environment (obstacles are assumed not moving), this seems like a reasonable resolution. Therefore an update rate of 60 times per second seems adequate for the ultrasonic distance measuring process.

Ultrasonic reception

Each time an ultrasonic burst occurs, the ultrasonic receiver and the microcontroller must then periodically sample the received signal. When the first such sample is above a pre-established threshold, the first obstacle is found. Determining the distance to the object is simply a matter of multiplying this time measurement by the speed of sound, which is 13,500 inches per second at room temperature.

For our purposes, 3 inches is probably a reasonable resolution. The fact that an object is 1.75 feet from the robot rather than 2 feet from the robot will make little difference in terms of the robot's response. Therefore, a good sampling rate for the ultrasonic reception is of the order of:

$$\text{Sampling Rate} = \frac{3}{13500}\ \text{sec} = 220\ \mu\text{sec}. \qquad (23.1)$$

Ultrasonic transmission/reception timing summary

Based on these calculations, we should transmit an ultrasonic burst every 1/60 seconds = 16.67 milliseconds and then sample, at 220 μsec intervals, the received signal reflected back from this burst.

There is one more thing. What if we are still receiving reflected pulses more than 16.67 milliseconds after a burst? In that case, the reflected pulse will look like a near event for the next burst. To avoid this situation, there should be no significant energy reflected at

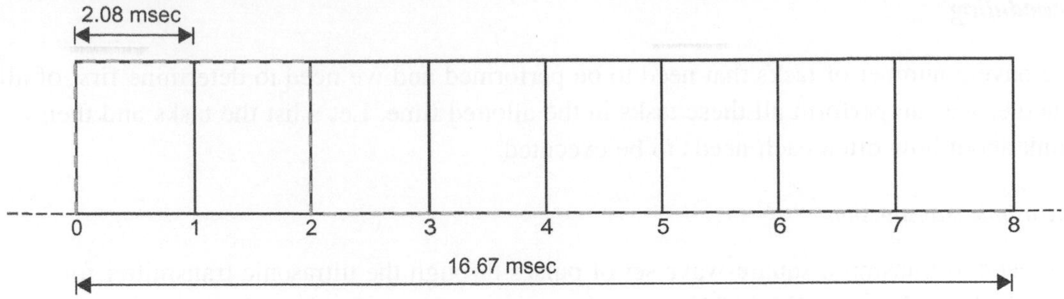

Figure 23.1
PWM motor drive options for increments of 1/8 duty cycle.

distances corresponding to times greater than 16.67 milliseconds. This translates to a distance of about 19 feet. In the event that there are sometimes reflection events of obstacles at distances this great, the high voltage to the ultrasonic transmitter would need to be reduced.

Motor drive

A PWM drive signal, based on a 16.67 msec period, produces a smooth drive to the vehicle's motor. The combination of the motor's response lag plus the inertia of the vehicle itself is the source of this smoothing. At low speeds there may be a slight jerking of the vehicle at this update rate, but this should pose no performance problem.

Allowing speeds of 1/8 duty cycle, 2/8, 3/8, etc., up to 100% drive gives reasonable speed selection to accommodate different drive scenarios. Thus, a good way to adjust the PWM drive is to turn it on at the beginning of the 16.67 millisecond period, then check the speed setting at equally spaced intervals during the 16.67 milliseconds, as shown in Figure 23.1.

As the figure shows, the PWM drive starts at the beginning of the 16.67 msec period. The intended speed of the number will be stored as a number between 0 and 8 at some memory location within the duty cycle. At 2.08 msec intervals, a counter is incremented and this count is compared to the memory location containing the intended speed. A decision can then be made about whether to turn the drive off or not.

The dashed line of the figure shows a 25% duty cycle motor drive signal. In that case, the memory location would have contained a 2 and, after the second 2.08 msec interval and the counter containing a 2, the program would have determined that the PWM signal should be changed to zero volts.

Steering

The steering on an inexpensive radio-controlled car is of the "bang-bang" variety. That is, the front wheels are slammed all the way to the left or to the right when a turn is made – there is no such thing as a slight turn. Also, a relatively long signal is needed to cause the wheels to move. A turn signal of 100 millisecond length works well.

High voltage generation

For the load presented by the LED drive circuits and the ultrasonic transmitter and for the inductor and other component choices used in this project, the boost circuit MOSFET switch will need to be turned on for a time of between 100 and 300 μsec.

Scheduling

Given these requirements, a convenient way to set up the schedule of events is to create an infinite loop that repeats every 16.67 msec. We will refer to this interval as a *frame*. Now, let's partition this frame into *time slices*. A convenient choice for the number of time slices in a frame is 256. So each time slice will be (16.67 msec/256) = 65.1 μsec.

Most of the functions just described can be performed in just one time frame. Two functions, High Voltage Generation and Ultrasonic Transmission, will need multiple time slices. There are two functions, Motor Drive and Ultrasonic Reception, which will need to be executed repeatedly within the frame.

In the case of the Motor Drive, the function is executed eight times within the frame, ideally every 2.08 milliseconds (32 time slices). In the case of the Ultrasonic Reception, sampling the received ultrasonic signal every four time slices results in a sampling rate of once per 260 μsec.

Let's assign a code to each type of task. These codes are arbitrarily chosen and are as shown in Figure 23.2.

The codes do a couple of things for us. First, they allow for a very concise way of identifying the task. We will create a 256-element table in the microcontroller's memory which will show what task should be performed at each time slice. Having a number between 0 and 15 requires only a single byte (in theory, it could be just 4 bits) per time slice to store the task identification for that time slice, whereas storing a string of characters for each time slice, such as "Ultrasonic Transmission", would require more than 20 bytes per time slice. Using these short codes to identify the tasks is a very simple example of what is known as data compression. As we will see shortly, the codes also provide an easy

Task	Code
Reserved	0
Reserved	1
High voltage generation	2
Steering	3
Motor drive	4
LED left	5
LED right	6
Ultrasonic transmission	7
Ultrasonic reception	8
Decision making	9
Idle	10
Reserved	11
Reserved	12
Reserved	13
Reserved	14
Reserved	15

Figure 23.2
Tasks and corresponding codes.

way to jump to the particular program section corresponding to the task to be performed in that time slice.

The schedule

Now, let's create a scheduling list. Here, the 256 tasks are listed in eight columns, each column containing 32 time slices. We can populate the time slices based on the constraints that we've already discussed for each type of function. Figure 23.3 shows such a table.

At the beginning of each frame, the microcontroller performs the Motor PWM task. Assuming the vehicle is to have any speed at all, the motor setting is turned on during this first occurrence of Motor PWM. Note that the Motor PWM task is repeated every 32 time slices (2.08 msec).

The Motor PWM task is followed by several frames of High Voltage Generation, then a time slice period performing the Steering task, then pulsing the left and right banks of LEDs, then Ultrasonic Transmission.

Time Slice	Task Description	Code	Time Slice	Task Description	Code	Time Slice	Task Description	Code	Time Slice	Task Description	Code	Time Slice	Task Description	Code	Time Slice	Task Description	Code	Time Slice	Task Description	Code	Time Slice	Task Description	Code
0	Motor PWM	4	32	Motor PWM	4	64	Motor PWM	4	96	Motor PWM	4	128	Motor PWM	4	160	Motor PWM	4	192	Motor PWM	4	224	Motor PWM	4
1	High Voltage	2	33		10	65		10	97		10	129		10	161		10	193		10	225		10
2	High Voltage	2	34		10	66		10	98		10	130		10	162		10	194		10	226		10
3	High Voltage	2	35	Ultrasonic RX	8	67	Ultrasonic RX	8	99	Ultrasonic RX	8	131	Ultrasonic RX	8	163	Ultrasonic RX	8	195	Ultrasonic RX	8	227	Ultrasonic RX	8
4	High Voltage	2	36		10	68		10	100		10	132		10	164		10	196		10	228		10
5	High Voltage	2	37		10	69		10	101		10	133		10	165		10	197		10	229		10
6	High Voltage	2	38		10	70		10	102		10	134		10	166		10	198		10	230		10
7	Steering	3	39	Ultrasonic RX	8	71	Ultrasonic RX	8	103	Ultrasonic RX	8	135	Ultrasonic RX	8	167	Ultrasonic RX	8	199	Ultrasonic RX	8	231	Ultrasonic RX	8
8	Pulse L LED	5	40		10	72		10	104		10	136		10	168		10	200		10	232		10
9	Pulse R LED	6	41		10	73		10	105		10	137		10	169		10	201		10	233		10
10	Ultrasonic TX	7	42		10	74		10	106		10	138		10	170		10	202		10	234		10
11	Ultrasonic TX	7	43	Ultrasonic RX	8	75	Ultrasonic RX	8	107	Ultrasonic RX	8	139	Ultrasonic RX	8	171	Ultrasonic RX	8	203	Ultrasonic RX	8	235	Ultrasonic RX	8
12	Ultrasonic TX	7	44		10	76		10	108		10	140		10	172		10	204		10	236		10
13	Ultrasonic TX	7	45		10	77		10	109		10	141		10	173		10	205		10	237		10
14	Ultrasonic TX	7	46		10	78		10	110		10	142		10	174		10	206		10	238		10
15	Ultrasonic TX	7	47	Ultrasonic RX	8	79	Ultrasonic RX	8	111	Ultrasonic RX	8	143	Ultrasonic RX	8	175	Ultrasonic RX	8	207	Ultrasonic RX	8	239	Ultrasonic RX	8
16	Ultrasonic TX	7	48		10	80		10	112		10	144		10	176		10	208		10	240		10
17		10	49		10	81		10	113		10	145		10	177		10	209		10	241		10
18		10	50		10	82		10	114		10	146		10	178		10	210		10	242		10
19		10	51	Ultrasonic RX	8	83	Ultrasonic RX	8	115	Ultrasonic RX	8	147	Ultrasonic RX	8	179	Ultrasonic RX	8	211	Ultrasonic RX	8	243	Ultrasonic RX	8
20		10	52		10	84		10	116		10	148		10	180		10	212		10	244		10
21		10	53		10	85		10	117		10	149		10	181		10	213		10	245		10
22		10	54		10	86		10	118		10	150		10	182		10	214		10	246		10
23	Ultrasonic RX	8	55	Ultrasonic RX	8	87	Ultrasonic RX	8	119	Ultrasonic RX	8	151	Ultrasonic RX	8	183	Ultrasonic RX	8	215	Ultrasonic RX	8	247	Ultrasonic RX	8
24		10	56		10	88		10	120		10	152		10	184		10	216		10	248		10
25		10	57		10	89		10	121		10	153		10	185		10	217		10	249		10
26		10	58		10	90		10	122		10	154		10	186		10	218		10	250		10
27	Ultrasonic RX	8	59	Ultrasonic RX	8	91	Ultrasonic RX	8	123	Ultrasonic RX	8	155	Ultrasonic RX	8	187	Ultrasonic RX	8	219	Ultrasonic RX	8	251	Ultrasonic RX	8
28		10	60		10	92		10	124		10	156		10	188		10	220		10	252		10
29		10	61		10	93		10	125		10	157		10	189		10	221		10	253		10
30		10	62		10	94		10	126		10	158		10	190		10	222		10	254		10
31	Ultrasonic RX	8	63	Ultrasonic RX	8	95	Ultrasonic RX	8	127	Ultrasonic RX	8	159	Ultrasonic RX	8	191	Ultrasonic RX	8	223	Ultrasonic RX	8	255	Make Decision	9

Figure 23.3
Time slice schedule table.

After this is a task called Make Decision. This is a program segment that takes the inputs gathered from the Ultrasonic receiver and the LED pulsing during the frame and makes decisions about speed, direction, and steering. It stores the results of these decisions so that they can be used during the following frame.

The additional tasks during the frame consist of the Ultrasonic Reception sampling that takes place every fourth time slice. The time slices where the microcontroller does nothing (Idle) are left blank in Figure 23.3.

To finish the table, let's insert the codes from Figure 23.2. The finished schedule is shown in Figure 23.4. Time slices are color-coded so that tasks of a particular type have the same color.

So Figure 23.4 shows the microcontroller's activity during the 256 65 μsec time slices. So where is this information stored? A microcontroller like the MSP430G2452 has 8 K bytes of program memory, so there should be plenty of room for this table, in addition to the program.

The MSP430G2452's program memory starts at $E000_{16}$, so let's start the scheduling table at $F800_{16}$. There are only 256 bytes to the scheduling table, so the last table entry will be at $F8FF_{16}$. The program memory, starting at $F800_{16}$, will just look like Figure 23.5. Note that both address and data are shown as hexadecimal.

Time Slice	Task Description	Code	Time Slice	Task Description	Code	Time Slice	Task Description	Code	Time Slice	Task Description	Code	Time Slice	Task Description	Code	Time Slice	Task Description	Code	Time Slice	Task Description	Code	Time Slice	Task Description	Code
0	Motor PWM	4	32	Motor PWM	4	64	Motor PWM	4	96	Motor PWM	4	128	Motor PWM	4	160	Motor PWM	4	192	Motor PWM	4	224	Motor PWM	4
1	High Voltage	2	33		10	65		10	97		10	129		10	161		10	193		10	225		10
2	High Voltage	2	34		10	66		10	98		10	130		10	162		10	194		10	226		10
3	High Voltage	2	35	Ultrasonic RX	8	67	Ultrasonic RX	8	99	Ultrasonic RX	8	131	Ultrasonic RX	8	163	Ultrasonic RX	8	195	Ultrasonic RX	8	227	Ultrasonic RX	8
4	High Voltage	2	36		10	68		10	100		10	132		10	164		10	196		10	228		10
5	High Voltage	2	37		10	69		10	101		10	133		10	165		10	197		10	229		10
6	High Voltage	2	38		10	70		10	102		10	134		10	166		10	198		10	230		10
7	Steering	3	39	Ultrasonic RX	8	71	Ultrasonic RX	8	103	Ultrasonic RX	8	135	Ultrasonic RX	8	167	Ultrasonic RX	8	199	Ultrasonic RX	8	231	Ultrasonic RX	8
8	Pulse L LED	5	40		10	72		10	104		10	136		10	168		10	200		10	232		10
9	Pulse R LED	6	41		10	73		10	105		10	137		10	169		10	201		10	233		10
10	Ultrasonic TX	7	42		10	74		10	106		10	138		10	170		10	202		10	234		10
11	Ultrasonic TX	7	43	Ultrasonic RX	8	75	Ultrasonic RX	8	107	Ultrasonic RX	8	139	Ultrasonic RX	8	171	Ultrasonic RX	8	203	Ultrasonic RX	8	235	Ultrasonic RX	8
12	Ultrasonic TX	7	44		10	76		10	108		10	140		10	172		10	204		10	236		10
13	Ultrasonic TX	7	45		10	77		10	109		10	141		10	173		10	205		10	237		10
14	Ultrasonic TX	7	46		10	78		10	110		10	142		10	174		10	206		10	238		10
15	Ultrasonic TX	7	47	Ultrasonic RX	8	79	Ultrasonic RX	8	111	Ultrasonic RX	8	143	Ultrasonic RX	8	175	Ultrasonic RX	8	207	Ultrasonic RX	8	239	Ultrasonic RX	8
16	Ultrasonic TX	7	48		10	80		10	112		10	144		10	176		10	208		10	240		10
17		10	49		10	81		10	113		10	145		10	177		10	209		10	241		10
18		10	50		10	82		10	114		10	146		10	178		10	210		10	242		10
19		10	51	Ultrasonic RX	8	83	Ultrasonic RX	8	115	Ultrasonic RX	8	147	Ultrasonic RX	8	179	Ultrasonic RX	8	211	Ultrasonic RX	8	243	Ultrasonic RX	8
20		10	52		10	84		10	116		10	148		10	180		10	212		10	244		10
21		10	53		10	85		10	117		10	149		10	181		10	213		10	245		10
22		10	54		10	86		10	118		10	150		10	182		10	214		10	246		10
23	Ultrasonic RX	8	55	Ultrasonic RX	8	87	Ultrasonic RX	8	119	Ultrasonic RX	8	151	Ultrasonic RX	8	183	Ultrasonic RX	8	215	Ultrasonic RX	8	247	Ultrasonic RX	8
24		10	56		10	88		10	120		10	152		10	184		10	216		10	248		10
25		10	57		10	89		10	121		10	153		10	185		10	217		10	249		10
26		10	58		10	90		10	122		10	154		10	186		10	218		10	250		10
27	Ultrasonic RX	8	59	Ultrasonic RX	8	91	Ultrasonic RX	8	123	Ultrasonic RX	8	155	Ultrasonic RX	8	187	Ultrasonic RX	8	219	Ultrasonic RX	8	251	Ultrasonic RX	8
28		10	60		10	92		10	124		10	156		10	188		10	220		10	252		10
29		10	61		10	93		10	125		10	157		10	189		10	221		10	253		10
30		10	62		10	94		10	126		10	158		10	190		10	222		10	254		10
31	Ultrasonic RX	8	63	Ultrasonic RX	8	95	Ultrasonic RX	8	127	Ultrasonic RX	8	159	Ultrasonic RX	8	191	Ultrasonic RX	8	223	Ultrasonic RX	8	255	Make Decision	9

Figure 23.4

Completed schedule with color coding for like tasks.

Finding the task associated with a particular time slice is easily performed as a lookup from this table. Let's say that the time slice that the program is executing is the 18th one (since we start counting at 0, this is actually the time slice labeled 17). By adding 11_{16} (that is, 17_{10}) to the table start location of $F800_{16}$, we have the address $F811_{16}$. We simply read the byte of data at that location ($0A_{16}$) and we know that we are to perform the task associated with this number (from the table in Figure 23.2, this is Idle).

The program executing the schedule keeps track of what time slice should be executed by simply incrementing an *index*. The index in the example just given had the value 17. After executing the Idle program segment (which is nothing more than just looping until 65 μsec has gone by) the index is incremented, to 18. This new index is added to the scheduling table start location, $F800_{16}$, and the contents of the byte pointed to by this new address, $F812_{16}$, contain the new task to be performed.

Handling the steering command

As mentioned earlier in the chapter, the Steering command will need to be held in its Left or Right state for a period of perhaps 100 msec. This means that the Steering command, either Left or Right, will need to be held over several 16.67 msec frames. We will therefore implement a frame counter that is incremented by 1 in each new frame. This frame counter is nothing more than a 16-bit word in RAM that we will use for the purpose of storing this free-running count.

Program memory location	Contents
F800	04
F801	02
F802	02
F803	02
F804	02
F805	02
F806	02
F807	03
F808	05
F809	06
F80A	07
F80B	07
F80C	07
F80D	07
F80E	07
F80F	07
F810	07
F811	0A

•
•
•

Figure 23.5
Scheduling table in program memory, starting at location $F800_{16}$.

If we wish to hold the Steering command for, say, six frames, then, if the frame count location currently holds the value N, we will hold the Steering command until the frame count is $N + 6$.

Dispatching the tasks

Okay, so we've got a straightforward way to figure out what task to perform at each new time slice – by reading the byte location at the address pointed to by the sum of the scheduling table start location and the index. But once we have this task number, how do we efficiently figure out where to jump to, in order to execute the appropriate task?

The jump table

Let's say that, for example purposes, we have just three different tasks. Each time we read a byte location in the scheduling table, the result is 0, 1, or 2. In assembly language, we can make use of something called a jump table.

Assume here that the task number, read from the scheduling table, is stored in a RAM location called Task. The jump table would look like this:

```
    rla      &Task
    add      &Task,PC
        jmp      Task0
        jmp      Task1
        jmp      Task2
Task0
    .
    .
    .
    jmp      NextTimeSlice
Task1
    .
    .
    .
    jmp      NextTimeSlice
Task2
    .
    .
    .
NextTimeSlice
```

What on earth is this doing? The first thing to realize is that jump instructions always occupy one word in program memory. The most significant 3 bits of the instruction identify the instruction as a jump instruction, the next 3 bits identify what type of jump (for example, JGE, JZ, JMP, etc.), and the least significant 10 bits are the offset from the location of the jump instruction to the location corresponding to the label used in the jump instruction. Actually, the information in the previous sentence is superfluous. All you really need to know is that the JMP instruction (as well as any of the conditional jump instructions) takes up exactly one word of program memory.

Note that the first thing the program does is to shift the contents of Task left one bit. We'll talk about why that's done shortly. At this point, just make note that a shift one bit left is equivalent to multiplying the contents of Task by 2. So, after executing the shift left, Task contains 0, 2, or 4.

Now, let's talk about that ADD instruction, in which the contents of Task (shifted left one bit) are added to the microcontroller's program counter (PC). That might seem unorthodox, but it's perfectly legitimate. The program counter is already pointing to the next instruction when the contents of Task (shifted left one bit) are added to it. The program counter, upon executing the ADD instruction, will point to the JMP Task0 instruction (if Task contains 0), to the JMP Task1 instruction (if Task contains 2), or to the JMP Task2 instruction (if Task contains 4).

Do you see why, now, the Task variable was shifted by 2? This was simply to account for the fact that there are two, not one, bytes in the jump instructions.

The real utility of the jump table comes about when we have a long list of tasks, as we have in this robot project. Rather than having to have a large number of compare and jump statements we just take the task number, multiply by 2, and add it to the program counter.

If you're familiar with C or C++ programming, you may have noticed how the jump table behaves very similarly to the Switch Case statements in those higher-level languages. Both the jump table in assembly language and the Switch Case in C or C++ are very efficient ways to dispatch a program to the correct section of code when a large number of choices are possible.

Intertask communications

At the beginning of the chapter, the three functions of an RTOS were listed as scheduling, dispatching the tasks, and intertask communications. In fact, in our very simple version of an RTOS, the intertask communication is trivial. All tasks have access to all of the RAM and exchange values that way.

RTOS wrap-up

As you can see, the RTOS provides the programming structure to control the timing of the robot program. Real-time operating systems can be very involved and are the subject of considerable research and literature. Fortunately, in the case of the robot program, the RTOS is a relatively simple structure.

Bibliography

[1] An Introduction to Real-Time Operating Systems, Quadros Systems Inc. <http://www.quadros.com/resources/white-papers/copy_of_introduction-to-real-time-operating-systems>.

[2] MSP430 Real Time Operating Systems Overview, Texas Instruments. <http://processors.wiki.ti.com/index.php/MSP430_Real_Time_Operating_Systems_Overview#PowerPac_for_MSP430.E2.84.A2_MCU>.

CHAPTER 24

Putting it all Together

Chapter Outline

If you've read all the chapters of this book, it probably seems like ages ago since we took the R/C car apart, removed the body, and prepared the wires to the motor and steering for connection to the board. We're finally ready to build the circuits and make the final connections.

We'll also talk about the complete program for the robot. This will include all the program segments that have been developed in previous chapters.

Final hardware

The schematic for all the electronics of the MSP430-based robot is shown in Figure 24.1. The schematic consists of six sheets and we'll briefly go through the functions contained on each sheet.

MSP430-based Robot Applications.
DOI: http://dx.doi.org/10.1016/B978-0-12-397012-1.00024-2

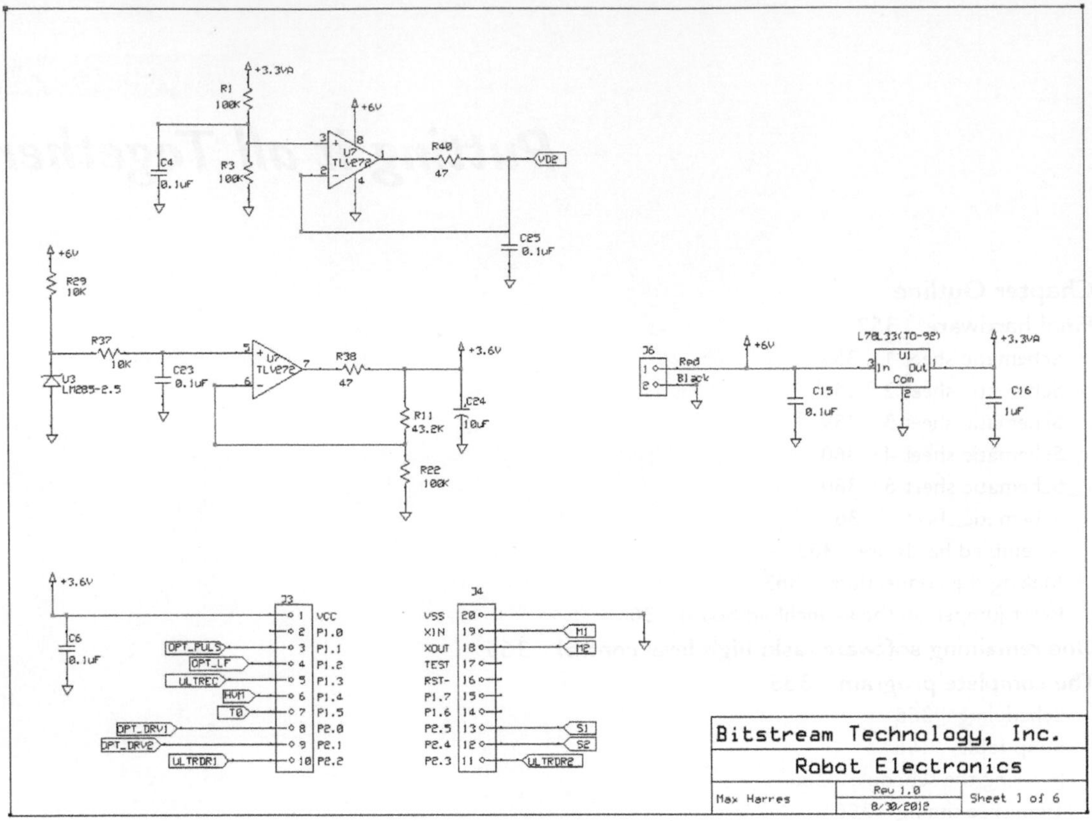

Figure 24.1a
Sheet 1, the LaunchPad connections and voltage regulation.

Schematic sheet 1

The first sheet (Figure 24.1a) shows the connections to the LaunchPad board (J3 and J4) as well as the voltage regulators. U1 provides the 3.3 V for the analog electronics and the U7 dual op amp provides the 3.6 V for the microcontroller as well as a reference voltage, *VD2*, that is half of the 3.3 V.

Schematic sheet 2

Sheet 2 (Figure 24.1b) contains two identical motor drive integrated circuits, based on the Toshiba TA7291. The left-hand IC is for the main robot motor and the right-hand IC is for the steering motor. These ICs can handle the several hundred milliamperes that the motors require.

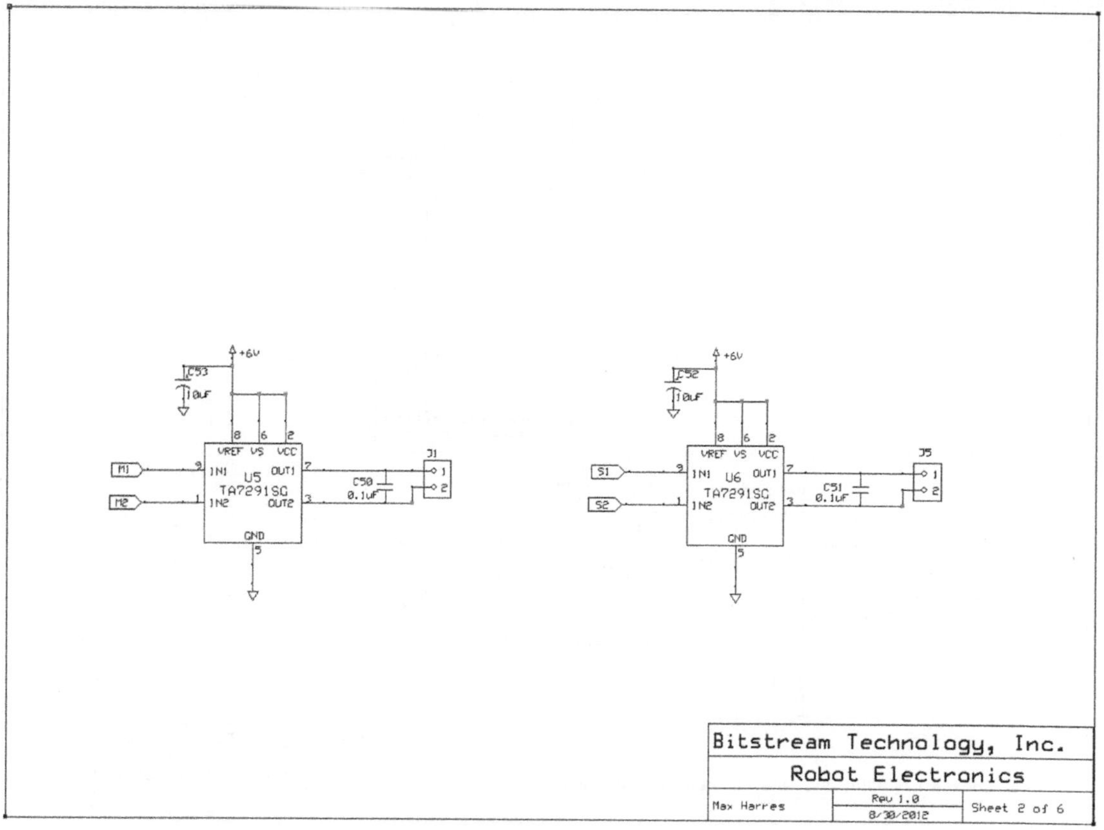

Figure 24.1b

Sheet 2, the steering and main motor drive circuits.

Schematic sheet 3

Sheet 3 (Figure 24.1c) includes the ultrasonic transmitter and its driver as well as the high voltage generator that provides the voltage needed for ultrasonic transmitter and LED drive operation. The ultrasonic transmitter is driven by one complete 40109 IC, which translates the 3.6 V signals from the microcontroller into the high voltage (approximately 15 V) that the ultrasonic transducer needs.

The high voltage generator circuit is a classic boost circuit that provides this 15 V. The control for this high voltage generation is provided by the MSP430 program, which monitors the voltage at C13 (the *HVM* signal) and adjusts the *T0* pulse width accordingly.

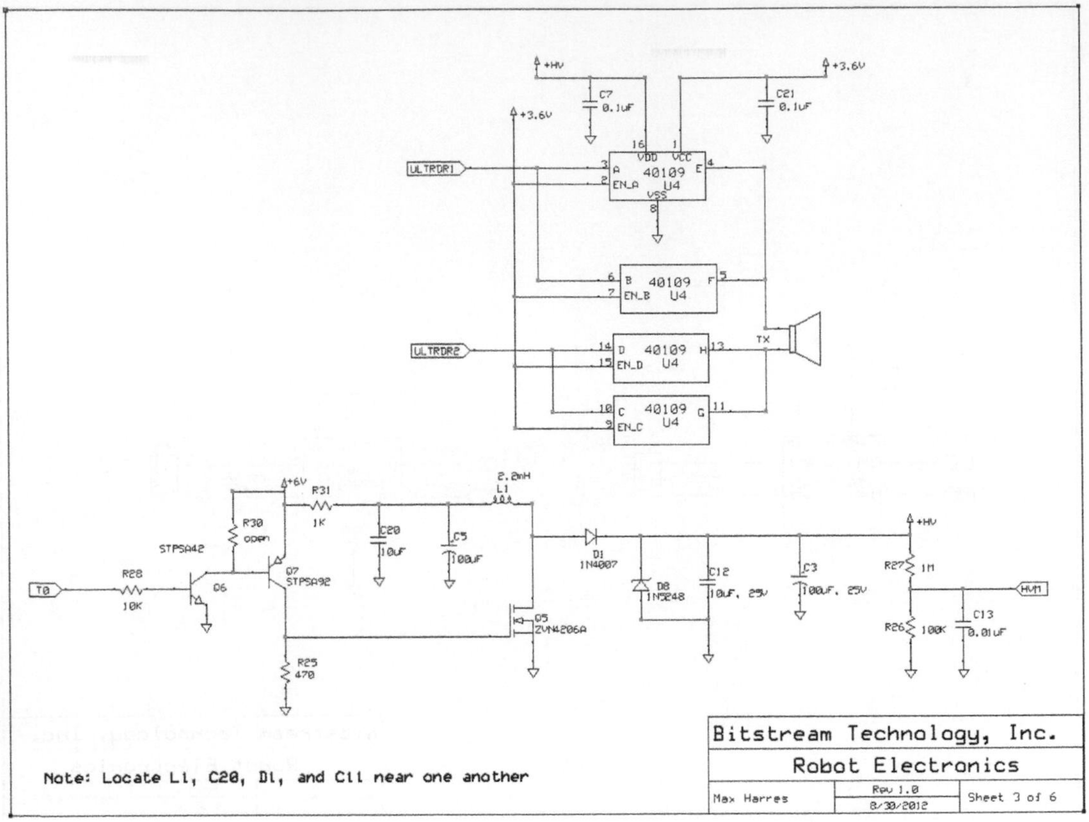

Figure 24.1c

Sheet 3, the ultrasonic transmitter and the high voltage generator.

Schematic sheet 4

Sheet 4 (Figure 24.1d) is the ultrasonic receiver circuit. It consists of two stages of gain for the ultrasonic reflected signal, followed by a rectifier (D6/R39/C8). Note that the gain stages are referenced to *VD2*, which is halfway between ground and the 3.3 V analog circuit power supply voltage.

Schematic sheet 5

Sheet 5 (Figure 24.1e) is the left/right LED "headlight" circuit. Recall that the purpose of this circuit is to flash light either from the left side of the board (which happens to be the left-hand circuit on sheet 5) or from the right side (right-hand circuit). The side producing the greater reflection is the side to avoid in the case of an obstacle being detected by the ultrasonic circuits.

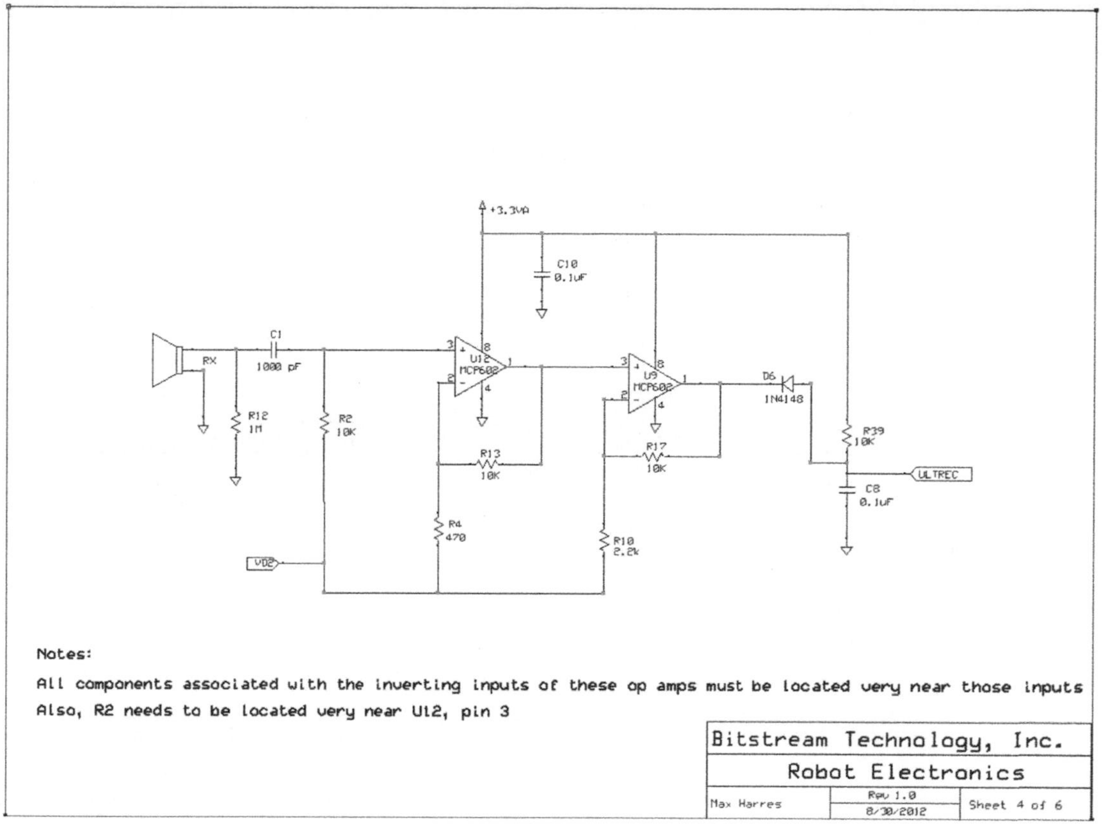

Figure 24.1d

Sheet 4, the ultrasonic receiver circuit.

These two LED driver circuits are able to produce nearly identical current for the LED drive, due to the op amp/transistor current driver configuration. Since each LED has a fairly large forward voltage of about 1.7 V and since the transistor and 10-ohm resistor require additional voltage, the LEDs derive their voltage from the high voltage on-board.

Schematic sheet 6

The last sheet (Figure 24.1f) of the schematic is the photodiode receiver for the LED "headlight" flash. This single circuit receives the light produced by either the left bank of LEDs on sheet 5 or the right bank of LEDs. The first stage is a transimpedance amplifier that converts the photodiode current into a voltage. The second stage is a gain-of-11 amplifier that is AC-coupled to the transimpedance amplifier (to eliminate

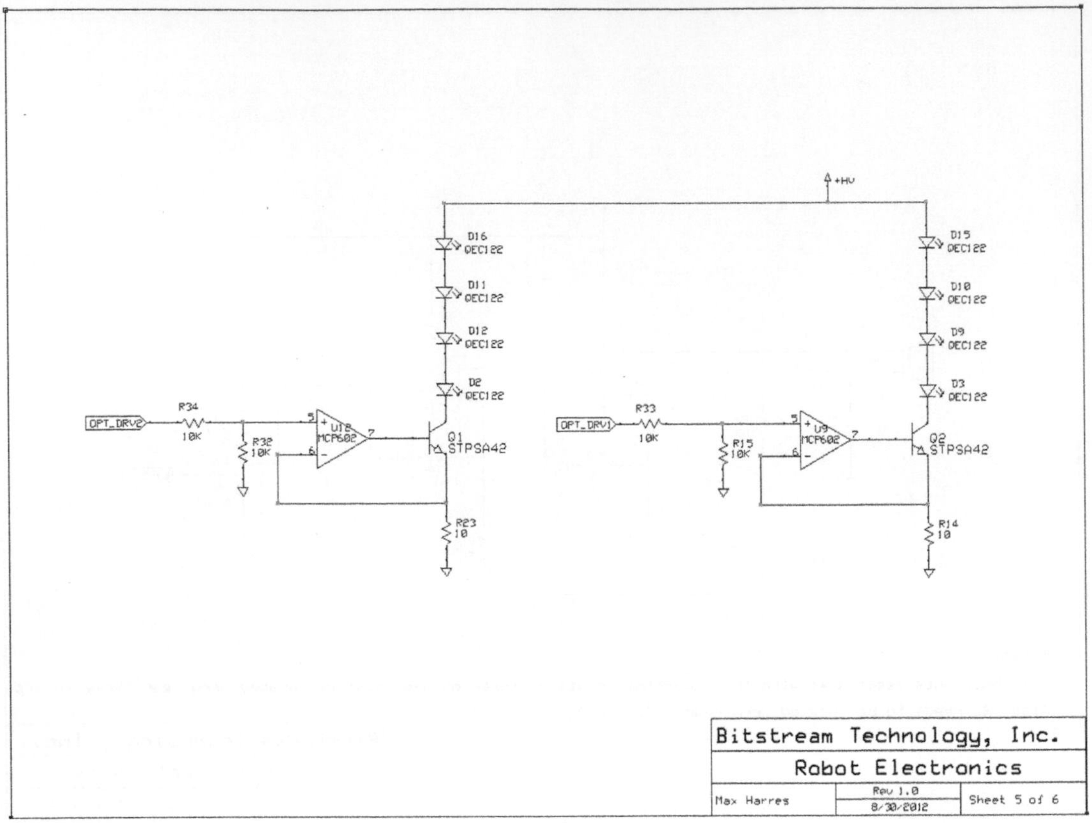

Figure 24.1e

Sheet 5, the LED drivers for the "headlight" flash.

as much of the ambient light as possible). Note that the reference voltage for this function is *VD2*.

Assembled hardware

Figure 24.2 shows the assembled electronics board. This was made from a printed circuit board sold by Bitstream Technology (www.bitstreamtechnology.com). Of course, as described in the chapter on circuit building, you may choose to build this board from vector board or some other construction technique.

Note that the board consists of the main electronics board (in front) plus the "piggyback" TI LaunchPad board (the red board) behind it. Figure 24.3 is a rear view of the main electronics board. The LaunchPad board has been removed in this photo to show the connectors, J3 and J4, that connect to the pins of the LaunchPad board.

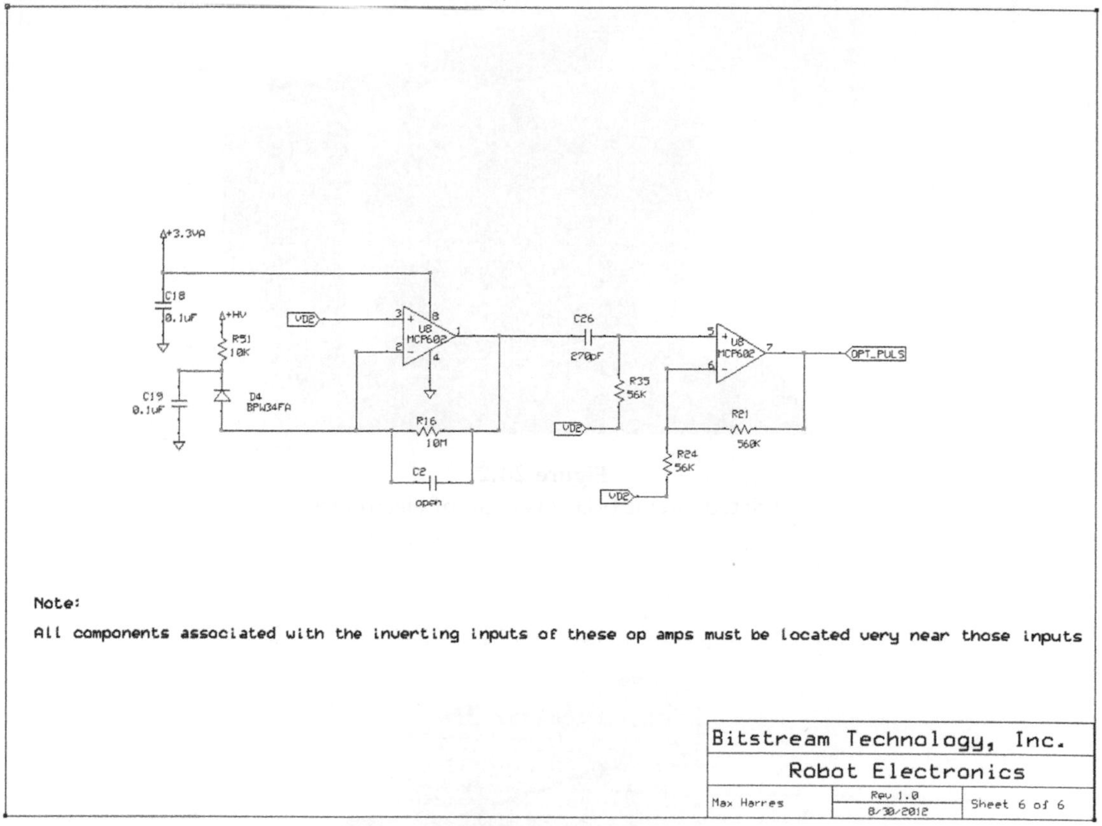

Figure 24.1f

The photodiode receiver circuit for detecting the LED "headlight" flashes.

Making the connections

Sheet 2 of the schematic showed the robot main drive motor and the steering motors. Note that each TA7291 driver IC had its output signals connected to a 2-pin connector (J1 for the main motor and J5 for the steering motor). In the Figure 24.2 photo these are two of the small blue connectors at the bottom of the main electronics board (the third blue connector brings in the power from the batteries). These connectors are where the wires from the radio-controlled cars drive motor and steering motor are connected.

Reset jumper on the LaunchPad board

We've mentioned jumper settings on the LaunchPad board but not really discussed the Reset jumper on that board. It turns out that when you use the LaunchPad as an extension

Figure 24.2
Printed circuit board version of electronics.

Figure 24.3
Back of electronics board with LaunchPad board removed (showing the mating connectors to the LaunchPad).

of your project board, as we are doing here, and not powering the auxiliary circuits on the LaunchPad board, just the MSP430 microcontroller, then you will need to remove the Reset jumper. There's no reason to power the auxiliary circuits when you're running your robot, so leave them unpowered (which you do by simply removing the VCC jumper). *So before running your robot, you should have both the VCC and Reset jumpers removed from J3 on the LaunchPad board.*

One remaining software task: high-level control

There is one more task to talk about before we get around to putting everything together. That is the high-level control. This is the function that takes the inputs from the LED/photodiode sensors and the ultrasonic sensor (plus any sensor that you might add in the future) and then processes this information to make decisions about speed, direction (forward or reverse) and steering.

There are two main conditions that the robotic vehicle will operate in – no obstructions and obstructions. When obstructions are "seen" by the vehicle's sensors, there are further decisions that will need to be made about turns, speed reduction, and direction.

For the purposes of getting started, the strategy will be a fairly simple one, but you'll be able to give your vehicle its own unique personality and optimize its performance by experimenting with this initial strategy.

- No obstructions
  For this condition, the vehicle will, of course, be run at maximum speed.
- Obstructions moderately far away on the left
  For this condition, speed needs to be reduced, since there is an obstruction ahead. The vehicle should turn right, to avoid hitting the obstruction. Speed is set to about half maximum.
- Obstructions moderately far away on the right
  The same comments apply as in the case of the obstruction moderately far away on the left, except that the vehicle turns left to avoid the obstruction.
- Obstructions nearby on the left
  In this case, the vehicle should reverse and turn the wheels toward the obstruction (left)
- Obstructions nearby on the right
  Turn the front wheels right and reverse.

The complete program

The complete program listing is given in the Appendix that follows this chapter. The program is also available for download from the Bitstream Technology website (www.bitstreamtechnology.com). New versions of the programs will also be available at that website as improvements to the program are made.

The following sections discuss the details of the program.

Scheduling

The first thing to notice is that the schedule for the time slices is contained in 256 program memory locations. This is at the beginning of the program and address-wise starts at location $F800_{16}$. The scheduling is set up as was discussed in the previous chapter.

Jump table

Next, note the jump table in the program. There are 12 tasks used (counting two reserved tasks, 0 and 1, that are treated as error conditions). The Task, once read from the scheduling table, is shifted left one bit, thus doubling the value. This is because each jump in the jump table is two bytes in length.

Indexing the schedule

The schedule is simply run through in a linear fashion, beginning with the first time slice in the schedule and proceeding, one-by-one, to each successive time slice. The process for controlling this is the variable Index. In most tasks, the index is incremented by one, so that Index is pointing to the next time slice. For the case of multiple-time-slice tasks, the index is incremented by more than one, as explained in the multiple-time-slice tasks section.

Time slice timing

Each time slice lasts (16.67 msec/256) = 65.1 μsec. This is accomplished by resetting the timer/counter, TAR, at the beginning of each frame and then letting it run continuously throughout the frame. After the execution of each time slice, a delay is added to produce a period of 65 μsec. The proper delay time is produced by comparing the TAR contents to a number corresponding to this 65 μsec.

The variable containing this number is NextStop. For example, when the State_LED1 task is executed, after driving the LED and checking for optical reflections, as required, the program instruction says to add the number 122 to NextStop. Since the MSP430's clocks are set to run at 2 MHz, this corresponds to only 61 μsec. However, there are some extra instructions that are executed after each time slice that account for the extra 4 μsec, bringing the total time for each time slice to roughly 65 μsec.

Multiple-time-slice tasks

There are several tasks, including High Voltage and Ultrasonic Transmission, which require more than one time slice. For example, the High Voltage Generation can require several hundred microseconds to fully charge the high-voltage capacitor, so six time slices are allocated to it. The High Voltage segment of the program does not check the timing at the end of each 65 μsec time slice. Instead, it executes the High Voltage segment of the program in its entirety, then increments NextStop by 772, which is slightly less than the 780 that is needed to achieve a 6 × 65 μsec timing interval with a 2 MHz clock speed. The difference between the 772 and 780 is accounted for, again, by a few overhead instructions not included in the count.

The Ultrasonic Transmission task is allocated seven time slices. However, for much of the time, the clock is run at 12 MHz rather than 2 MHz (to improve the ultrasonic frequency accuracy), so the number added for NextStop is actually larger than 7 × 130.

The multiple-time-slice, multiple-frame steering

Steering presents a different approach than other functions of the robot control. Whereas all the other functions can be contained within a 16.67 msec frame and most within a 65 μsec time slice, the robot's R/C car steering requires longer on-times. Typical on-times of 100 msec are common for these vehicles. Less than that may result in an anemic turn or no turn at all.

As a result, a left or right Steering output must be maintained over several frames. To make this happen, a frame count variable, SteerCount, is used. SteerCount is initially set to 6 when steering starts and is decremented by one with each new frame. When it reaches zero, the steering ends. The choice of 6 for SteerCount results in an electronic steering signal to the steering motor that is 100 msec long (6 × 16.67 msec).

Programming and circuit wrap-up

As you've probably realized, the real focus for this project has been on the individual circuits and MSP430 peripherals, which we've individually covered in all the previous chapters. This chapter ties up the loose ends.

I hope, at this point, that you understand the discussions about how things work that went on in each of the chapters. If there are things that still don't make sense, don't worry. You can still build the circuits according to the schematic (either breadboarding the circuit or using the printed circuit board from Bitstream Technology) and you can use the provided MSP430 program and get things working without understanding 100% of why these things are the way they are.

What you will find, if you pursue this type of field, either as work or hobby, is that there is lots to know and there will always be that one more thing that you want to learn and don't quite understand.

Figure 24.4
The MSP430-based robot in action.

Future robot peripherals

The functionality of the robot described in this book is pretty basic, despite the considerable theory and explanation required to get things to this point. Electronics and microcontroller programming are like that – there's a huge amount to learn just to get to the stage where you can do something relatively simple. Fortunately, all of that learning needed to do the simple stuff creates the foundation for you to do the innovative, creative, and cool things later on.

Nevertheless, it's true that our robot really just goes relatively fast (Figure 24.4) while avoiding collisions. The question is: what else can it do?

Before answering that question, it's reasonable to ask how the robotic controller developed in this book can be expanded. Not shown on the schematic of Figure 24.1 but actually present on the Bitstream Technology printed circuit board is a connector that implements the I^2C serial port pins (P1.6 and P1.7 on the MSP430 LaunchPad). These pins, along with the +3.6 V and ground, are available to create an expansion port for future peripherals that either you or Bitstream Technology, our robotics company, may come up with.

So what might these future products be? We've already described a couple of functions that would be useful: an optical remote control to turn the robot on and off; and a speed sensor. Another function that could be useful is a line tracker – an optical sensor that allows the robot to follow lines. If you start thinking about other types of vehicles that could be built – for example, air vehicles, surface water vehicles, and submersible water vehicles – there are many other sensors and actuators that could be useful to add.

Learn and enjoy!

It's my hope that you've learned not just how to make this particular robot but the basics of electronics and programming from this book. Electronics and microcontrollers are important topics in technology and can lead to a rewarding and successful career as well as a satisfying pastime. If you find that the things discussed in this book appeal to you, dig in and start learning and doing!

Appendix — Program Listing

```
;*****************************************************************************
;
; The way this version works is that the intermediate state will vary between
;   coasting one frame and full speed the next
; There will only be one Motor PWM routine per frame
;
;*****************************************************************************
#include  "msp430G2452.h"
;-----------------------------------------------------------------------------

     ORG    0F800h    ; Beginning of Look-up Table
      DB     4    ; Motor PWM
      DB            2              ; High Voltage Generation
      DB            2              ; High Voltage Generation
      DB            2              ; High Voltage Generation
      DB            2              ; High Voltage Generation
      DB            2              ; High Voltage Generation
      DB            2              ; High Voltage Generation
      DB            3              ; Steering
      DB            5              ; Pulse L LED
      DB            6              ; Pulse R LED
      DB            7              ; Ultrasonic TX
      DB            7              ; Ultrasonic TX
      DB            7              ; Ultrasonic TX
      DB            7              ; Ultrasonic TX
      DB            7              ; Ultrasonic TX
      DB            7              ; Ultrasonic TX
      DB            7              ; Ultrasonic TX
      DB            10             ; Do Nothing
      DB            10             ; Do Nothing
      DB            10             ; Do Nothing
      DB            10             ; Do Nothing
      DB            10             ; Do Nothing
      DB            10             ; Do Nothing
      DB            8              ; Ultrasonic RX
      DB            10             ; Do Nothing
      DB            10             ; Do Nothing
      DB            10             ; Do Nothing
      DB            8              ; Ultrasonic RX
      DB            10             ; Do Nothing
      DB            10             ; Do Nothing
      DB            10             ; Do Nothing
      DB            8              ; Ultrasonic RX
      DB            10             ; Do Nothing
      DB            10             ; Do Nothing
      DB            10             ; Do Nothing
      DB            8              ; Ultrasonic RX
      DB            10             ; Do Nothing
      DB            10             ; Do Nothing
      DB            10             ; Do Nothing
      DB            8              ; Ultrasonic RX
      DB            10             ; Do Nothing
```

```
DB      10      ; Do Nothing
DB      10      ; Do Nothing
DB      8       ; Ultrasonic RX
DB      10      ; Do Nothing
DB      10      ; Do Nothing
DB      10      ; Do Nothing
DB      8       ; Ultrasonic RX
DB      10      ; Do Nothing
DB      10      ; Do Nothing
DB      10      ; Do Nothing
DB      8       ; Ultrasonic RX
DB      10      ; Do Nothing
DB      10      ; Do Nothing
DB      10      ; Do Nothing
DB      8       ; Ultrasonic RX
DB      10      ; Do Nothing
DB      10      ; Do Nothing
DB      10      ; Do Nothing
DB      8       ; Ultrasonic RX
DB      10      ; Do Nothing
DB      10      ; Do Nothing
DB      10      ; Do Nothing
DB      8       ; Ultrasonic RX
DB      10      ; Do Nothing
DB      10      ; Do Nothing
DB      10      ; Do Nothing
DB      8       ; Ultrasonic RX
DB      10      ; Do Nothing
DB      10      ; Do Nothing
DB      10      ; Do Nothing
DB      8       ; Ultrasonic RX
DB      10      ; Do Nothing
DB      10      ; Do Nothing
DB      10      ; Do Nothing
DB      8       ; Ultrasonic RX
DB      10      ; Do Nothing
DB      10      ; Do Nothing
DB      10      ; Do Nothing
DB      8       ; Ultrasonic RX
DB      10      ; Do Nothing
DB      10      ; Do Nothing
DB      10      ; Do Nothing
DB      8       ; Ultrasonic RX
DB      10      ; Do Nothing
DB      10      ; Do Nothing
DB      10      ; Do Nothing
DB      8       ; Ultrasonic RX
DB      10      ; Do Nothing
DB      10      ; Do Nothing
DB      10      ; Do Nothing
DB      8       ; Ultrasonic RX
DB      10      ; Do Nothing
DB      10      ; Do Nothing
DB      10      ; Do Nothing
DB      8       ; Ultrasonic RX
DB      10      ; Do Nothing
DB      10      ; Do Nothing
DB      10      ; Do Nothing
```

```
DB      8       ; Ultrasonic RX
DB      10      ; Do Nothing
DB      10      ; Do Nothing
DB      10      ; Do Nothing
DB      8       ; Ultrasonic RX
DB      10      ; Do Nothing
DB      10      ; Do Nothing
DB      10      ; Do Nothing
DB      8       ; Ultrasonic RX
DB      10      ; Do Nothing
DB      10      ; Do Nothing
DB      10      ; Do Nothing
DB      8       ; Ultrasonic RX
DB      10      ; Do Nothing
DB      10      ; Do Nothing
DB      10      ; Do Nothing
DB      8       ; Ultrasonic RX
DB      10      ; Do Nothing
DB      10      ; Do Nothing
DB      10      ; Do Nothing
DB      8       ; Ultrasonic RX
DB      10      ; Do Nothing
DB      10      ; Do Nothing
DB      10      ; Do Nothing
DB      8       ; Ultrasonic RX
DB      10      ; Do Nothing
DB      10      ; Do Nothing
DB      10      ; Do Nothing
DB      8       ; Ultrasonic RX
DB      10      ; Do Nothing
DB      10      ; Do Nothing
DB      10      ; Do Nothing
DB      8       ; Ultrasonic RX
DB      10      ; Do Nothing
DB      10      ; Do Nothing
DB      10      ; Do Nothing
DB      8       ; Ultrasonic RX
DB      10      ; Do Nothing
DB      10      ; Do Nothing
DB      10      ; Do Nothing
DB      8       ; Ultrasonic RX
DB      10      ; Do Nothing
DB      10      ; Do Nothing
DB      10      ; Do Nothing
DB      8       ; Ultrasonic RX
DB      10      ; Do Nothing
DB      10      ; Do Nothing
DB      10      ; Do Nothing
DB      8       ; Ultrasonic RX
DB      10      ; Do Nothing
DB      10      ; Do Nothing
DB      10      ; Do Nothing
DB      8       ; Ultrasonic RX
DB      10      ; Do Nothing
```

```
DB          10          ; Do Nothing
DB          10          ; Do Nothing
DB          8           ; Ultrasonic RX
DB          10          ; Do Nothing
DB          10          ; Do Nothing
DB          10          ; Do Nothing
DB          8           ; Ultrasonic RX
DB          10          ; Do Nothing
DB          10          ; Do Nothing
DB          10          ; Do Nothing
DB          8           ; Ultrasonic RX
DB          10          ; Do Nothing
DB          10          ; Do Nothing
DB          10          ; Do Nothing
DB          8           ; Ultrasonic RX
DB          10          ; Do Nothing
DB          10          ; Do Nothing
DB          10          ; Do Nothing
DB          8           ; Ultrasonic RX
DB          10          ; Do Nothing
DB          10          ; Do Nothing
DB          10          ; Do Nothing
DB          8           ; Ultrasonic RX
DB          10          ; Do Nothing
DB          10          ; Do Nothing
DB          10          ; Do Nothing
DB          8           ; Ultrasonic RX
DB          10          ; Do Nothing
DB          10          ; Do Nothing
DB          10          ; Do Nothing
DB          8           ; Ultrasonic RX
DB          10          ; Do Nothing
DB          10          ; Do Nothing
DB          10          ; Do Nothing
DB          8           ; Ultrasonic RX
DB          10          ; Do Nothing
DB          10          ; Do Nothing
DB          10          ; Do Nothing
DB          8           ; Ultrasonic RX
DB          10          ; Do Nothing
DB          10          ; Do Nothing
DB          10          ; Do Nothing
DB          8           ; Ultrasonic RX
DB          10          ; Do Nothing
DB          10          ; Do Nothing
DB          10          ; Do Nothing
DB          8           ; Ultrasonic RX
DB          10          ; Do Nothing
DB          10          ; Do Nothing
DB          10          ; Do Nothing
DB          8           ; Ultrasonic RX
DB          10          ; Do Nothing
DB          10          ; Do Nothing
DB          10          ; Do Nothing
```

```
DB    8     ; Ultrasonic RX
DB    10    ; Do Nothing
DB    10    ; Do Nothing
DB    10    ; Do Nothing
DB    8     ; Ultrasonic RX
DB    10    ; Do Nothing
DB    10    ; Do Nothing
DB    10    ; Do Nothing
DB    8     ; Ultrasonic RX
DB    10    ; Do Nothing
DB    10    ; Do Nothing
DB    10    ; Do Nothing
DB    8     ; Ultrasonic RX
DB    10    ; Do Nothing
DB    10    ; Do Nothing
DB    10    ; Do Nothing
DB    8     ; Ultrasonic RX
DB    10    ; Do Nothing
DB    10    ; Do Nothing
DB    10    ; Do Nothing
DB    8     ; Ultrasonic RX
DB    10    ; Do Nothing
DB    10    ; Do Nothing
DB    10    ; Do Nothing
DB    8     ; Ultrasonic RX
DB    10    ; Do Nothing
DB    10    ; Do Nothing
DB    10    ; Do Nothing
DB    8     ; Ultrasonic RX
DB    10    ; Do Nothing
DB    10    ; Do Nothing
DB    10    ; Do Nothing
DB    8     ; Ultrasonic RX
DB    11    ; Coast Initial State
DB    10    ; Do Nothing
DB    10    ; Do Nothing
DB    8     ; Ultrasonic RX
DB    9     ; Make Decision
DB    10    ; Do Nothing
DB    10    ; Do Nothing
DB    10    ; Do Nothing
DB    10    ; Do Nothing
DB    10    ; Do Nothing
DB    10    ; Do Nothing
DB    10    ; Do Nothing

DB    10    ; Do Nothing
DB    10    ; Do Nothing
DB    10    ; Do Nothing
DB    10    ; Do Nothing
DB    10    ; Do Nothing
DB    10    ; Do Nothing
DB    10    ; Do Nothing
DB    10    ; Do Nothing
```

```
;------------------------------------------------------------------------------
        ORG     0E000h
;------------------------------------------------------------------------------
PWMWidth   EQU   0200h
Index    EQU   0202h
State     EQU   0204h
NextStop  EQU    0206h
Distance   EQU    0208h
Direction  EQU    020Ah
ReadLED1A  EQU    020Ch
ReadLED1B  EQU    020Eh
LED1Diff   EQU    0210h
ReadLED2A  EQU    0212h
ReadLED2B  EQU    0214h
LED2Diff   EQU    0216h
SteerState  EQU    0218h
MotorState  EQU    021Ah
SteerCount  EQU    021Ch
MotorCount  EQU    021Eh
HVReading  EQU    0220h
FoundObstacle EQU  0222h
RxCount    EQU    0224h
MotorDirection EQU 0226h
MotorSpeed  EQU    0228h
InterState  EQU    022Ah
MotorSpinning EQU  022Ch
TABLE     EQU    022Eh

;------------------------------------------------------------------------------
RESET
        mov     #0280h,SP            ; Initialize stackpointer

; Start of section to set input/output directions and initial values for Port1
; and Port2 along with CLK frequency and variable initial values
        call    #StopWDT
        call    #SetupP1
        call    #SetupP2
        call    #P2BitSet
        call    #SetupClk_2Mhz
        mov     #0,&Index
        mov     #1,&PWMWidth
        clr     &TAR
        clr     &NextStop
        clr     &SteerState
        clr     &SteerCount
        clr     &MotorCount
        clr     &InterState
        clr     &MotorSpeed
        clr     &MotorDirection
Mainloop
        mov     #64,&Distance        ; Make the initial distance large
        clr     &State
```

```
        clr     &RxCount
        clr     &FoundObstacle
        clr     &MotorSpinning
        mov     #TABLE,R10
Schedule
        mov     #0F800h,R5
        add     &Index,R5      ; consulting lookup table for Schedule values
        mov.b   0(R5),&State
        rla     &State
        add     &State,PC

; Start of jump tables for the subroutine tasks
        jmp     State_HV_USTx
        jmp     State_HV_LED1
        jmp     State_HV_LED2
        jmp     State_Steering
        jmp     State_Motor
        jmp     State_LED1
        jmp     State_LED2
        jmp     State_USTx
        jmp     State_USRx
        jmp     State_Decision
        jmp     State_Nothing
        jmp     State_Initial_Coast
Trap
        jmp     Trap
State_HV_USTx
State_HV_LED1
State_HV_LED2
        call    #High_Voltage
        ; Add of 772 to NextStop due to 6 time slices being reserved for HV
        add     #772,&NextStop
        call    #delay
        ; There are 6 increment instructions due to the back to back
        ;    table listings for HV
        inc     &Index
        inc     &Index
        inc     &Index
        inc     &Index
        inc     &Index
        inc     &Index
        jmp     Next
;-----------------------------------------------------------------------------
State_Steering
        call    #Steering
        ; 122 is used instead of 130 (65usec*2Mhz)
        ;   due to extra instructions happening after delay subroutine
        add     #122,&NextStop
        call    #delay
        inc     &Index
        jmp     Next
;-----------------------------------------------------------------------------
State_Motor
        call    #Motor
```

```
        add   #122,&NextStop
        call  #delay
        inc   &Index
        jmp   Next
;-------------------------------------------------------------------------------
State_LED1
        call  #opt_drv1
        add   #122,&NextStop
        call  #delay
        inc   &Index
        jmp   Next
;-------------------------------------------------------------------------------
State_LED2
        call  #opt_drv2
        add   #122,&NextStop
        call  #delay
        inc   &Index
        jmp   Next
;-------------------------------------------------------------------------------
State_USTx
        ; 12Mhz clock needed to accurately transmit US at 40kHz
        call  #SetupClk_12Mhz
        call  #Transmit
        call  #SetupClk_2Mhz
        ; 5452 (5460 minus usual 8) is used due to a 12Mhz clk freq
        ;   for the USTx and there are 7 timeslices allocated
        ;   for USTx ([130*6]*7=5460)
        add   #5452,&NextStop
        call  #delay
        add   #7,&Index
        jmp   Next
;-------------------------------------------------------------------------------
State_USRx
        call  #Receiver
        add   #122,&NextStop
        call  #delay
        inc   &Index
        jmp   Next
;-------------------------------------------------------------------------------
State_Decision
        call  #Decision
        add   #122,&NextStop
        call  #delay
        inc   &Index
        jmp   Next
;-------------------------------------------------------------------------------
State_Nothing
        add   #122,&NextStop
        call  #delay
        inc   &Index
        jmp   Next
;-------------------------------------------------------------------------------
State_Initial_Coast
        call  #Initial_Coast
```

```
        add    #122,&NextStop
        call   #delay
        inc    &Index
        jmp    Next
;------------------------------------------------------------------------------
Next
        ; 37890 is used as a total sum of all 256 time slices
        ;  with allowance of 5460 added to the other 130 values
        cmp    #37890,&TAR
        jc     NewPeriod
        cmp    #256,&Index
        ; jump to holding pattern if 256 time slices completed before 16.67 msec
        jge    StartOver
        jmp    Schedule
StartOver
        cmp    #37890,&TAR
        jc     StartOver
NewPeriod
        ; Setup certain variables for next 16.67 msec block
        clr    &TAR
        clr    &Index
        clr    &NextStop
        mov    #TABLE,R9
        clr    R6
        jmp    Mainloop

;------------------------------------------------------------------------------
StopWDT
        mov    #WDTPW+WDTHOLD,&WDTCTL  ; Stop watchdog timer
        ret

;------------------------------------------------------------------------------
SetupP1
        bic.b  #01Eh,&P1DIR           ; P1.1-P1.4 input direction
        bis.b  #0E1h,&P1DIR           ; P1.0,P1.5-P1.7 output direction
        ret

;------------------------------------------------------------------------------
SetupP2
        bis.b  #0FFh,&P2DIR           ; P2.7-P2.0 output direction
        bic.b  #0FFh,&P2SEL           ;
        bic.b  #0F0h,&P2OUT
        ret

;------------------------------------------------------------------------------
SetupADC
        mov
#SREF_1+ADC10SHT_3+ADC10SR+REF2_5V+REFON+ADC10ON,&ADC10CTL0
        mov.b  #010h,&ADC10AE0
        ret

;------------------------------------------------------------------------------
P2BitSet
        bic.b  #0Ch,&P2OUT  ; Clear and set P2.2 and P2.3 for use with USTx
```

```
        bis.b  #08h,&P2OUT
        ret

;------------------------------------------------------------------------------
SetupClk_2Mhz
        clr.b  &DCOCTL  ;  Select lowest DCOx and MODx settings
        mov.b  &CALBC1_8MHZ,&BCSCTL1 ; Set range
        mov.b  &CALDCO_8MHZ,&DCOCTL ; Set DCO step + modulation
        mov.b  #DIVS_2+DIVM_2,&BCSCTL2
        mov    #TASSEL_2+MC_2,&TACTL
        ret

;------------------------------------------------------------------------------
SetupClk_12Mhz
        clr.b  &DCOCTL  ;  Select lowest DCOx and MODx settings
        mov.b  &CALBC1_12MHZ,&BCSCTL1  ; Set range
        mov.b  &CALDCO_12MHZ,&DCOCTL ; Set DCO step + modulation
        clr.b  &BCSCTL2
        mov    #TASSEL_2+MC_2,&TACTL
        ret

;------------------------------------------------------------------------------
High_Voltage
        call   #SetupADC
        mov    &PWMWidth,R8
Read_V
        bic    #ADC10IFG,&ADC10CTL0
        mov    #INCH_4,&ADC10CTL1
        bis    #ENC+ADC10SC,&ADC10CTL0
TestADC
        bit    #ADC10IFG,&ADC10CTL0
        jz     TestADC
        mov    &ADC10MEM,&HVReading
        ; with a Vref = 2.5V, 1.36V = 022Eh and this is our value
        ;   for A2D reading of 15V
        cmp    #022Eh,&ADC10MEM
        jge    Dec_R8
        inc    R8
        ; gives T0 pulse a chance to be wide enough to reach 15V
        cmp    #170,R8
        jge    Dec_R8
        jmp    hold_high
Dec_R8
        cmp    #2,R8            ; Safeguard against unwanted state of R8
        jl     again
        dec    R8
hold_high
        bis.b  #020h,&P1OUT         ; Set P1.5 output
        mov    R8,&PWMWidth
again
        dec    R8               ; Decrement R8
        jnz    again
        bic.b  #020h,&P1OUT         ; Clear P1.5 bit
        ret
```

```
;-------------------------------------------------------------------------------
Steering
; Decision subroutine will have set parameters for steering
; Initial test is to determine if steering is necessary

        tst     &SteerState
        jeq     No_steering
        tst     &SteerCount
        jne     Cont_Steer
        jmp     Start_steer
        jl      End_routine
Start_steer
        mov     #6,&SteerCount          ; Any steering will last 100 msec
        dec     &SteerCount
        jmp     Steering_direction
Cont_Steer
        dec     &SteerCount
        jmp     End_routine
Steering_direction
        cmp     #1,&SteerState  ; Left or Right steering is implemented here
        jeq     Turn_left
        cmp     #2,&SteerState
        jeq     Turn_right
Turn_left
        ; S1 and S2 are first cleared to avoid unstable situation
        bic.b   #030h,&P2OUT
        bis.b   #010h,&P2OUT            ; S1 = 1 and S2 = 0 for left turn
        jmp     End_routine
Turn_right
        bic.b   #030h,&P2OUT
        bis.b   #020h,&P2OUT            ; S1 = 0 and S2 = 1 for left turn
        jmp     End_routine
No_steering
        bic.b   #030h,&P2OUT
End_routine
        ret

;-------------------------------------------------------------------------------
Motor
; For Far:
;   MotorDirection is 0 (forward) and MotorSpeed is 1
; For Intermediate:
;   MotorDirection is 0 (forward) and MotorSpeed is 0
; For Near
;   MotorDirection is 1 (reverse) and MotorSpeed is 1
        tst     &MotorDirection
        jnz     Reverse
        tst     &MotorSpeed
        jnz     Fwd
        ; If the program got to here it is in intermediate
        ; If InterState=0 then Coast
        ; If InterState=1 then Fwd
        ; Toggle InterState before returning
```

```
        xor    #1,&InterState
        jnz    Fwd
Coast
        bic.b  #0C0h,&P2OUT
        jmp    done
Brake
        bic.b  #0C0h,&P2OUT
        bis.b  #0C0h,&P2OUT
        jmp    done
Reverse
        bic.b  #0C0h,&P2OUT
        bis.b  #040h,&P2OUT
        jmp    done
Fwd
        bic.b  #0C0h,&P2OUT
        bis.b  #080h,&P2OUT
done
        ret

;-------------------------------------------------------------------------------
opt_drv1
A2DLoop    bic    #ENC,&ADC10CTL0
        bic    #ADC10IFG,&ADC10CTL0
        mov    #INCH_1,&ADC10CTL1
        call   #StartA2D
        call   #FlagCheck
        mov    &ADC10MEM,&ReadLED1A
        bis.b  #1,&P2OUT
        call   #StartA2D
        call   #FlagCheck
        mov    &ADC10MEM,&ReadLED1B
        bic    #ENC+ADC10SC,&ADC10CTL0
        bic.b  #1,&P2OUT
        mov    &ReadLED1A,R6
        mov    &ReadLED1B,R7
        sub    R7,R6
        mov    R6,&LED1Diff
        ret

;-------------------------------------------------------------------------------
opt_drv2
A2DLoop1   bic    #ENC,&ADC10CTL0
        bic    #ADC10IFG,&ADC10CTL0
        mov    #INCH_1,&ADC10CTL1
        call   #StartA2D
        call   #FlagCheck
        mov    &ADC10MEM,ReadLED2A
        bis.b  #2,&P2OUT
        call   #StartA2D
        call   #FlagCheck
        mov    &ADC10MEM,ReadLED2B
        bic    #ENC+ADC10SC,&ADC10CTL0
        bic.b  #2,&P2OUT
        mov    &ReadLED2A,R6
```

```
        mov   &ReadLED2B,R7
        sub   R7,R6
        mov   R6,&LED2Diff
        ret

;-----------------------------------------------------------------------------
Transmit
        call  #SetupClk_12Mhz
        mov   #18,R6
forty_k
        mov   #47,R7
        nop
        nop
hold_1
        dec   R7
        jnz   hold_1
        xor.b #0Ch,&P2OUT
        mov   #47,R7
        nop
        nop
hold_2
        dec   R7
        jnz   hold_2
        xor.b #0Ch,&P2OUT
        dec   R6
        jnz   forty_k
        ret

;-----------------------------------------------------------------------------
Receiver
        inc   &RxCount            ; Keep track of receiver count
        cmp   #1,&FoundObstacle
        jeq   finish_loop
        mov   #INCH_3,&ADC10CTL1
        call  #StartA2D
        call  #FlagCheck
        cmp   #01ABh,&ADC10MEM
        jge   finish_loop
set_decision
        mov   &RxCount,&Distance
        mov   #1,&FoundObstacle
finish_loop
        ret

;-----------------------------------------------------------------------------
; Direction = 0 for left turn
; Direction = 1 for right turn
; For Far:
;    MotorDirection is 0 (forward) and MotorSpeed is 1
; For Intermediate:
;    MotorDirection is 0 (forward) and MotorSpeed is 0
; For Near
;    MotorDirection is 1 (reverse) and MotorSpeed is 1
; When the motor is put into reverse, it stays in that state for 32 frames
```

```
; It does this using MotorCount
; The first frame during which the motor is reversed, MotorCount is set to 32
; Thereafter, when the previous motor state is Reverse and MotorCount is not
;   equal to 0, the Decision routine just decrements MotorCount and returns
Decision
        tst   &MotorDirection   ; Was the motor already reversing?
        jz    ContinueDecision
        dec   &MotorCount
        jnz   end_loop          ; If MotorCount is zero start reversing again
ContinueDecision
        clr   &Direction
        cmp   &LED1Diff,&LED2Diff
        jge   TurnLeft
        inc   &Direction
TurnLeft
        cmp   #30,&Distance
        jge   far_steer
        cmp   #6,&Distance
        jl    short_steer
; Intermediate Speed
        clr   &MotorDirection     ; Forward
        clr   &MotorSpeed         ; Slow
        mov   #01h,&SteerState    ; Left
        tst   &Direction
        jeq   end_loop
        rla   &SteerState         ; Right
        jmp   end_loop
far_steer
        clr   &SteerState         ; Straight
        clr   &MotorDirection     ; Forward
        mov   #1,&MotorSpeed      ; Fast
        jmp   end_loop
short_steer
        mov   #32,&MotorCount
        mov   #1,&MotorDirection ; Reverse
        mov   #1,&MotorSpeed      ; Fast
        mov   #1,&SteerState      ; Left
        tst   &Direction
        jne   end_loop            ; reverse direction of steering for reverse
        rla   &SteerState         ; Right
end_loop
        ret

;------------------------------------------------------------------------------
Initial_Coast
        bic.b  #0C0h,&P2OUT
        ret

;------------------------------------------------------------------------------
StartA2D
        clr   &ADC10CTL0
        bis   #ADC10ON,&ADC10CTL0
        bis   #ENC+ADC10SC,&ADC10CTL0
        ret
```

```
;-------------------------------------------------------------------------------
FlagCheck
        bit    #ADC10IFG,&ADC10CTL0
        jz     FlagCheck
        bic    #ENC+ADC10IFG,&ADC10CTL0
        ret

;-------------------------------------------------------------------------------
delay
        cmp   &NextStop,&TAR
        jnc    delay
        mov    &TAR,&NextStop
        ret

;-------------------------------------------------------------------------------
;       Interrupt Vectors Used MSP430x2013
;-------------------------------------------------------------------------------
        ORG     0FFFEh              ; MSP430 RESET Vector
        DW      RESET               ;
        END
```

Index

Note: Page numbers followed by "*f*" and "*t*" refer to figures and tables, respectively.

S

W

X

Z

Printed and bound by CPI Group (UK) Ltd, Croydon, CR0 4YY
03/10/2024
01040319-0004